Praise for George Gilder's *Microcosm*

"A thought-provoking journey through the microchip-driven world of modern technology, highlighting the people and ideas that made it happen. I recommend it highly."
—Arno Penzias,
Nobel Laureate in Physics, and Director, Bell Laboratories

"By 'listening to the technology,' George Gilder has produced a stunning narrative of the men and women who are supplying much of the intellectual capital of America. More than that, he has demonstrated that the new technology is on the side of freedom and antithetical to centralized power of all kinds."
—Walter Wriston,
former Chairman, Citicorp/Citibank

"Gilder's brilliant, rigorous analysis of the coming death of television alone is more than enough reason to read this book."
—Jim Rogers,
Professor, Columbia Business School

"If you liked *The Bonfire of the Vanities,* you will like *Microcosm* even better because it is written with just as much flair, but it is about the real people and pioneering ideas of the Silicon Valley. The chapter describing the AMD annual Hawaiian sales meeting is as accurate as a Polaroid picure . . . I was there."
—T. J. Rodgers,
President, Cypress Semiconductors

Also by George Gilder

The Spirit of Enterprise
Wealth and Poverty
Visible Man
Men and Marriage

GEORGE GILDER

The Quantum Revolution in

MICROCOSM

Economics and Technology

A TOUCHSTONE BOOK
Published by Simon & Schuster Inc.
New York   London   Toronto   Sydney   Tokyo   Singapore

Touchstone
Simon & Schuster Building
Rockefeller Center
1230 Avenue of the Americas
New York, New York 10020

Copyright © 1989 by George Gilder

1   3   5   7   9   10   8   6   4   2

1   3   5   7   9   10   8   6   4   2   Pbk.

Library of Congress Cataloging in Publication Data
Gilder, George F., date
Microcosm: the quantum revolution in economics and technology.
Bibliography: p.
Includes index.
1. Microelectronics industry.   2. Microelectronics—
Social aspects.   I. Title.
HD9696.A2G55   1989   338.4'7621381   89-11476
ISBN 0-671-50969-1
ISBN 0-671-70592-X Pbk.

To Richard Vigilante

CONTENTS

PREFACE

Listen to the technology and find out what it is telling you.
—CARVER MEAD

This book is an exploration of the meaning and future of modern technology. With its origins in quantum physics and its embodiment in the microchip, the exemplary product of this technology is the computer. Still in the infancy of its influence and power, the computer in its current form is the product of some twenty-five years of development. In all its manifestations—reaching from the microprocessor in a car's fuel injection system to the microprocessor at the heart of a personal computer, from a supercomputer modeling the weather to a worldwide ganglion of fiber-optic threads of glass and light—computer technology epitomizes the fruits of the microcosm of quantum physics. It was quantum theory early this century that revealed the inner structure of matter for the first time and made modern computers possible.

Broadly considered, the computer is the most important product of the quantum era. By exploring this central machine of the age, however, we discover not the centrality of machines and things but the primacy of human thought and creativity.

The quantum era is still unfolding in a fourfold transformation of the world—in science, technology, business, politics—and even in philosophy. But all the changes converge in one epochal event: the overthrow of matter.

The change originates in the microcosm of quantum theory itself —the new physics launched in Europe early this century—which overthrew matter in the physical sciences. At the foundation of the universe, Isaac Newton's hard, inert, and indivisible solids gave way to a rich panoply of paradoxical sparks, comprising waves and particles that violate every principle of Newtonian solidity. At the root of all the cascading changes of modern economic life—devaluing material resources in technology, business, and geopolitics—is this original overthrow of material solidity in the science of matter itself.

The second step in the overthrow of matter came in the use of quantum theory to overcome the material limits of weight, heat, and force in the creation of new machines. The industrial age essentially managed and manipulated matter from the outside, lifting it against gravity, moving it against friction, melting or burning it to change its form. The quantum era manipulates matter from the inside, adapting its inner structure to human purposes.

In the microchip, combining millions of components operating in billionths of seconds in a space the size of the wing of a fly, human beings built a machine that overcame all the conventional limits of mechanical time and space. Made essentially of the silicon in sand— one of the most common substances in earth—microchips find their value not in their substance but in their intellectual content: their design or software.

The third great manifestation of the overthrow of matter is the impact of this technology on the world of business. By overcoming the constraints of material resources, the microchip has devalued most large accumulations of physical capital and made possible the launching of global economic enterprises by one entrepreneur at a workstation. With the overthrow of the constraints of material scarcity, gravity, and friction, large bureaucracies in government and business lose their power over individual creators and entrepreneurs.

The fourth phase of the overthrow of matter is the collapse of the value of natural resources and territory in determining the distribution of power among nations. The microcosms of science, technology, and enterprise have converged in a global quantum economy that transcends all the usual measures of national power and wealth. The most valuable capital is now the capital of human mind and spirit. Intellectual capital can transform any physical environment, even a few small, cold, stony islands off the eastern coast of Asia, into a center of production and wealth. And the lack of such capital, or its abuse, can turn the greatest empire into a hollow shell reverberating with the frustration of tyrants.

In overthrowing the thrones of matter, this new epoch—the quantum era—also overthrows the great superstitions of materialism. Worship of things—whether in a Marxist dialectic or a Midas's hoard —collapses in a world in which thought is paramount even at the heart of matter itself.

PART

The Overthrow
of Matter

I

CHAPTER

1

The Message from the Microcosm

However much discussed and however promising this atomic theory might appear, it was, until recently, regarded merely as a brilliant hypothesis, since it appeared to many far sighted workers too risky to take the enormous step from the visible and directly controllable to the invisible sphere, from the macrocosm to the microcosm.

—MAX PLANCK, inventor of quantum theory

The central event of the twentieth century is the overthrow of matter. In technology, economics, and the politics of nations, wealth in the form of physical resources is steadily declining in value and significance. The powers of mind are everywhere ascendant over the brute force of things.

This change marks a great historic divide. Dominating previous human history was the movement and manipulation of massive objects against friction and gravity. In the classic image of humanity, Atlas bears the globe on stooped shoulders, or Sisyphus wrestles a huge rock up an endless slope. For long centuries, humans grew rich chiefly by winning control over territory and treasure, slaves and armies. Even the Industrial Revolution depended on regimented physical labor, natural resources, crude energy sources, and massive transport facilities. Wealth and power came mainly to the possessor of material things or to the ruler of military forces capable of conquering the physical means of production: land, labor, and capital.

Today, the ascendant nations and corporations are masters not of land and material resources but of ideas and technologies. Japan and other barren Asian islands have become the world's fastest-growing economies. Electronics is the world's fastest-growing major industry.

Computer software, a pure product of mind, is the chief source of
added value in world commerce. The global network of telecommunications carries more valuable goods than all the world's supertank-
ers. Today, wealth comes not to the rulers of slave labor but to the
liberators of human creativity, not to the conquerors of land but to
the emancipators of mind.

Impelled by an accelerating surge of innovation, this trend will
transform man's relations with nature in the twenty-first century. The
overthrow of matter will reach beyond technology and impel the
overthrow of matter in business organization. Devaluing large accumulations of fixed physical capital, the change will favor entrepreneurs
over large bureaucracies of all kinds. The overthrow of matter in
business will reverberate through geopolitics and exalt the nations in
command of creative minds over the nations in command over land
and resources. Military power will accrue more and more to the masters of information technology. Finally, the overthrow of matter will
stultify all materialist philosophy and open new vistas of human imagination and moral revival.

The exemplary technology of this era is the microchip—the computer inscribed on a tiny piece of processed material. More than any
other invention, this device epitomizes the overthrow of matter. Consider a parable of the microchip once told by Gordon Moore, chairman of Intel and a founding father of Silicon Valley:

"We needed a substrate for our chip. So we looked at the substrate
of the earth itself. It was mostly sand. So we used that.

"We needed a metal conductor for the wires and switches on the
chip. We looked at all the metals in the earth and found aluminum
was the most abundant. So we used that.

"We needed an insulator and we saw that the silicon in sand mixed
with the oxygen in the air to form silicon dioxide—a kind of glass.
The perfect insulator to protect the chip. So we used that."

The result was a technology—metal oxide silicon (MOS), made
from metal, sand, and air—in which materials costs are less than 1
percent of total expense. Combining millions of components on a
single chip, operating in billionths of seconds, these devices transcend
most of the previous constraints of matter. The most valuable substance in this, the fundamental product of the era, is the idea for the
design.

The microchip not only epitomizes but also impels the worldwide
shift of the worth of goods from materials to ideas. These transvalua-
tions are not mere luck, to be annulled by some new scarcity. Nor do

they reflect only the foresight of one industry, summed up in Moore's parable. The rise of mind as the source of wealth spans all industries and reflects the most profound findings of modern physics and philosophy. The overthrow of matter in economics is made possible by the previous overthrow of matter in physics. All the cascading devaluations of matter in the global economy and society originate with the fundamental transfiguration of matter in quantum science.

Max Planck, the discoverer of the quantum, offered the key when he asserted that the new science entailed a movement from the "visible and directly controllable to the invisible sphere, from the macrocosm to the microcosm." The macrocosm may be defined as the visible domain of matter, seen from the outside and ruled by the laws of classical physics. The microcosm is the invisible domain, ruled and revealed by the laws of modern physics.

The borders of these rival domains are not set by size or location. Some of the largest phenomena of the age—from global telecommunications to nuclear explosions—follow the quantum laws of the microcosm. Indeed, because atoms observe the microcosmic rules, the microcosm is ubiquitous and comprises all objects large and small. But because the microcosm is invisible, it has been relatively uninfluential in shaping public views and philosophies.

It is understandable that humans resist the microcosm and even rebel against it. Quantum theory is an abstruse and difficult set of ideas. It baffles many of its leading exponents and it perplexed Albert Einstein to his grave. Defying the testimony of the human senses, the new physics is contrary to all human intuition and metaphor. In the quantum domain, all conventional analogies of physics—such as tops, springs, and billiard balls—are radically misleading. Therefore, we cannot "understand" quantum theory in the way we can comprehend classical physics. Quantum theory simply does not make sense.

The reason the new physics does not make sense to most humans, however, is that prevailing common sense is wrong. Common sense serves the materialist superstition: the belief that we live in a world of solid phenomena, mechanically interconnected in chains of cause and effect. The common wisdom of mankind has yet to absorb the simple truth that in proportion to the size of its nucleus the average atom in one of our cherished solids is as empty as the solar system. Few ponder the fact that an electron—a key to physical solidity—does not occupy any specific position in space; in a famous experiment, a single electron passes simultaneously through two separate holes in a screen. Such quantum images are difficult for human beings to grasp or be-

lieve. Humans balk at the basic paradox of a physical theory that defies
the testimony of human senses and overthrows matter in the very
science of matter itself.

In various grudging ways, physical scientists all recognize this overthrow of matter. As the author of the quantum theory of chemistry,
Linus Pauling developed the fundamental model of what we call matter. He knows as much about it as any other man. But at the beginning of his popular text, *Chemistry,* he concedes that "no one really
knows how to define matter. . . ." Like virtually all physical scientists,
he then proceeds to use the term anyway. Yet the paramount theme
of the new physics is that what scientists now call matter is totally
unlike what classical physicists used to call matter and completely alien
to the solid objects that the term calls to mind.

A key reason that quantum theory is so difficult to understand, so
apparently riddled with paradox, is this continued use of the vocabulary of materialism to discuss phenomena that in the usual sense lack
all material qualities, such as solidity, location, continuity, and inertia.
On the other hand, it is this very transcendance of material limits that
explains why quantum technologies far outperform mechanical devices and promise a new era of technology and economics.

Separating the old and new sciences is a nearly unbridgeable gulf.
Compare Pauling's agnosticism about the definition of matter with
Isaac Newton's materialist confidence. Newton described matter as
"solid, massy, hard, impenetrable, movable particles . . . even so very
hard, as never to wear or break in pieces; no ordinary power being
able to divide what God himself made one in the first creation . . .
[to] compose bodies of one and the same nature and texture in all
ages." Newton's matter at heart was inert, opaque, and changeless.

Just as important, Newton had no problems with common sense.
His science was anthropomorphic. He assumed that matter at its
fundamental levels behaved like the material objects that we perceive.
"We no other way know [the characteristics of matter] than by our
senses. . . ." Its impenetrability, for example, "we gather not from
reason but from sensation."

For some two hundred years nearly all leading scientists shared
these materialist assumptions, based on sensory models and determinist logic. Newton's great eighteenth-century countryman Adam Smith
extended the materialist metaphor to society, contending that the
economy itself is a clockwork, a "great machine." Later Karl Marx
applied materialism to the very fabric of political ideas, which were
deemed mere figments of ownership in physical capital, or of aliena-

tion from it. Sigmund Freud and his followers developed a psychological theory of forces and pressures, inhibited or released, built up or fed back, much like the classical mechanics of steam engines.

Today most sophisticated people imagine that they have transcended Newton and have come to terms with the findings of modern science. But they have not. As an intellectual faith, materialist logic still prevails. Even theologians and philosophers who spurn materialism in defining the meaning of life accept it as the lesson of science. They still believe that the solid world they see and feel—governed by determinate chains of cause and effect, rooted in Newtonian masses and forces—is real and in some sense definitive.

The atom may not be ultimate, but they assume some other particle is, perhaps the quark. It has become a cliché to call the quark "the fundamental building block of nature." At the foundations of the physical world, so it is supposed, are physical solids—"building blocks"—that resemble in some way the solids we see. They link together in causal chains of mechanical logic like a set of cogs and levers. These solids are deemed to comprise all matter, from atoms and billiard balls to bricks and the human brain.

Such beliefs are manifest in how we think. MIT mathematician Gian Carlo Rota has written: "The naive prejudice that physical objects are somehow more 'real' than ideal objects remains one of the most deeply rooted in Western culture. . . . A consequence of this belief—which until recently was not even perceived as such—is that our logic is patterned exclusively on the structure of physical objects." A logical argument, for example, works very much like a material machine. Many of us still assume that like material things, causes and effects are directly and mechanically related in time and space.

Since most visible objects and machinery seem to observe the macrocosmic rules most of the time, most human beings are comfortable in a macrocosmic world. Like primitive tribes, they worship things they can see and feel. They think in logic based on the behavior of these things. In all these attitudes they cling to the materialist superstition: the belief that mechanical and mindless interactions of inert and impenetrable matter are the ultimate foundation of reality. They resist the apparently murky and paradoxical message from the microcosm: the overthrow of matter in a quantum world.

Understood as episodes in the dismantling of Newtonian matter, however, the historic discoveries of the new physics follow a coherent and powerful logic. The move from macrocosm to microcosm can be seen as a progress from a material world composed of blank and inert particles to a radiant domain rich with sparks of informative energy. This overthrow of matter in science leads plausibly to the ascendancy of information and mind in contemporary technology, and hence in economics as well.

The first crisis of classical physics revolved around the theory of light. In a famous experiment in 1801—beaming light through two holes at a screen equipped with a photodetector—Thomas Young had conclusively demonstrated that light is a wave. Like the interplay of the wakes of two stones dropped in a pool, light rippled across the screen in spirals of brighter and darker rings. Called an interference pattern, it is the signature of all wave phenomena.

In the 1860s, a Scotsman, James Clerk Maxwell, brought the wave theory of light to a new pinnacle. Integrating electricity with magnetism, Maxwell offered a full and definitive description of all electromagnetic phenomena, including light, in just four simple equations that still stand as a towering synthesis of physical knowledge. For example, using his equations, Maxwell could compute exactly the speed of light long before experiments of that precision. One hundred years later Nobel Laureate Richard P. Feynman would describe Maxwell's achievement as the most important event of the nineteenth century, relegating the American Civil War of the same decade to "provincial insignificance" by comparison.

Nonetheless, Maxwell's concept was incomplete because it failed to dispose of the issue of a medium for the waves to wave through. Traveling at a fixed speed through a continuous and infinitely divisible space, Maxwell's electromagnetic flux seemed to require a fixed frame of reference. Maxwell's successors assumed that this medium was "ether," an "elastic solid" with fixed coordinates filling all cosmic space. Waves of light were believed to propagate through ether somewhat as sea waves through water.

The first great disaster to befall the Newtonian theory occurred in 1887 when two American researchers, Albert Michelson and E. W. Morley, tried to measure the impact of this ubiquitous solid. Raiding the fortune of Michelson's wife, the two men contrived not one but several of the most exacting experiments in the history of physics to that day. But they never found any inkling of material ether.

This effort by Michelson and Morley is usually depicted as an important stage in developing the quantum theory of light. And so it was. Light emerged as an esoteric paradox of waves without substance traveling at a fixed speed in relation to a medium without substance. But the real meaning of Michelson-Morley was its role as the first step in the overthrow of matter: the intellectual annihilation of the ether previously supposed to comprise most of the matter in the universe.

Similarly, Max Planck's discovery of the quantum is usually treated as an episode in the study of certain paradoxes in black body radiation —beams emitted from heated substances such as coal or charcoal that essentially absorb all incident rays. He discovered, in essence, that the waves emanating from any given atom of such a body were not a continuous function of the energy it received. Rather, the atom would not radiate at all until it had absorbed a precise minimal dose of energy, its quantum, defined as the frequency of its waves times a number called Planck's constant. This idea seemed to enhance the concept of quanta as material particles. But the true meaning of Planck's breakthrough was the insufficiency of Maxwell's model of continuously fluctuating waves. In any usual sense, quantum matter was proving to be neither waves nor particles.

In 1905, in a Nobel Prize-winning paper, Albert Einstein used Planck's insight to take another giant step in dethroning matter. He declared that if there was no ether, light could not be a wave. It had nothing to wave through. Extending Planck's insight, he said that light consisted of quanta—packets of energy—that he called photons. Although said to be "particles," photons observed the wave equations of Maxwell. They seemed a cross between a wave and a particle.

In terms both of classical physics and of human observation, this wave-particle cross was an impossible contradiction. Waves ripple infinitely forth through a material medium; particles are single points of matter. At once definite and infinite, a particle that is also a wave defies our sensory experience. Newton's universe built of "solid, massy, hard, impenetrable, movable" particles was foundering. But by clinging verbally to the material concept of a particle, scientists obscured the true meaning of this collapse of physical solidity in the science of matter.

The overthrow of matter afflicted the theory of physics with contradiction and paradox. But eschewing the materialist superstition allowed physicists at last to address the real world and unleashed a golden age of practical creativity. If Maxwell's equations were the paramount event of the nineteenth century, the paramount event of the twentieth century may well prove to be the atomic theory of Niels Bohr and his colleagues. Mapping the microcosm for the first time, Bohr made possible all the cascading advances of the overthrow of matter in technology.

Experimenters for years had been noticing that electrically or ther-

mally disturbed materials release distinctive and identifiable wave patterns. Every solid has a specific frequency. In the prevailing theory of
a kind of subatomic solar system, the frequencies of these waves were
assumed to signify the rate at which the electrons were rotating
around the nucleus. But like the waves shed from a furnace in Planck's
theories, these radiations persistently violated classical predictions.

An eminent Danish physicist, Bohr suggested that rather than measuring the frequency of electron orbits, these waves signified the energy emitted by electrons in jumping from one orbit to another.
Dropping from a higher to a lower energy state, an electron released
a photon of light of its resonant frequency. Absorbing a photon of its
resonant frequency, an electron would move to a higher energy level.
In quantum theory, an atom was not a system of continuous orbits
but a hierarchy of distinct energy states.

Announced in 1913 and proved for the single electron of the hy-
drogen atom, the Bohr model was the first great vindication of quantum theory. One test of scientific advance is whether it extends the
realms of human understanding and control. Bohr's breakthrough
ultimately opened the microcosm as an industrial site.

The established physics could not explain the effectiveness of chemistry, let alone extend it to atoms. Unlike a solar system, atoms do not
exist in majestic isolation. Ceaselessly in movement, they endlessly
jiggle together in what is called Brownian motion. People even step
on them. In a world of Newtonian continuities, electron orbits would
vary continually as atoms collided with one another. Constantly
knocked loose in these collisions, electrons in a conductor should flow
far more copiously and respond to heat more massively than experiments showed.

In this realm of readily flowing and colliding particles, chemical
elements would not be stable. Elements would ooze together rather
than assume the fixed properties of the periodic table, the basis of all
chemistry. Negative electrons, in fact, would rapidly spin into the
positive nucleus, collapsing the matter of the universe. Far from en-
abling the vastly precise functions of microelectronics, Newtonian
physics could not even explain why you do not fall through the floor,
and then through the world.

Reunifying chemistry and physics in the microcosm, the new model
of the atom explained the apparent solidity of the physical world.
Establishing a gap, called a band gap, between an electron in its
ground state and an electron excited to a higher energy level, the new
physics showed why the constant collisions of atoms do not cause the

atomic structure to collapse. A small collision will not affect an atom. Just as a musical string can resound only in its fundamental and resonant frequencies (all others canceled out by the rules of wave interference), an electron will not respond to any small disturbance. It will react only if it receives its necessary quantum of energy, defined by its resonant frequency times Planck's constant.

In this way the quantum concept specified precisely the energy needed to change an atom or to connect it with another atom. Thus the new physics defined a grammar of chemical combinations. By restricting electron energies to specific quanta and by restricting the number of electrons in each energy state through the laws of quantum wave interference, Bohr's model brought an intelligible order to the atom. Revealed as a fixed mathematical construct, one atom could connect to another by fixed laws.

Bohr and his followers thus opened the way to understand the forces of attraction and repulsion among atoms. Only atoms with a missing electron in their outermost energy level are chemically active. Electrons tied to their atoms create stability; electrons freed of their atoms create electricity; electrons reaching out resonantly to other atoms or switching back and forth between atoms tie together the elements of chemistry, such as silicon, oxygen, and aluminum. By regulating the atomic binding and unbinding of electrons, the new theory illumined the laws of the chemical bond and the nature of electricity.

Without this new understanding, we could never adequately comprehend the flow of electricity through a solid. Nor could we see how to make a kind of semi-conducting sand become a conductor under certain precise conditions—thus rendering it an excellent pathway for information. These and thousands of other precise manipulations would be impossible without quantum mechanics.

Knowledge is power. Quantum physics gave humans access to the inner structure of matter. As Mitchell Feigenbaum, a pioneer of chaos theory, put it, quantum theory "tells you how you can make computers from dirt." But in overcoming Newtonian solidity, quantum scientists failed to offer a coherent or intelligible alternative image of matter.

As a result, the most profound discoveries of the twentieth century failed to penetrate the public consciousness, intellectual life, or natural philosophy. Most people continued in the old materialist idiom, forcing the new concepts into the old frame. Others used the quantum as a warrant for a murky subjectivism totally alien to the spirit of the science, which is rigorously objective and supremely practical. Like all fundamental shifts in the scientific view of reality, the quantum perspective has been slow in working its way through the culture. Bound by the necessary conservatism of their trade, scientists hesitate to traffic in the unprovable domains of meaning and non-scientists demur at entering the abstruse domains of science.

Such problems of the microcosm are evident in the history of the theory of the electron, the basic entity of electronics. Most interested people understand much of what electrons do. But very few have any clear idea of what an electron actually is, or its implications for the concept of matter and its overthrow in the world economy.

From the telephone to the human brain, from the television set to the computer, information mostly flows in the form of electrons. This function of electrons has quantum roots. As in Planck's black body radiation, electrons do not respond to applied energy in a continuous, proportional, or linear way. They are non-linear; they have quantum thresholds and resonances. These quantum functions shape their electrical properties. In order to move through a solid, electrons must be freed from their atoms, jumping from one energy state to a free state across measurable energy "band gaps" in strict accordance with quantum rules. These rules give electrons identifiable and controllable features that can be used to convey information.

With controlled pulses of electrons down wires, computers could be interconnected around the world. With controlled flows of electrons in and out of tiny capacitors, computer memories could be constantly read, written, and restored. With minute charges of electrons in silicon, computer transistors could be switched on and off. The most studied phenomenon in physics, electrons are constantly measured, manipulated, traced, aimed, and projected. Yet throughout the history of science, the electron has remained a humbling perplexity. Let us listen, and find out what it is telling us about the bizarre abundance of the domains beyond matter.

At the time of Bohr's breakthrough, the electron was assumed to be matter. By contrast with the photon, the source of its energy, it showed measurable mass. But in 1923 Louis de Broglie developed a wave theory of the electron and Einstein endorsed it. In 1926, Erwin Schrödinger contrived a famous equation, still widely used by engineers and physicists today, that describes the electron's wave behavior. The following year, Werner Heisenberg offered a set of strange mathematical matrices derived from particle concepts that explained electron behavior as well as Schrödinger's wave equation did. Schrödinger then strengthened the theory immeasurably by showing that Heisenberg's particle paradoxes were mathematically convertible into his own wave equation. Like light, electricity proved to be a flow of wave-particles.

As if to banish any chance of a revival of Newtonian theory at the heart of the atom, Heisenberg then presented a radical new concept called the Uncertainty Principle. He asserted the intrinsic impossibility of ascertaining at once both the momentum and location of an electron. Thus Heisenberg showed that the necessary parameters of Newtonian equations were necessarily unmeasurable at the atomic level.

In subsequent years, the hugely more massive protons and neutrons of the atomic nucleus were also shown to exhibit wave action. Wave behavior was clearly essential to the quantum domain. Many investigators began to see the world as consisting of simple waves of energy that could consolidate into mass. But this precarious trench of materialism also became untenable.

Crossing decisively into the microcosm, Heisenberg declared that the waves which Bohr had examined in recreating the atom were not conventional waves at all. Designated "probability amplitudes," they were waves or fields that defined the statistical likelihood of finding an electron at any particular location.

This was a climactic step in the overthrow of materialism in physics. With the electron itself depicted as a wave and the wave depicted as a probability field, the specific particle in this theory had disappeared into a cloud. With it disappeared the last shreds of Newtonian logic and mechanistic solidity. As Bohr put it, quantum theory required "a final renunciation of the classical idea of causality and a radical revision of our attitude toward the problem of physical reality."

Sixty years later, the quantum cloud has still not been dispersed. Nobel Laureate Steven Weinberg recently summed up the current state of the argument:

> The inhabitants of the universe were conceived to be a set of fields
> —an electron field, a proton field, an electromagnetic field—and
> particles were reduced to mere epiphenomena. In its essentials, this
> point of view has survived to the present day and forms the central
> dogma of the quantum field theory: *the essential reality is a set of fields*
> subject to the rules of special relativity and quantum mechanics; all

else is derived as a consequence of the quantum dynamics of those fields.

Einstein made the point even more strongly: "There is no place in this new kind of physics both for the field and matter, for the field is the only reality."

What is the field? As Einstein explains, "Instead of a model of actual space-time events, it gives the probability distributions for possible measurements as functions of time." Although actual waves measured by physical apparatus, the fields of atoms are less fields of force than fields of information.

This idea of information fields cannot be translated into a physical analogy. A probability wave is a wave—a detectable force—but it is also an idea, an index of the likelihood an electron will appear at one place rather than another. It is a field of information.

The information remains full of paradox. But the paradoxes all derive from the materialist superstition. All analysts want to retain the last purchase of solidity, the concept of a particle. But at the most fundamental level, there is no such thing as a particle.

What are called elementary particles are neither elementary nor particles. If elementary is taken in the Newtonian sense of indestructible matter, Heisenberg points out, electrons are not elementary because they dissolve on contact with their antimatter twin, the positron. Envisaged as more intense points moving through a wave of probability, electrons do not even retain any particular substance.

The most famous of many efforts to resolve such perplexities came in experiments suggested by Niels Bohr, based on Thomas Young's original two-hole scheme for showing the wave nature of light. Once again, researchers shoot an electronic "gun" at two tiny holes in a screen in front of an electrically sensitive target. Scientists first assumed that wave interference arose when the "waves" of electrons collided in some way. But the assemblage of hits on the target can show Young's wave interference pattern even when electrons are shot one by one minutes or hours apart. The waves are not "crowd effects"; they are intrinsic to each quantum particle.

In some experiments, a single quantum particle passes through both holes at the same time, thus exhibiting wavelike non-locality. In some crucial microelectronic products, from diodes to non-volatile memories, electrons use probability waves to "tunnel" through barriers that they are incapable of either penetrating or surmounting as physical particles.

Richard Feynman—Caltech's inventor of the current consummation of the theory, quantum electrodynamics (QED)—believed that electrons can jump momentarily backward in time. "Things on a very small scale," he wrote, "behave like nothing that you have any direct experience about. They do not behave like waves, they do not behave like particles, they do not behave like clouds, or billiard balls, or weights on springs, or like anything you have ever seen."

Such results apply to all quantum entities, from quarks to photons. In explaining them, the consensus of quantum theorists, wrought over fifty years of intense debate and experiment, excludes all materialist logic. The common popularizations of the theory are all wrong. The paradoxical duality of wave-particles is *not* an effect of the measuring apparatus. Probability waves are *not* a manifestation of hidden variables that researchers have failed to detect. The paradoxical results are *not* a phenomenon reducible to simpler and more intelligible materialist terms. There is no concealed clockwork in the quantum. In the usual way, quantum physics does not make materialist sense.

Quantum physics can make sense, however, if it is treated in part as a domain of ideas, governed less by the laws of matter than by the laws of mind. The words used to describe atomic phenomena offer a clue. "Reflection," "absorption," "non-locality," and "probability" are concepts that accord as much with thoughts as with things. While things have mass that is always conserved, opposing thoughts can cancel each other out; sympathetic ideas can resonate together in the mind. The paradoxical stuff of the microcosm, atomic phenomena seem to represent the still mysterious domain between matter and mind, where matter evanesces into probability fields of information and mind assumes the physical forms of waves and particles.

Even paradox, a perplexity for things, is a relatively routine property of thoughts. Conceiving of the quantum world as a domain of ideas, we make it accessible to our minds. The quantum atom is largely an atom of information.

The shift between Newton's materialist clockwork and Feynman's quantum enigmas—between Newton's sure embrace of sensory data and Feynman's dismissal of common experience—is a Copernican break in modern history. In the early sixteenth century, Copernicus displaced the earth from the center of the universe. Early this century, quantum physics displaced the human senses as the central test of reality.

This sacrifice of the sensory knowledge of matter, however, yielded a bonanza of technological opportunity. Newtonian physics was intelligible to human beings; but it rendered the solid state unintelligible and opaque. The paradoxical propositions of quantum theory are utterly unintelligible to the human sensory system; but they render the inner space of matter both intelligible and manipulable.

The "informativeness" of subatomic matter is the key to modern electronics. Because the world of the microcosm consists not of inert and opaque solids but of vibrant, complex, and comprehensible fields, it constitutes a useful arena for modern information technology. It is because there is so much analogical information in the microcosm that the microcosm is a uniquely powerful medium for information. The quantum-regulated movement of electrons across quantum-mapped crystalline paths epitomizes information technology, and information technology epitomizes the quantum era.

In the atom of information, this era acquires its definitive symbol. What was once a blank solid is now revealed in part as information, what was once an inert particle now shines with patterns and probabilities, what was once opaque and concrete is now a transparent tracery of physical laws. Far from plunging reality into clouds, quantum theory makes the universe radically more intelligible.

The new science does not estrange human beings from their environment. Since thought is the most distinctly human power, the quantum world is actually more anthropomorphic than the world of Newtonian masses and forces.

A more intelligible universe, penetrable by the human mind, endows people with greater power to create wealth. But it also radically changes the way wealth is created. Throughout previous human history, the creation of wealth depended chiefly upon the extraction, transport, combination, and modification of heavy materials against the resistance of gravity, the constraints of entropy, and the constrictions of time and space. When things are large and approached outside-in, it is expensive to move and manipulate them. Their costs derive from the weight, rarity, entropy, and resistance of their matter. But small things, virtually devoid of matter, move less like weights than like thoughts. In the microcosm, the costs of fuel and materials decline drastically; the expense devolves from matter to mind. Just as quantum science overthrew Newtonian matter in the explanation of the universe, the quantum economy overthrows Newtonian matter in the creation of new wealth.

From the immateriality of the microcosm comes a further great benefit: the awesome precision of microelectronic technology. Although probabilistic in nature, quantum waves are incomparably more predictable than the waves in your bathtub. Such macrocosmic phenomena as turbulence in the sky or in a furnace or in a tear running down your cheek defy measurement and prediction by present human powers. The new approaches of chaos science at last are opening useful avenues of understanding. But the quantum physics of microcosmic phenomena is far more accurate and reliable than any macrocosmic process.

Thoughts can be incomparably more accurate than things. Quantum physics, as Feynman points out, achieves a greater correspondence between theory and experiment than any previous science. He estimates that such quantum computations as the magnetic charge of an electron are comparable in accuracy to a measurement of the distance between New York and Los Angeles correct to the width of a human hair. Indeed, it is the immense precision of microcosmic data compared to macrocosmic chaos that accounts for the increasing efficiency of electronic components as they become even smaller and more densely packed together, moving ever deeper into the quantum domain. It is this precision of the microcosm that makes possible the overthrow of matter in technology and economics.

That microcosmic devices would be faster and better and more reliable than macrocosmic machines was little understood in the early years of the electronic era. The unreasonable effectiveness of microcosmic technology was asserted and largely explained in the 1960s by Carver Mead, a man who deserves the title of prophet of the microcosm.

CHAPTER

2

The Prophet

The information industry entered the microcosm almost as an accident. The seeds were sown in 1959 in Pasadena, at California Institute of Technology, home of Feynman and Pauling . . . and Carver Mead.

In the 1950s and 1960s, as Mead remembers, "Computers were still big ugly things." Rococo cathedrals of refrigerated wires and tubes, costing millions of dollars, they demanded the fealty of thousands of servants to the most minute whims of their machine languages. Using this technology, human beings had first to line up for access, then bow to a rigid mechanism, following the intricate liturgy of a priesthood of experts. Rather than tools or chips, computers were shrines.

During the 1960s, protests against this priesthood shook Berkeley, the computerized campus to the north. Lampooning the IBM punchcards then used to feed computers, students wore pins declaring: "I am a human being. Do not spindle, fold, or mutilate." But student protests were more successful against the Vietnam War than against the mainframe computer. For decades giant computers seemed a permanent fact of life.

Still based on the architecture first propounded in a memo by John von Neumann in 1945, the computer's central processing units

(CPUs) worked on programs and data kept in a separate storage system. Proceeding through its tasks step by step, fetching data and instructions one at a time from relatively remote memories, the entire system was bound together with miles of copper wire.

Through the 1960s and 1970s, the makers of new computers would slowly replace the Von Neumann components, one for one, with solid state switches. But the Von Neumann architecture remained supreme, on its pedestal in the central processing room, still mostly an electromechanical device, wrapped in wire and refrigerated, and still essentially a part of the macrocosm.

Nonetheless, working as surely as Planck's constant on the foundations of nineteenth-century physics, microcosmic ideas were infiltrating the fabric of twentieth-century technology. A Japanese physicist named Leo Esaki passed through Pasadena in 1959 and left behind him a quantum concept. It led to a flawed product, a failed research project, and a new vision for microelectronics. Through a long struggle, still under way, that vision would transform the industry, then the global economy, and finally the politics of nations.

The flawed product was called a tunnel diode. Invented at Sony in Japan in the mid-1950s, the tunnel diode won a Nobel Prize for Esaki, its inventor. The prize was for demonstrating a rare quantum effect in a seemingly practical device. As it turned out, the device was not practical. But it dramatically announced to the electronics world that the quantum era had begun.

The failed research project was Carver Mead's. In response to Esaki's Pasadena visit, Mead resolved to perfect the tunnel diode. He failed. But his failure bore fruit more important than any prize. Still in his early twenties, Mead found in this flawed device the secrets of the quantum era and led the way into it.

Named from the Greek words meaning two roads, an ordinary diode is one of the simplest and most useful of tools. It is a tiny block of silicon made positive on one side and negative on the other. At each end it has a terminal or electrode (route for electrons). In the middle of the silicon block, the positive side meets the negative side in an electrically complex zone called a positive-negative, or p-n, junction.

Because a diode is positive on one side and negative on the other, it normally conducts current only in one direction. Thus diodes play an indispensable role as rectifiers. That is, they can take alternating current (AC) from your wall and convert it into direct current (DC) to run your computer.

In this role, diodes demonstrate a prime law of electrons. Negatively charged, electrons flow only toward a positive voltage. They cannot flow back against the grain. Like water pressure, which impels current only in the direction of the pressure, voltage impels electrical current only in the direction of the voltage. To attempt to run current against a voltage is a little like attaching a gushing hose to a running faucet.

It had long been known, however, that if you apply a strong enough voltage against the grain of a diode, the p-n wall or junction will burst. Under this contrary pressure, or reverse bias, the diode will eventually suffer what is called avalanche breakdown. Negative electrons will overcome the p-n barrier by brute force of numbers and flood "uphill" from the positive side to the negative side. In erasable programmable read-only memories (EPROMs), this effect is used in programming computer chips used to store permanent software, such as the Microsoft operating system in your personal computer (MS-DOS). Avalanche breakdown is also used in Zener diodes to provide a stable source of voltage unaffected by changes in current.

Esaki, however, took diode breakdown into the quantum domain. Applying less than one fifth of the avalanche voltage, he made electrons move through the junction at some five times the speed. In an appropriately prepared device, the Esaki electrons would violate all the rules of electricity and simply bore through the barrier, tunneling from the positive to the negative side. Relatively large currents would flow well before any avalanche breakdown began.

Indeed, running the diode through all voltages up to the point of breakdown, researchers showed that the Esaki current was entirely separate from the avalanche current. Following the burst of Esaki current and before the avalanche, the voltage could double and triple. But the tunneling flow would *drop* some 80 percent. Flouting Ohm's Law, the most basic rule of electricity, this drop in current with a rise in voltage added yet a further mystery to Esaki's magic tunneling.

Magic or not, however, the tunnel diode was both scientifically intriguing and commercially exciting. Because electrons tunnel far more quickly than they normally flow in semiconductors, the Esaki burst effect promised extremely fast switches, approaching the speed of light. All things being equal, the faster the switch the better the computer, which uses vast arrays of on-off switches to perform its high-speed calculations. For a long period in the fifties and sixties, Esaki was the toast of the industry.

Defying the usual rules of matter, tunneling was also a prime interest of quantum physicists. It was as if a high jumper with a normal seven-foot limit had a certain restricted but statistically reliable probability of being able to leap over a twenty-foot barrier, provided you didn't get him too excited. Or perhaps a better analogy is a football player with a statistically dependable likelihood of being able to bull through a brick wall without breaking it. But in the quantum domain, all such analogies founder.

In essence, the electrons tunneled through the wall in the form of quantum probability waves before they slowed down, assumed mass, and began their flow as particles which the barrier could stop. Showing that the probability theory of quantum effects could countervail the most sanctified rules of electricity—and create a superfast device to boot—Esaki had achieved an important vindication of the new physics. It was the kind of surprising phenomenon that fascinated the physicist in Carver Mead.

Despite Esaki's brilliant papers, the taciturn foreigner had little to say at Caltech. He left Mead's laboratory suffused with the fumes from his chain-smoked cigarettes and spurned Mead's effort to recruit him. Instead, he took a fellowship in the research labs at IBM, where he remained until his retirement more than a quarter century later. But his visit in 1959 had a profound effect on Mead, prompting him to launch an intense investigation of tunneling effects.

Mead pursued these efforts for close to a decade and wrote some important papers. At first his interest focused on the possibility of making more effective diodes or other devices. He tried to determine how far an electron could be made to tunnel. But soon this work led him to a more general line of exploration.

Tunneling only occurs at dimensions measured in angstroms. An angstrom is truly a quantum measure: one ten thousandth of a micron. The usual yardstick of the microcosm, a micron is one millionth of a meter, or between one seventy-fifth and one hundredth of the width of a human hair. An angstrom thus is close to one millionth the width of a human hair.

Plunged so deep into the microcosm, Mead gained an intuitive sense of the quantum domain and reached an amazing conclusion for all semiconductor electronics. The industrial world might be telling him to invent a faster diode. But the technology was telling him a way to transform the industrial world.

During that period in the mid-1960s, most of the leading electronic laboratories—from Bell and RCA in New Jersey to IBM in Yorktown Heights and the research centers of Silicon Valley—were seeking new ways to accelerate switching speeds. Most analysts also recognized the benefits of integrated circuits, joining many transistor switches on one piece of silicon. With more components on each chip, each component became cheaper to build. But as the 1970s approached and transistor dimensions dropped below 10 microns, the industry concluded that the process of miniaturizing electronic circuitry was approaching its limits. Further gains would have to come chiefly from accelerating the speed of existing components.

In the prevailing view, not only would smaller switches be more difficult to build, but as they approached micron size, they also would suffer all kinds of transient effects and defects, particles and distortions, that were not significant in larger devices. A notable paper from Sweden accurately predicted that below 10 microns cosmic rays or alpha particles from the package itself would disrupt the circuits.

Smaller things, crowded more closely together, it seemed obvious to the industry, would be more delicate and problematic to make and to operate. The usual notion was that with higher speeds and greater densities, the silicon chips in which the switches were embedded would eventually heat up and melt. This belief was not foolish. Judging from the heat-emission levels of chips at that time, silicon devices with just ten times more transistors would seethe like a frying pan over an open fire.

Carver Mead, however, in his years in pursuit of the elusive tunnel diode, had plunged deep into the microcosm, listened to what the technology was telling him, and emerged with a radically different message: a set of Mead's laws—contrary to the laws of the world— that would order the future of microelectronics. As you move down into the microcosm, he revealed, everything gets better as it gets smaller, cooler as it gets faster, cheaper as it gets more valuable. As the traffic of electrons becomes denser, speedier, more complex, and more plentiful, the number of accidents drops, defects decline, and nothing ever wears out.

Mead found that none of the expected threats to miniaturization— neither heat, nor reliability, nor low-threshold voltages, nor silicon defects—actually posed a serious problem. Instead, in a great irony, the chief obstacle to smaller devices turned out to be tunneling itself, long cherished as the key to a fast switch. Spontaneous tunneling occurred if devices were shrunk beyond a certain point—with dimensions a small fraction of a micron and far smaller than all previously conceived limits. Mead came to see this spontaneous quantum tunneling as the *only* prohibitive limit to scaling down the features on these microchips. Once the p-n junctions become so thin that spontaneous tunneling occurs, the device will no longer turn off; its barriers will be permeable to electrons. But his researches revealed no other fundamental limits to miniaturizing transistors.

Mead estimated that transistors could operate safely with dimensions as small as a quarter micron, or 2,500 angstroms. One four hundredth the width of a human hair, this would be forty times smaller than the industry then believed possible.

This unexpected bonanza stemmed from the basic differences between microcosm and macrocosm. Chips consisted of complex patterns of wires and switches. In the macrocosm of electromechanics, wires had been simple, cool, reliable, and virtually free; switches were vacuum tubes, complex, delicate, hot, and expensive. Wires were routinely mass-produced; vacuum tubes were intricately crafted. In the macrocosm the rule was, the fewer and faster the tubes, and the more wire to serve them, the better the circuit. But in the microcosm, all the rules changed.

In the microcosm, switches are almost free—microscopic transistors inscribed in silicon. Wires become the major problem. Longer wires laid down on the chip and more wires connected to it translated directly into greater resistance and capacitance and more needed power and resulting heat. The shorter the wires, the purer the signal and the smaller the resistance, capacitance, and heat. As electron movements approach their mean free path—the distance they can travel without bouncing off the internal atomic structure of the silicon —they get faster, cheaper, and cooler. As Mead showed, tunneling electrons, the fastest of all, emit virtually no heat at all.

The overall results seemed a new quantum paradox. The smaller the space, the more the room. Space and time seem to expand to meet the demands put upon them. The microcosm overthrew the usual constraints of the material world.

Mead presented his initial findings in 1968, nine long years after Esaki's visit, to an industry workshop at a resort in Missouri called the Lake of the Ozarks. As the young researcher made his arguments to a dubious audience, his excitement increased. Flying back to Pasadena, he computed some of the effects in greater detail on the back of envelopes. Eventually he was to show that scaling down the web of switches on a chip by a factor of 10, you get no increase in power usage or heat dissipation, and thus no silicon meltdowns. But you get ten times more speed and one hundred times more transistors per unit area. Meanwhile the power-delay product—the key index of semiconductor performance, combining switching delays with heat emission —drops a thousandfold. With the power-delay product collapsing, reliability soars. With transisitor size dropping, production yields mount and costs plummet.

All these benefits were available with just a tenfold reduction in the size of the features on the chip. But it was clear to Mead that a fortyfold drop could be achieved, meaning a drop by a factor of hundreds of thousands in the power-delay product. The conclusion was monstrous and irresistible. Electronic circuits made of untold thousands or even millions of transistors on single silicon chips would rule the world. All the development efforts on tunnel diodes and other exotic devices became suddenly irrelevant. Other acclaimed technologies, such as magnetic bubble memories and Josephson Junctions, celebrated at IBM and Bell Laboratories, would never achieve any widespread markets.

The industry would thrive with soaring improvements in cost-performance ratios. The problem would not be heat or speed or reliability or cost; it would be designs. How would it be possible to design chips with hundreds of thousands or even millions of transistors on them? What would they do? One thing they would do, it was clear to Mead, was transform forever the computer industry.

As Mead foresaw in 1968—three years before it was done— it would soon be possible to put an entire computer on a chip and sell it for a few dollars. The semiconductor industry would have to convert itself from a manufacturer of components to a maker of entire systems on single chips. This would mean an entirely new kind of microchip, moving the industry to new domains beyond matter, into complex challenges of systems design and microcosmic manufacturing.

Born in 1934, near Bakersfield in Southern California, Mead is a descendant of the first settlers to run cattle into the Kern River Valley. Son of the manager of a remote electrical power station for Los Angeles, Mead as a boy earned money fishing and trapping in the nearby rivers and forests. But as a teenager, afflicted with a bad complexion, disdainful toward the sports and enthusiasms of his peers, spurned by the svelte California girls in his classes, he retreated to the microcosm. For modern pioneers it was a frontier as spacious and intriguing as the West had been for earlier American pioneers.

As a high school freshman he learned calculus and began befriending members of the electronics club at the local community college.

Eventually he entered Caltech and studied with Pauling, Feynman, Max Delbrück, and other giants of the quantum era. As a Caltech freshman, Mead learned from Pauling the elegant permutations of the chemical bond, the quantum sources of solid state, the foundations of his career. But his most important personal lesson came from Feynman.

At the time, Mead was suffering a small but persistent canker of doubt about his scientific prowess. Surrounded on all sides at Caltech by mathematical wizards—making chalk cluck and squeak and leaving chicken tracks all over the blackboard—he wondered whether there was room in science for less poultry, more poetry. Early in Feynman's own career, Enrico Fermi and Robert Oppenheimer had provoked the same doubts in the young man from the Bronx. But as a Caltech professor, Feynman had found that his own intuitive method was often indispensable to achieving new truth.

Mead says simply today: "Feynman gave me confidence to go ahead and do science. . . . He taught me that the first and most crucial step was usually to think about what the problem is trying to tell you, rather than tell it what you know already about the math."

This approach impelled Mead to a lifelong effort to escape the momentary claims and crises of his field and to see the thing whole: to transcend the common sense of the day, the dense traffic of convention, the ways of the wealthy wisemen of the Valley, and uncover the deeper meanings of the technology: "to sort out what the thing was really trying to tell me from the things that the experts were telling me."

In 1968, when Mead got back to Pasadena from his presentation at Lake of the Ozarks, he immediately assigned his most brilliant available graduate student to get to work calculating more precisely all these marvels of the microcosm. Mead himself began to enrich his vision.

In the decade since Esaki's visit, Mead had done more than investigate tunnel diodes. He had invented the first commercially viable transistor made of gallium arsenide, a compound material that offered speed advantages over silicon. He also had been introduced to Fairchild Semiconductor by Gordon Moore and had become a noted consultant in Silicon Valley. There he gained a valuable perspective on the commercial aspects of the technology.

Broadening his academic horizons as well, he had studied biology with Max Delbrück. Intrigued by the recognition that the nerve membrane works like a transistor, Mead spent a summer at Coblenz in

Germany exploring the links between physics and biological systems. In 1968 he had also embarked on a study of computer systems design, a field most other solid state theorists had regarded to be beyond their domain.

From all these experiences, he began to weave together a powerful prophecy for the future of the computer industry. A gnome of quiet voice, pointed beard, and kindly smile, he would bring his vision— tenaciously, trenchantly—to prestigious men who disdained it, to old friends who bitterly opposed it, to allies who betrayed it. His role is recognized in no current history of the field. But carried by hundreds of his students throughout the industry, by thousands of his readers, and by many unconscious followers of his logic, his word became law. In the end the microelectronics industry would transform itself— against the resistance of some of its most powerful leaders in both the United States and Japan—to conform with his message.

Writing in 1968, he described "the dilemma of the computer industry": "It has an enormous investment in big machines and big software programs, and the only thing the industry can do right now is to use the new microelectronics as it fits into the existing system. . . ." To use chips, that is, as replacement parts that add power and speed to the existing big and ugly machines.

Mead continued, "We have computer power coming out of our ears. What we need is the kind of systems we would like to have in our automobiles, in our telephones, in our typewriters—where people now spend vast amounts of time on the repetitive and mundane operations involved in keeping track of a lot of little things. . . ."

Mead contrasted the lordly computer to the car. "A person doesn't feel dehumanized by such a machine—one that frees him from routine tasks and is under his control. . . . An automobile is a machine that gives you a lot of power, yet it leaves you as much in control as if you were walking." He predicted that the new silicon technologies would allow creation of small computers that brought power to the people as automobiles did, rather than bringing it to large institutions and sophisticated programmers, as mainframe computers did.

Mead offered the analogy of the electric motor, "invented at a time when most industries had a big steam engine out in back driving a big shaft the length of the factory. . . . Even though it was perfectly clear that the way this innovation should have been used was to put electric motors on each machine . . . the most that could be done economically was to replace the big steam engine with a big electric motor."

Small electric motors eventually came to inject power everywhere into the world's tools, its appliances, and even its toys. Mead predicted that small computers could similarly inject human intelligence into the entire environment of human beings.

Beginning his argument with an analysis of the angstrom dimensions of electron tunneling, Mead had proceeded to a radical case for the transformation of the computer industry. In the autumn of 1968, he was summoned to the living room of his friend Gordon Moore to offer advice to Moore and Robert Noyce, who were then preparing to form Intel Corporation. To them, too, he made his argument about the obsolescence of the large computer and the need to escape the predicament of design complexity.

As a portent of the future, Mead pointed to Hewlett Packard's new programmable calculator—the HP 9100. A machine the size of a typewriter, it would soon find its way onto nearly every engineer's desk and later as a hand-held device into the pocket of every student of engineering. It both performed an array of scientific calculations and could be used to control HP's line of scientific instruments. Mead urged Noyce and Moore to develop chips and design techniques for the coming era of dispersed computing power.

The two men agreed with Mead about the future of computing. But they were initially wedded to a contrary strategy for their new company, replacing the current electronics of the mainframe with faster and cheaper semiconductor devices operating within the old design. The two entrepreneurs wanted to begin by exploiting a clear existing market. But the future would not wait. Within three years, Intel would become the leading instrument of the new market predicted by Mead, providing all the key components of the personal computer.

As a result, in most companies today, all information workers now have computers on their desks. To accomplish a computing task, no longer is the user forced to join a queue outside the doors of the data-processing center waiting for access to the lordly mainframe on a pedestal inside. In the end, the 1960s Berkeley revolt against the mainframe largely triumphed. The microcomputer revolution has redistributed computing power onto the desks of nearly every student, office worker, and corporate manager.

Mead celebrates this achievement. But he believes it represents only a first step in a more far-reaching agenda for information technology. For while computer power is now dispersed throughout most companies and around the world, there still thrives in miniature, in the inner structure of the machine itself, the old tyranny of the data-processing center.

It is a secret scandal of computing. Enshrined inside nearly every computer, large or small, is its own central processing room. Data and instructions queue up endlessly in registers, buffers, caches, and stacks waiting for access to a single lordly central processing unit (CPU). Most of every computer, most of the time, is in a wait state, scuttlebutting and flipflopping, taking turns at access to the CPU. Modern CPUs are extremely fast but nevertheless constitute a drastic bottleneck.

It is the original computer architecture of John von Neumann, far more pervasive today than ever before. In 1947, when he proposed it, and in the 1960s when it provoked protest at Berkeley, the bottle-neck was a necessary economy. The industry operated in the macro-cosm of electromechanics where switches were expensive vacuum tubes and wires were nearly free. Von Neumann's computer architecture reflected these relative costs of switches and wires in the macrocosm. He economized on vacuum tube switches for processing and was lavish with interconnections.

Moving to the microcosm, however, the Von Neumann compo-nents—storage, interconnect, and processor—all can be made of the same sliver of silicon. The computer on a chip outperforms the com-puter on a pedestal and costs one millionth as much. Switches are now virtually free—microscopic transistors inscribed on a chip. Today, as Mead declares, it is wire—once free—that has become costly in every way, clogging the chip with complex metals hard to lay down, subject to deterioration from heat, and difficult and expen-sive to link to the world.

The microcosm of free switches and expensive wires dictates a re-versal of current practices. Rather than wiring up one superfast pro-cessor to lots of supporting chips, the new technology dictates combining lots of processors on each chip. The microcosm favors innovation designed to channel data to large numbers of switches processing in parallel.

Yet the Von Neumann architecture still widely prevails. The com-puter remains the same lopsided technology it was when information workers lined up in long queues outside the central processing rooms of large corporations. Nearly all the processing power is devoted to manipulating data rather than producing or acquiring it and getting it into the machine. Yet the effectiveness of any system is determined by the speed and accuracy of input and output as much as by its millions of instructions per second (MIPS). And input and output cannot drastically improve as long as the computer interfaces with the world one bit or even one byte (eight bits) at a time.

To overcome this problem, Mead urges the industry to listen to the technology. For computer theorists seeking the secrets of interfacing —seeing, hearing, recognizing, and sensing the world—the only relevant technology is the human brain.

The computer industry, however, has long been blinded to this truth by its own cherished set of materialist superstitions. What MIT's Gian Carlo Rota called the "naive prejudice" of a system of logic "patterned exclusively on the structure of physical objects" has long been the only logic of computer science. Computer theorists have long imagined that by creating a logical machine they could simulate the brain.

From John von Neumann and Herbert Simon to John McCarthy and Marvin Minsky, many of the pioneers in the field believed that the computer's capacity to perform logical functions—based on its already stunning superiority in calculating speed—would lead to a machine more powerful than the brain in every critical way. And in the course of a passage through the domains of higher thought, from chess and mathematics to the diagnosis of disease and the composition of poetry, the computer would relatively easily attain such subordinate goals as sight, hearing, speech, and learning.

Seeing logic as the main role of the brain, leading theorists supposed that sensory powers were by-products of logical powers. Yet the computer has already surpassed the brain in logic without making major progress toward sensory competence. Both direct investigation of the brain itself and the evidence of the evolutionary paradigm suggest the clear reason: the brain is only incidentally a logic machine. The senses—hearing, sight, smell, touch—all came first, long before thought, and the brain is primarily set up to service them. It is increasingly clear that the brain's logical capacity grew in response to the complexity of its sensory apparatus.

To become fully effective tools, computers must undergo the same evolution. The key problems coming to the forefront of the computer industry, from speech recognition to missile defense, are chiefly perceptual rather than computational. Rather than focusing chiefly on logical functions, computers and their masters should become humble students of the brain's perceptual, associative, and adaptive powers.

The essential agenda was given by Mead's teacher, Max Delbrück. A particle physicist who began his career working with Niels Bohr in

Copenhagen, Delbrück in 1969 went on to win the Nobel Prize for Physiology. By then already a professor of biology at Caltech, Delbrück declared in his Nobel lecture: "Sensory physiology in a broad sense contains hidden at its kernel an as yet totally undeveloped but absolutely central science: transducer physiology, the study of the conversion of the outside signal to its first 'interesting' output."

Transducer physiology is the domain of the sensory faculties. It is the way the microcosm of the brain interfaces with the macrocosm of matter. It is the prime function of the brain itself, the world's most effective quantum technology.

The most amazing of all quantum two-hole experiments, after all, is the one that is performed routinely by two human eyes. The webs of wetware convert the onrush of photons and other sensory data into the absorptive, reflective, superposing, and interfering flux of mind and memory we call thought. Computer theorists have been humbled to discover that one human being can perform more image processing than all the supercomputers in the world put together. They have found that a baby in its mother's arms performs feats of learning and pattern recognition that defy large arrays of today's computers. They described computer operation as faster than thought. They were right. Computers *are* faster than thought. But they still must wait for humans to do the most crucial thinking.

The age of the microcosm began by overthrowing the human senses as a test of material reality. But it can be consummated only by giving the computer a rudimentary sensory system, partaking of the quantum powers of human perception. If human beings are to consummate the overthrow of matter and pursue their own higher creative and intuitive roles, the computer must be taught to perform routine sensory functions, identifying and learning patterns, recognizing speech, structuring information, and inputing data. Releasing humans from bondage to the keyboard and the screen, the computer can help restore the primacy of the human mind in both science and industry.

The brain could never function using the Von Neumann architecture. But Mead's early discoveries in the microcosm enabled a serious drive to imitate the architecture of the brain in massively parallel machines. Then by collapsing the computer to invisibility and imbedding it in the matter of everyday life, man may impregnate the world with his mind and waken it to the sound of its master's voice. Such in essence was the prophecy that was emerging in the mind of Carver Mead.

The history of the quantum era would chiefly unfold as the fulfillment of this essential prophecy. But its roots were in the past, at Bell Laboratories, where executive vice president Mervin Kelley decided after World War II to "take advantage of the understanding of solids made possible by quantum mechanics," and launched a huge quantum web of wires and switches.

CHAPTER

3

Wires and Switches

William Shockley worked in the midst of the world's widest mesh of wires. They reached across the continent, spanned the Atlantic Ocean, and spread in trunks and branches to nearly all points of the industrial world.

The intellectual center of this miraculous web, spread out to catch the voices of nations, was Shockley's employer, Bell Laboratories. Moving intelligible signals through these wires and around the globe represented an immense feat of engineering science.

In 1947, however, most of the switching of these signals was still done by human operators plugging patch cords into switchboards. The global telephone network still depended for intelligence on the wet wires and switches, axons and neurons of the human brain.

Like the first computers emerging in that era, telephones in 1947 were still mostly electromechanical systems, entrenched in an industrial macrocosm of heavy machinery attended by hundreds of thousands of regimented workers, reciting by rote: "Number please." Statisticians calculated that at the current pace of expansion of telephone use, all the women in the country would soon be telephone switchers.

By the early 1950s, the mechanical switches were rapidly giving way to electromechanical crossbar and vacuum tube systems that could automatically respond to dialed numbers. Sure enough, all the nation's citizens did become operators . . . of their own phones' dialing devices. But as the number of telephones increased additively, the number of possible connections multiplied. The problem shifted out from taking calls to routing them to their destinations.

The challenge of finding the best routes resembled the classic traveling salesman problem. Not only did determining the shortest among increasing millions or even billions of possible paths confound human beings but even an approximate solution strained the computational resources of computers. Soon AT&T's labyrinth of wires was served by the world's most formidable computer-switching systems.

The computers were themselves new, more compact ganglia of wires and switches. In the telephone firm, central switching systems captured the signals from dialed numbers, sought their addresses, defined the routes between them, reconciled conflicts, checked line performance, and collected billing data, in much the way that a business computer might capture the signals from a keyboard, search memory addresses for data, process it in accordance with spreadsheet formulas, and transmit results to screens, printers, and remote memory addresses, using a variety of networks, including probably the phone network itself.

Both systems are logic machines. In the telephone network, each switch is a decision point that sums up all previous decisions in the chain that links one caller to another. Like a syllogism in logic, it reaches the one right answer through a concatenation of other right answers. Is the handset off the hook, yes or no? If yes, connect switch for dial tone. Then follow the number, digit by digit, each opening a new level of the network as it branches through the logic chain, summing up previous switching choices in the form of an open channel and actuating new choices or switching decisions that open a further channel, step by step to the destination.

Similarly, the millions of switches inside a computer are decision points for the complex calculations that it performs. In a computer the on-off positions of these switches represent one and zero (and thus on a base-two system all numbers), or true and false and yes and no in the binary logic of the machine. Although these are minimal bits of information, each can be assigned a specific meaning of any needed complexity by a consistent software language. Then the accumulation in a logical code of millions of such decisions can figure your sales tax and estimate it for the next five years, pore through *War and*

Peace to find how often Andrei talks to Natasha or Pierre talks to God, or manage your voice through thousands of switches from your phone in Queens to the Queen's phone in London.

The more numerous, the faster, and more efficient the switches, the more powerful the logic of the system. In 1947, however, the only available switches were electromechanical gadgets or vacuum tubes that collectively needed the power of a locomotive to pull a voice around the world or tug a differential equation to its solution. The size, expense, and power requirements of such switches placed an oppressive upper limit on the capacity of all logical systems, from telecommunications to computers.

Switches are the substance of the artificial mind. As long as the switches were macrocosmic, the millions required in any complex logical process, from word processing to telephone routing, would bulk too large and allow too many compounding defects to function efficiently. In overthrowing matter and creating a new technology of mind, engineers would have to dissolve this substance into the microcosm.

Today, across the entire telephone labyrinth, from its supercomputer central switches to its microcomputer desktop autodialers, voice mailers, modems, and facsimile boards, nearly all the switching is performed by sophisticated computer systems consisting chiefly of integrated circuits and advanced microprocessors. Each of these integrated circuits and microprocessors, moreover, is itself a compact and intricate microsystem of tiny wires and switches.

From a far-flung but finite geometrical system of wires, interconnected by human operators, the telephone system has exploded into a gigantic fractal array. That is, it is self-similar at every level. It repeats the same essential pattern of wires and switches, lines and nodes, on every scale of size and operation from continental waveguides to microchip connections.

The worldwide pattern of cables and huge ESS-5 central switching computers is repeated in the pattern of open wires and smaller switches at local end offices, and again in the twisted wires and Private Branch Exchanges in commercial buildings, and again on the printed circuit boards inside the PBXs, and again in the repeated patterns of logic gates on the microprocessors on the printed circuit boards, and finally in the complex vias, contacts, gates, and channels of the physical chip layout.

All the way down the hierarchy of scale, the system becomes more dense with repeated structural details of wires and switches. In accord with the rule of such fractals and the law of the microcosm, the richness of informative detail increases far faster than the drop in size as the system moves toward the quantum domain. Thus the power of the logic and the features of the system improve far faster than the switches multiply. And nearly every one of the infinitesimal switches inscribed by the many trillions throughout the telephone system—on integrated circuits, microprocessors, memory chips, signal processors, controllers, converters, multiplexers, demultiplexers, clocks, counters, and thousands of other devices—is a *transistor*. The transistor is the microcosmic switch. Crucial to the overthrow of matter in both computers and telecommunications, it was defined and impelled by the researches of William Shockley at Bell Labs in the late 1940s. Now informing a global system of wires and switches, transistors control an endless quantum interplay of electrons and photons comprising the essential fabric of modern information technology.

Combined in branching nets of logic spread across minute slivers of silicon, millions of these wires and switches comprise a computer. Stretched across the mostly silicon surfaces of continents, the wires and switches comprise a telecommunications system. Fused into a global ganglion of interconnected tools, the wires and switches of both computers and telecommunications join to form the central nervous system for a new world economy. Eventually, in the form of fiber-optic cables, it will become chiefly a web of light and glass, in which messages flash around the world with no material embodiment at all.

The invention of the transistor was the critical first step in the invention of this new technology of mind. Understanding this historic achievement, you will doubtless have no trouble identifying the man, with a last name beginning with "*Sh*" and ending "*ley*," who was featured in *The New York Times* in its issue of July 1, 1948, the day after the invention of the transistor was announced by Bell Laboratories.

The answer, of course, is George Shackley, composer of "Anthem for Brotherhood," which was to be performed the following Sunday on the New York television station WPIX.

Below the Shackley story, the *Times* did tell of the transistor announcement. But it failed to mention the inventors of the device, William Shockley, Walter Brattain, and John Bardeen. This oversight was not peculiar to the *Times*. Most of the American press failed to cover the Bell announcement at all.

Thirty years later, the microcosm was still a small matter to the *Times* and much of the rest of the American media. In 1978, as part of a thirteen-volume encyclopedia of Great Contemporary Issues, the *Times* published a volume of stories from the paper entitled *Science in the Twentieth Century*. The editor of the volume was Walter Sullivan, the *Times*'s distinguished science editor, who later became a fervent follower of the microcosm. The book included entries on Neil Armstrong, the atom bomb, the nuclear submarine, the atom smasher, Jane Goodall, the Great Crab Nebula, Richard M. Nixon, Mars, J. Robert Oppenheimer (ten pages), the Piltdown Man, P. C. Steptoe, Werner von Braun, the yeast fly, Sputnik, Women Astronauts, Lewis Thomas, solar energy, marmots, Mariners 2, 4, 5, 6, 7, 9, 10, the Glomas Challenger sea derrick, Yuri Gagarin, Gerald R. Ford, Dmitri Balyayev, Carl Sagan, Trofim Lysenko, and yes, indeed, William Shockley, the inventor of the junction transistor and the pioneer of solid state physics.

But wait. There is something wrong here. Shockley's entry is devoted entirely to his cranky theories on the genetics of intelligence. There is no mention of the transistor or of its other inventors. In the late 1970s, the fabulous tales of the microcosm were still considered less fit to print than the saga of nuclear-powered steam engines and protests against the archaic pretense of white racial superiority.

Science in the Twentieth Century is merely a symbol, and a relatively unimportant one at that, of a strange ignorance of the most important feats of the modern age. When Bell Laboratories announced the invention of the transistor, it expected a general uproar of interest and the threat of classification from the Pentagon. But the Pentagon was indifferent and George Shackley took precedence at the *Times*. History would take some thirty-five years fully to grasp the importance of Shockley's incredible shrinking switch.

The transistor is a switch. Yet it may be defined as simply two wires, crossing. A small force in one of the wires controls a large force in the other. Rather than requiring the intervention of a cumbersome mechanical device to direct the current and reset the switch, the transistor uses electrons to control electrons. It attains this capability solely by operating in the very microstructure of the solid state materials from which it is constructed. In the microcosm, the wires are the switches.

Necessary to building switches for information technology— whether neural nets in the brain, computer systems, or telephone networks—are two key requirements. One is gain, or amplification. Gain is crucial because power is constantly lost as the signals proceed through the wires and switches. Without amplifiers with controllable gain, power would steadily dwindle by resistance and capacitance in the wires and the result could never be computed or the call completed.

In a transistor, gain may be defined as the ratio between the controlling force and the larger working current. In a faucet, an analogous switch or valve, the gain would be the ratio between the force on the handle and the flow of water from the spout.

The second key requirement for information technology, from the life-giving molecules of DNA to the solid state of the transistor, is crystalline regularity in the structure of the medium. Microcosmic devices, which cannot be seen or investigated directly, must operate with absolutely dependable regularity in time and space. The movement to the microcosm thus entailed manufacture of large amounts of pure and reliable substances of a regular microstructure. Information technology would require a crystalline medium.

In practice, the crystalline medium would have to be a semiconductor, for only in the microstructure of semiconductors is it possible to build a solid state switch with gain, that will set and reset itself without moving parts.

Semiconductors got their name from a peculiar skittishness in conducting electricity. As the quantum physicists showed, the conductivity of a substance springs from the number of free or detachable electrons in it. Rubber and glass, for example, are both insulators; all their electrons are so tightly bound to the nuclei of their atoms that the materials will break down or dissolve before conducting. These substances have no free electrons at all; their "conduction band," as it is called, is empty. Most metals, on the other hand, are full of free electrons and conduct easily under nearly all circumstances.

Semiconductors are unique in that they can be made to switch between insulating and conducting modes. Moreover, because a semiconductor contains relatively few electrons free to carry charge, proportionately small changes in its condition—adding or subtracting relatively few charge carriers—can produce large effects in its conductivity. Thus semiconductors offered the promise of gain or amplification: a large and manageable increase in electrical output resulting from a small change in a controlling circuit. But until 1948, no one could figure out how to make a reliable amplifier from a semiconductor. When it happened, among all the wires and switches at Bell, William Shockley was not altogether pleased.

. . .

The imperious young scientist with the stern black spectacles and the spare handsome features beat his fist on his desk at Bell and declared: "I am a better physicist than anyone working for me."

Well, yes, Dr. Shockley, that may be true. But it is John Bardeen, the quiet and shy theoretician, and Walter Brattain, the kindly silver-haired veteran at Bell, who have invented the transistor. Moreover, it is a design that has virtually nothing in common with the field effect (or voltage-controlled) transistor you have been trying to build for many years, following the inspiration of Julius Lilienfeld, who patented such a device in 1926.

True, as far as anyone knows, Lilienfeld never built one that worked. But neither have you, Dr. Shockley. In fact, Dr. Bardeen has shown that contrary to your theory, it is impossible to switch or amplify a current by applying a voltage or field effect to a semiconductor. Electrical pressure will not do it alone; you need an actual current. The surface states—all the dangling chemical bonds on the interface between the voltage source and the semiconductor—prevent the voltage from controlling the flow of electricity.

Shockley knew all that. He had ordered the program to investigate the surface phenomena. The invention of the transistor, he wrote, was an example of "creative failure methodology." The failure of the underlings in their effort to create a field effect transistor would lead to the creation of some other splendid device by Shockley, the brilliant manager of the research program. In fact, Shockley would go on to create a grown junction transistor.

But Bardeen and Brattain were supposed to be creatively failing in field effect devices, not sneaking off to make epochal inventions unbeknown to their brilliant young boss. For a while it looked as if Bardeen and Brattain might receive most of the credit—even the Nobel Prize—and Shockley remain an obscure manager at Bell Labs. In that case, the *Times* would not even notice if he came up with bizarre theories of white supremacy.

Indeed, John Bardeen, later willing to go back and start again in a new research program in superconductivity, would eventually become the only man to receive two Nobel Prizes in the same field. Some observers might reasonably question whether Shockley was in fact a better physicist than any of his subordinates in 1948. Superconductivity, moreover, might prove as important in the end as all these little switches. History can surprise you. Better get to work, Bill.

Shockley did get to work and history would notice. In addition, he had a legitimate complaint. The prize-winning point contact transistor invented by Bardeen and Brattain was to a considerable degree a retrograde development.

Two of the three electrodes (routes for electrons) on the point contact transistor were conventional metal wires, their points delicately in contact with a semiconductor base; it was only partly a solid state device. In fact, it hearkened back to the whisker diode, used in crystal sets for crude radios, more than forward into the age of the integrated circuit.

Shockley was partly right that this transistor, announced by Bell in 1948 and responsible for Brattain's and Bardeen's Nobel Prize, set back the cause of electronics. Difficult to control, hard to explain, and balky to manufacture, the point contact device was just menacing enough to prompt a massive new effort to improve vacuum tubes but not sufficiently superior to pull the industry into the new era.

Shockley's concept of a field effect device was simpler and easier to miniaturize. It was unipolar; it would operate entirely on one semiconductor block. Thus for most purposes it was preferable even to subsequent Shockley bipolar transistors. Rather than running two flows of *current* into a complex arrangement of three semiconductor materials joined in two p-n junctions, the switch worked by applying *voltage* to an insulated gate on top of the block. The electrical pressure or voltage—the field effect—opened a channel for electrons below the insulated gate.

Years later, engineers necessarily turned to Shockley's field effect concept when they rid themselves of all excess baggage for their descent deep into the microcosm. Today, Shockley's dream dominates the industry. But in the late 1940s, Shockley could not build it. So he turned to second best: an effort to create what was called a "grown junction" transistor.

The grown junction transistor was a bipolar device. This meant—in Shockley's new and entirely solid state concept—that the base was of a different polarity (positive or negative) than the emitter and collector on each side of it. All these elements would have to be grown on or "doped" into the crystal. This would take some doing. Although the materials specialists at Bell had made great gains in growing crystals, the process was still unreliable; essentially they lined up the suspects, tested them, and tried to find substances with the appropriate "dirt." If Shockley's grown junction device was to supplant the point contact transistor as he intended—and make him once again the solid state champion of Bell Labs—he would have to come up with a way to mass-produce the grown junctions. But he was making little progress. It appeared that Shockley, this intellectual titan widely regarded as the most brilliant scientist in solid state electronics, would fail to get a product out the door and would be eclipsed in history by two of his subordinates.

Almost two years later, however, in 1950, Shockley's project was saved by a stubborn young solid state chemist from Texas named Gordon Teal. Working with molten tubs of semiconductor glop, Teal had persisted doggedly in a long effort to create pure single crystals of material. Most of his colleagues, including Shockley, wondered at first why it was worth the trouble, since semiconductor devices could be made perfectly well from polycrystalline or amorphous substances.

Yet Teal's effort was indispensable to Shockley's success with the grown junction transistor. Through a process of alternately doping the material with p- and n-type impurities as it was being pulled from the melt, Teal was able systematically to produce an ingot of germanium already appropriately layered for a junction transistor. Shockley took it from there.

By slicing the crystal across the junctions, attaching wire leads to the positive regions at both ends and a control circuit to the n-type base in the middle, Shockley showed the way to manufacture grown junction transistors. His first devices—the p-n-p's—were based partly on the use of positive "holes" as carriers of charge. In electronics, a hole is a positive charge left by the absence of an electron in an atom of the semiconductor crystal. Shockley showed that positive holes— moving like bubbles through water—could pass through a semiconductor nearly as fast as free electrons. Western Electric produced these grown junction devices for a decade; they completely eclipsed point contact transistors and laid the foundations for a prosperous and important semiconductor industry.

Also in 1950, Shockley published his book *Electrons and Holes in Semiconductors*, which stood for many years as the definitive work in the field and confirmed his credentials for the Nobel Prize that he shared with Brattain and Bardeen in 1956. The fact was that for his theory of the field effect transistor that later dominated the industry and for the junction transistor that was dominating it at the time, Shockley deserved the prize alone. He had at last made his point.

Yet Shockley was not satisfied. "Every physicist," he said at the time, "wants two things: glory and money. I have won the glory. Now I want the money." A superb scientist and inventor, with some ninety patents to his name, Shockley gained glory in a research setting. But like many intellectuals, he greatly underestimated the disciplines of money.

Shockley sought a sure thing. He began by simply demanding a million-dollar fee for his consulting services. That approach failed. Then he started the first semiconductor company in Silicon Valley and attracted to it from across the country what would prove to be the leading business and technical talent to emerge in the field over the next thirty years. Included were Robert Noyce and Gordon Moore, later to found both Fairchild Semiconductor and Intel. But like so many of the venturers who would follow him, Shockley wanted a proven market. Proven markets tend to be markets that are already being filled. For providing small improvements in proven markets, the entrepreneur gains small fees at best, not the glorious riches Shockley desired.

As Peter Drucker has pointed out, a new technology cannot displace an established technology—with its installed base of plant, equipment, training, personnel, and satisfied customers—unless the innovation is about ten times more cost-effective than its predecessor. The established technology will have far more than ten times the effort and momentum behind it and will continue to improve fast enough to prevent the success of any marginal invention.

As a result of this mandate for a tenfold gain, the world is full of bitter inventors with conspiracy theories about the hostility of corporations to their terrific ideas—at least "five times as good as the established product." Five times as good is no good if the new device requires drastic changes in an existing industry. As a discrete power device, the Shockley transistor was too small an improvement over existing technologies to succeed except in a few niches.

Although Shockley had worked at the center of the greatest information system in the world, he never clearly saw that the transistor could succeed as a bearer of information but not as a smaller and more efficient power device. At Shockley Semiconductor, he sought to build replacement parts such as four-layer thyristors to "control every light switch" or supplant every telephone relay, rather than to launch logic tools that could digitize entire systems and yield compounding gains of ten and more for every network of wires and switches in the world.

Shockley always wanted to use his unique knowledge of solid state materials to create clever discrete devices that were low in information content but could control ever larger amounts of physical force. The microcosm demanded numerous low-power devices integrated to control ever larger amounts of information.

Shockley and his colleagues had solved a key problem of information technology. The transistor could both represent information and send it to several destinations. It provided both a bit and a boost: a crucial switch and amplifier both to register the ones and zeroes, bits and branches of binary logic, and to transmit them to the world. To go with its binary uses as a switch, the transistor had gain and fan-out; the output of one transistor could activate several other transistors. Moreover, the transistor reset itself, returning to its original state when the power was removed. With gain, fan-out, and automatic reset, information could cascade through the system without complex replenishment or other outside intervention.

This was a major step. Lack of gain and resetting capabilities would ultimately doom many fast devices otherwise supremely attractive for computation. Tunnel diodes, Josephson Junctions, and optical logic devices, for example, all operate at or near the speed of light. Each can represent a bit. But they lack gain, so they can't boost it or send it out to many other points in the machine, and they often cannot reset themselves. Thus these devices cannot function alone in extended computations without awkward supporting systems. So information technology everywhere still relies on Shockley's far slower transistors.

Shockley and the Bell Labs leadership, however, had underestimated the second key requirement of information technology: a pure crystalline substance to be manufactured in volume. To overthrow matter, it would be necessary to convert matter into a medium, to create matter pure and regular enough to serve as a transparent vessel of data, matter so symmetrical that in itself it would constitute a form of information. Then it would be necessary to produce it cheaply in large volumes.

The impulse to achieve this breakthrough would be unlikely to come from laboratory science itself. Pure scientists can thrive on ingenious prototypes, winning fame, wealth, and prizes. Scientists could get by without a mass-produced crystalline medium. But the creation of a new technology of mind would require hundreds of trillions of perfectly replicated wires and switches. It would entail a manufacturing breakthrough.

To turn this elite science into the fractal fabric of a mass technical culture would requre an entrepreneurial compulsion. The transistor would have to reshape the business environment, and be reshaped by it. It would have to feed upon the creative interplay betwen science and society, technology and enterprise, that is the systole and diastole of economic growth.

Quantum science was vital to the fulfillment of Mead's vision. But equally important was the market, joining producers and consumers in a new interplay of learning. It turned out that the new information technology would emerge most powerfully not at the exalted centers of research where priorities were defined by the scientists such as Shockley, but at small firms struggling to find products to sell.

First it would be Shockley's Bell collaborator, Gordon Teal, defecting to Texas Instruments, who would create the essential elements of success. Finally it would be a group of defectors from Shockley Semiconductor, called by Shockley the "traitorous eight," who would pioneer logic devices at Fairchild and Intel and launch an entire industry to overthrow matter.

PART

The Technology of Mind

II

CHAPTER

4

The Silicon Imperative

The common sand that you tread underfoot, let it be cast into the furnace to boil and melt and it will become a crystal as splendid as that through which Galileo and Newton discovered the stars.

—VICTOR HUGO, *Les Misérables*

Using electrons to control electrons, Shockley's transistor was the first microcosmic switch. Making sense and syntax from the evanescent entities of the quantum realm, the new invention moved the world well beyond conventional matter. The transistor provided the first major bridge between the quantum revolution against matter in physics and a quantum revolution against matter in technology. Crossing this bridge would be thousands of inventors and entrepreneurs who would ultimately overthrow matter in business structure and in the world economy as well.

Governing this process of creation is a prime truth of classical economic theory called Say's Law: supply creates its own demand. Invented by the Frenchman Jean-Baptiste Say, the law usually refers to the fact that—including wages, rents, and profits—the payments made in producing a good or service create enough buying power to purchase it. As a theory of the cyclical flow of funds, Say's Law is a foundation of the equilibrium in most economic models and a source of the stability of capitalism. But in practice, Say's Law also unleashes powerful forces of disequilibrium. The prime disequilibrium is called economic growth. It is a process greatly enhanced in the microcosm.

As the driving force of economic growth, Say's Law exalts the creativity of suppliers over the wants and needs of demanders or consumers. As entrepreneurs invent new things and learn how to make them more efficiently, unit costs and prices drop and goods become more attractive. As goods become affordable to a wider public, more people work to acquire them by creating goods to exchange. These new suppliers both provide and acquire new wealth at ever lower expense.

The key is not wants and needs. Wants and needs are as ubiquitous as poverty, and as easy to produce. The key to wealth is always the creativity and diligence of ever-widening circles of suppliers, led by inventors and entrepreneurs.

This truth is difficult for many people to accept. An optical illusion leads people to believe that wealth inheres in physical things they can see rather than in ideas that are invisible. This materialist superstition seems especially plausible wherever scarce and immobile matter predominates in an industry, whether coal, land, gold, iron, or oil.

Under these circumstances, the power to seize or sequester material resources seems the route to wealth. Existing establishments of landowners or resource suppliers dominate society and change is realtively slow. Cost reductions seem to come chiefly from rising efficiency in extracting and manipulating materials and exploiting physical labor. As a scarce resource is used up, its price will even rise in accord with the law of diminishing returns.

It is human inventiveness that lends value to raw materials (even oil was useless gunk until 1855). But the rarity and weight of such inputs impose a clear limit on the productive process. People come to revere the substances themselves more than the inventiveness that gave them worth.

Until the end of World War I, support for the materialist view could even be found in the statistics of growth. Increases in labor productivity came chiefly from the substitution of energy for labor. U.S. output grew more slowly than growth in the use of energy. This meant that while labor productivity was increasing, energy productivity was actually declining; it took more and more energy to achieve a specific advance in labor output.

Materialists could claim that industrial progress was in a sense a sham: a zero-sum game, where the gains of the winners are precisely offset by the losses of the losers. The human race was paying for its economic progress by the exhaustion of irreplaceable fossil fuels. Entropy—the inexorable conversion of usable energy into waste and pollution—was the rule of both thermodynamics and economics.

From this point of view, the current generation of humans was merely stealing the real wealth of the earth from future generations. As socialists said, property was theft.

Since World War I, total output in capitalist nations has been rising far faster than the consumption of energy. But materialists could continue to claim that the gains were spurious. Irreplaceable fuels were being underpriced because future generations could not bid for them. Moreover, energy benefits were offset by pollution and other so-called externalities. In the ecological balance sheets, economic growth still seemed a Sisyphean quest.

In the microcosm, however, all these materialist claims collapse. Ideas are not used up as they are used. Where intellect is the decisive source of value, the economic burdens of matter decline and costs can follow Mead's laws of the microcosm. Space and time expand as size and power drop. In the age of the microcosm, the inventive inputs of producers launch a spiral of economic growth and productivity at steadily declining cost in every material domain: land, energy, pollution, and natural resources.

Say's Law returns with ever-increasing effect. Liberated from the burdens of matter, Atlas bears the globe not on stooped shoulders but on the crests of creative thought. Sisyphus hurls aside his boulder and plunges beyond gravity, entropy, and friction into the vast inner space of matter. Shifting from the manipulation of materials to the generation of ideas, entrepreneurs launch a spate of novelties, radically diminishing every material cost—from weight and space to toxic waste.

Entrepreneurship is a creative process, and by its very nature, creativity comes as a surprise to us. To foresee an innovation is in effect to make it. If creativity were not unexpected, customers could demand it and expert planners could supply it by rote. An economy could be run by demand. But an economy of mind is necessarily impelled by Say's Law, driven by the unforced surprises of human intellect.

It is a spontaneous process of discovery, of "creative failure," as Shockley put it, of "creative destruction," according to economist Joseph Schumpeter. Under any name, however, innovation tends to devalue the materials of the established system and create a new means of production with a higher content of intellect and ideas. This displacement of materials with ideas is the essence of all real economic growth.

On the manufacturing level, the growth of intellect and ideas in the means of production is often measured in terms of learning or experience. As producers enlarge output, they learn how to produce more cheaply and well. Impelled by these increases in the real value of the commodity, demand for it increases. Quantified as a "learning curve," this concept shows that efficiency in manufacturing any product increases some 30 percent with each doubling of accumulated volume. Not only physical economies of scale but also efficiencies of learning and experience, metaphysical and inexhaustible, lead to lower costs.

Just as important as the producer's experience in Say's dynamic of growth is the learning curve of the customers. As the price of a product drops, customers learn new ways to use it and improve it. The customers' knowledge may increase as fast as the producers'. Particularly in impelling new technologies, the experience of customers is as important as the experience of the producers.

Immured in expensive laboratories, academic and industrial researchers tend to eschew this entire interplay of learning between manufacturers and users. Thus, in an intellectual environment outside the marketplace, products miss the spiral of increasing knowledge that steadily increases the share of intellect in the output of goods and services.

As philosopher Michael Novak has written, capitalism is the system under which the key source of wealth is the *caput*, Latin for head. Capitalism is supremely the mind-centered system. Joined with microelectronics—the mind-centered technology—capitalism is reaching a new creative pinnacle. The microcosm, entirely a phenomenon of capitalist societies, is now eroding the significance of the world's physical capital and exalting this true source of capitalist success. Nowhere was this dynamic so clear as in the semiconductor industry itself.

The elite scientists at Bell never gave Gordon Teal full support in his efforts to open the way for mass production of transistors. Few noticed when in 1952 he answered the siren call of the entrepreneur. Responding to an advertisement in *The New York Times,* he decided to return to his native state to run an electronics lab for a small firm in Dallas, later to become Texas Instruments (TI).

With the possible exception of Sony Corporation in Japan, the Texas firm was the least promising purchaser of the rights to Bell Labs' transistors. But Teal's new boss, Patrick Haggerty, far inferior to Shockley as a scientist, commanded the intuitive sense of the microcosm that Shockley lacked. Haggerty was searching for a proprietary product that would give the company, then chiefly a vendor of oil services, an entree into what he saw as the coming era in electronics.

As Haggerty put it, "I was trying to conceive of how one could approach it in a way that would be fundamental . . . like being at the center of a sphere, where there were opportunities—and problems, of course!—in all directions.

"The heart of electronics," as he saw it, "was the valve that controlled the flow of electrons," and that valve was still the vacuum tube. After contemplating a number of specialized tube products, however, Haggerty began to sense that the sphere was subtly but irrevocably shifting. Although the spread of television was pushing vacuum tube revenues to new highs, Haggerty felt that this technology—along with firms that relied on it—was becoming peripheral.

When he heard the good news from Bell Labs at Murray Hill—the invention of the transistor—Haggerty immediately grasped the key idea of the microcosm: that the "future of electronics" lay in the manipulation of knowledge at the "level of the structure of matter." Recently a military man, in a country then mobilizing for war in Korea, Haggerty resolved at once to bring the U.S. defense industry into the era of solid state.

At that time a regulated monopoly, Bell was under pressure to share useful inventions. When it offered to license its semiconductor discoveries, TI immediately signed up. Then in 1952 at a conference for its licensees came the bad news from Murray Hill.

Bell's germanium transistors could not operate at the high temperatures or in the harsh environments required by military and oil services work. An ex-military man in an oil services company, Haggerty was crushed. But Gordon Teal suggested silicon as an alternative transistor material for high-temperature applications. The world's leading expert on germanium, Teal displayed a rare entrepreneurial willingness to inflict creative destruction on his own wealth of expertise.

With a larger band gap (a greater quantum distance between the valence band and the conduction band of its electrons), silicon would operate more slowly than germanium, and thus might be less useful for radios and other high-frequency functions. The high band gap of silicon also made it much more difficult to work with; it did not melt until 1200 degrees Celsius. But the larger band gap would give silicon greater stability under heat and other harsh conditions. Rejecting the consensus view that silicon transistors required a base too thin to be manufactured, Teal set to work on silicon in January 1953 in a former bowling alley on Lemon Avenue in Dallas.

Haggerty's vision reached beyond silicon transistors themselves. As he explained, "I was sure that this was a volume business. . . . The silicon transistor would take care of this military, environmental problem. We could probably get started that way, technically, in small quantities. . . . Still unless you found a way to do it in volume," to make it a commodity, "you weren't going to stay in business. It was a volume business, and you were going to be forced into volume processes, doing things at low cost."

In the Navy, Haggerty had encountered the concept of the learning curve. Lower costs, Haggerty reasoned, meant lower prices and expanding market share in a spiral of benefits. Haggerty was seeking a high-volume, low-cost product that could bring the firm down the learning curve in semiconductor electronics, gaining experience that would be invaluable wherever the sphere might move.

He named the product confidently: a portable radio that would use fast and tiny germanium transistors in place of vacuum tubes. Haggerty thus was committing his small squad of engineers, few boasting any credentials in solid state electronics, to mass-produce transistor devices from two radically different semiconductor substances, slow silicon and fast germanium. Moreover, one of his products, the transistor radio, would be TI's first consumer product.

To compound TI's problems, the military began giving financial aid to the competition. The Signal Corps inaugurated a grant program of several million dollars annually to the established firms to improve transistor manufacturing techniques. The amount seems small today but it was significant in the 1950s electronic business.

Yet TI overcame all obstacles. As the years passed, the vacuum tube giants—RCA, Sylvania, GE, Westinghouse, Philco-Ford, Raytheon —with their long head starts, their growing government subsidies, and their large military contracts, all would leave the semiconductor trade. In all the annals of business history, there are few triumphs so grand and improbable as TI's domination of this central electronic technology.

Mark Shepherd, the Dallas policeman's son who later took over the firm from Haggerty, offered as good an explanation as any: "Those companies all knew the things that weren't possible. We didn't. We were stupid."

In the wee hours of the morning of September 25, 1954, two of the heroes of the story, together with an increasingly distracted onlooker, joined in one of the more improbable scenes of the drama. On one side of the room was the mechanical hero. Six feet high, three feet wide at the girth, it looked a lot like a popcorn vending machine left over from the lobby of the Lemon Avenue Bowling Alley. This was Gordon Teal's crystal-growing device, a unique machine, perfected at TI, which Haggerty believes gave the firm a two- to three-year lead in process technology. It had a small motor on top running a pulley down the broad transparent neck into a quartz crucible, glowing like an illuminated jar of melted butter. From this glowing vessel were pulled the pure molten crystals of germanium or silicon at the heart of microcosmic technology.

Sitting nearby, also some three feet wide, and looking uncomfortable, was Mrs. Mark Shepherd, who the next day would bear her second child, Mary K. Crouching attentively was her husband, his face creased with a near-parental concern for the condition of the crystal-pulling machine. Many hours and some one thousand tries before—in fact, on his third attempt—he had managed a major break-through in crystal processing, crucially needed for the creation of an adequate radio. Shepherd had made what is termed a rate grown germanium transistor. But if he could not grow another one, it might be written off as a fluke. For the last several hours he had promised that at any moment he would pull another rate grown device, if he could "hit the right set of parameters. . . ."

Shepherd would have given up and gone home except that the future of the company in consumer markets was hanging in the balance, in that glowing womb of quartz, and he had had such a tantalizing success on his third try. At 3:00 A.M., on try 1,005, he finally found the clue to the combination, finally figured out the problem. Attempt 1,006 produced another perfect rate grown transistor. All the next day, Shepherd was as proud and happy and tired as any attentive young father.

Although TI eventually lost the transistor radio business to Sony, the Texas company sold several hundred thousand portable radios at a time when all the leading radio firms had decided that portables were unmarketable. Most important, TI honed its semiconductor fabrication skills and demonstrated its powerful technology. The radio had been designed as much to send as to receive a signal, and its key message, said Haggerty, was the announcement of a technology and a company: "That transistors were usable now, not some years in the future, and that we were ready, willing, and able to supply them."

Up in Armonk, New York, Tom Watson, Jr., leader of IBM, received the message, bought a hundred or so radios to distribute to his staff, and remarked to an assistant: "If that little outfit in Texas can make these radios work for this kind of money, they can make transistors that will run our computers too." IBM eventually became TI's biggest customer and the world's biggest producer of semiconductors.

For all the effect and portent of Haggerty's rash rush into germanium transistor radios, though, the most important event at TI in 1954 was Teal's success—also considered unlikely for several years—in producing dependable silicon transistors in volume. Like the radio, the silicon transistor gave TI an important monopoly position. Unlike transistor radios, however, silicon semiconductors would remain a TI bastion for decades.

Silicon was adopted because it was less sensitive to heat than germanium. But it ultimately prevailed because, unlike oxides of other substances, silicon dioxide was both electrically and chemically inert and thus could both protect and insulate the devices of near-micron dimensions that would appear within the next two decades. The substance of sand, silicon was made in large quantities in polysilicon form by firms such as DuPont and later TI. Then, through methods perfected by Teal, it could be rendered in perfectly regular and variable crystals, suitable to bear and process information.

Without silicon, the move to the large-scale integrated circuits and microprocessors that shape the information age and run your Macintosh or IBM personal computer would have lagged for decades. As DNA was the crystalline substance bearing the informative codes of life, silicon was the crystalline medium of the microcosm.

Bell Labs had worked with silicon, but it lacked the entrepreneurial compulsion to develop the tools to manufacture pure silicon crystals in volume. Metallurgists at the firm believed that such single crystals were unnecessary and their production much too expensive for manufacturing. Yet it was no accident that Teal had come from Bell Labs. He was a leading fundamental scientist from Brown who exemplifies the utter dependence of information technology on the quantum revolution.

Without the light shined into the atomic structure by the quantum physicists, the world would never have achieved the understanding of the microstructure of materials on which Haggerty based his business plan. Without Teal's deep command of the quantum properties of silicon and germanium single crystals, TI could not have mastered their mass production. Teal was following closely in the footsteps of Pauling and other giants who provided the rigorous quantum calculus at the heart of modern materials science and thus made possible the industrial overthrow of matter.

Teal, however, was not only a master of solid state chemistry, he also was mastered by a vision of mass production and profits. Over a year of frenzied effort at TI, he worked out the concepts for mixing selected impurities into the molten silicon and for growing crystals suitable for slicing into transistor blocks.

Gordon Teal launched this new medium in a famous announcement, offhandedly astonishing the crowd, in Dayton, Ohio, on May 10, 1954. The nominal event was the National Conference on Airborne Electronics, but the real event, so Teal resolved, was the sudden consummation of a silicon coup.

Gordon Teal was the next to last speaker on the program. A quiet man who speaks in a low voice, he followed several other reasonably soporific scientists. It is likely that any flickers of excitement that rippled through the crowd as the former Bell Labs luminary approached the podium were extinguished as he began a droning rendition of his thirty-one-page address. Teal was very tired, having stayed up until 5:00 A.M. rewriting the final sections to include his announcement. But the title, "Some Recent Developments in Silicon and Germanium Materials and Devices"—though mildly provocative in the precedence of silicon—gave no hint that the man with the rumpled suit and the bulge in his right pocket had anything extraordinary to divulge.

To some, it might seem strange that a man would travel to Dayton, Ohio, to declare the overthrow of the establishment. But Teal's audience, assembled at Dayton's Engineering Club Auditorium, comprised a group of the unacknowledged legislators of our age. The leading American media would not notice them unless they protested the military budget or defected to Russia, yet they constituted a perfect group for Teal's theatric purposes. Technical press, electronics engineers, Pentagon officials, all increasingly conscious of the rising promise of transistors, they were ready to nod knowingly at the conventional wisdom and then stiffen abruptly at the sudden announcement of revolution. They could be depended upon to publish it to the relevant world.

As the earlier speeches proceeded, Teal made a thoroughly exhilarating discovery about his audience. Significant parts of it were vocally impatient with the limitations of germanium devices. They had been taught to expect miracles from semiconductors, but miracles at room temperature alone would not suffice for anyone whose products had to perform under more extreme conditions.

For example, at this conference on airborne electronics, many of the engineers were dealing with airplane engines and avionics, and with missiles and other weapons that were used outside the office environment. There were also representatives and reporters from a spectrum of other industries. Yet germanium transistors would be useless in car motors, stoves, oil-logging equipment, and even computers used in harsh conditions. Several men on the program already had disappointed the crowd with predictions that silicon transistors could not be manufactured for four or five more years. After giving a long recitation of the marvels of his germanium products, one speaker bristled when his first questioner said, "Yes, fine, but when are you going to have the silicon transistor?"

"You are not paying attention," the speaker reproached him, "to our good germanium transistor."

Teal was duly grateful for such unwitting contributions to the drama of his silicon announcement. As he listened, he scrawled—and doubly underlined—a bold claim at the end of his speech: "Contrary [to] the opinions expressed in the morning session, this will begin immediately." In the margin he placed two exclamation points. But as he began to speak he managed to conceal his growing excitement.

Most of his first twenty-four pages gave a lulling account of the properties of germanium on which he was the world's leading savant. Then he began his discourse on the higher-priced spread: "Many laboratories have been very active in the study of silicon and investigating its possibilities as a transistor material. Substantial progress has been made at Texas Instruments. . . . The work to be reported is the result of collaborative investigation by W[illis] A. Adcock, Morton E. Jones, Jay W. Thornhill, and Edmund D. Jackson.

"They have successfully constructed NPN silicon grown junction transistors and have developed the process to a point such that our company now has three types of silicon transistors in production. . . . They forecast an exciting future for silicon materials and devices, and one which will strongly affect circuit and apparatus designs in the years to come. . . ."

After Teal's statement, he recalls, a member of the audience in the middle of the auditorium about eight or ten rows from the stage stood up as if on cue and asked incredulously, "Did you say you have silicon transistors *in production*?"

"Yes," Teal responded amiably, "we have three types of silicon transistors in production. In fact," he said, "I just happen to have some here in my coat pocket." He took several out and raised them up to be seen, a tired but exhilarated man, with arms thrust high like a wearied boxer who has just won a decision.

But the show was not over. Lest anyone in the audience miss the significance of his announcement or doubt its veracity, he went backstage to get some props for a demonstration. He brought out a small record player with some of its innards dangling from the side and set a beaker of boiling oil next to it. The dangling device, he said, was a germanium amplifier. When he placed the needle in the grooves, the swinging strings and clarinet of Artie Shaw's "Summit Ridge Drive" pealed through the room. Teal allowed the technically saturated crowd of engineers to enjoy a brief respite of melody. Thereupon he plunged the germanium amplifier into the boiling oil. The sound made a sickening swoon to silence. Then he replaced the germanium with a silicon transistor and repeated the experiment. The lilting music of TI's triumph blared through the hall.

Although there is no report of any dancing, *Fortune* magazine described "a stampede for copies" of the speech (Teal said a few were available "in case some of you are interested") and a rush to the phone booths. "One man from Raytheon," wrote *Fortune,* "put in a call to his executive vice president and was heard in the booth, croaking hoarsely: 'They got the silicon transistor down in Texas!' "

As it turned out, TI, with Teal's crystal puller, was the only place that had silicon transistors in effective production for nearly five years. Moreover, most of the companies that began to challenge TI after 1958 scarcely existed at the time of Teal's announcement; Raytheon and the other tube companies were never again a major force in the market. The Bastille had fallen. Silicon, the most common substance on earth (being earth itself), would inherit the world of semiconductors and hurtle the TI juggernaut up Summit Ridge Drive, casually taking up, *en passant,* the last man on the Dayton list of speakers. "He was in trouble," recalled then TI chairman, later Dallas mayor, Erik Jonsson, "he had no audience left, and to this day I don't know who he was." It was the kind of thing that happens with a decisive shift of the sphere.

Unit sales of germanium devices for mild environments still dwarfed the silicon numbers—as they would until the age of integrated circuits—but TI could sell its silicon devices for nine times the germanium price. Spearheaded by soaring sales of silicon transistors, TI's revenues rose from $24 million in 1954 to $232 million in 1960, and profits rose at a pace of 200 percent a year.

TI had provided the crystalline medium and the high-gain mechanism for the age of information and had rushed down the learning curve to turn them to gold. But the silicon transistor was still what

Carver Mead had called it as a teenager. It was a replacement part, looking back hungrily at vacuum tube slots rather than forward to the age of the microcosm. Individual discrete transistors were relevant to the ultimate promise of the technology chiefly because they provided training in the ways of silicon.

The silicon transistor was the key to the cosmos only because millions could ultimately be packed together on single chips and insulated by their own oxide. Packed so tightly together, slow silicon transistors would operate incomparably faster than any assemblage of speedier germanium devices. No one at TI at the time could even imagine such a thing.

CHAPTER
5

The Monolithic Idea

In the move into the microcosm—Planck's invisible sphere—the idea of combining many transistors and other circuit elements on a single piece of silicon, one chip, marked a point of no return. An engineer could see and handle single discrete transistors and connect them to other devices with ordinary copper wire. He could still imagine that he was working with ordinary materials in a visible world. He was still manipulating the solid stuff of his sensory experience. But putting an entire system of electrical components on one chip the size of a fingernail was a new challenge, which required the crossing of a great divide. The integrated circuit would take the industry down a slippery slope from the familiar shores of the senses into a quantum sea.

Yet this invention was not made at a major industrial or university laboratory full of quantum physicists and expensive equipment. Making the plunge instead were two engineers at two small companies, following the logic of information technology wherever it should lead.

One of these engineers arrived early in 1958 at the Dallas headquarters of Texas Instruments. A tall, quiet man named Jack Kilby, he had been designing small systems for hearing-aid companies at a firm in Wisconsin called Centralab. He would apply unpackaged transistors and other devices to a ceramic substrate and then connect them by depositing metal lines and resistors on the surface of the ceramic by a silk-screen process. Although far inferior to true integrated circuits—ICs on single chips—this hybrid approach was more efficient than wiring together separately packaged devices. Kilby's idea was appealing enough that as late as 1964, IBM, in a moment of typical conservatism, adopted it for the breakthrough 360 mainframe computer series.

Kilby's first assignment in Texas, however, was to develop "micromodules" that seemed actually inferior to the ceramic systems he had built in Wisconsin. The micromodule concept envisaged encasing each transistor or other component, together with all its wiring, in a separate plastic package. These identical modules could be plugged together like Lego blocks. The key appeal of the micromodule to TI was its sponsor, the U.S. Army. Its appeal to the Army was its secure footing in the known world of the macrocosm. With careful directions, even a small child could assemble any specified circuit from these fungible units.

The problem was, so it seemed to Kilby, that the small child would grow old and gray before any important project was finished. By the time Kilby arrived at TI, military systems for such functions as missile control, space travel, supercomputers, airplane avionics, and other complex uses entailed many millions of components. Built out of so-called micromodules, they would fill skyscrapers. The Army's proposal in fact illustrated the futility of macrocosmic approaches to microcosmic technology.

Still, the Army wanted it and was willing to pay, so Kilby set out to design appropriate modules . . . and if they were tiny enough, perhaps . . . if they were cheap enough, possibly . . . they could be dumped on the market to sell as miniature Lego blocks at Christmas. Then, in July 1958, TI dispersed for one of its mass vacations, leaving Jack Kilby—as a recent arrival unentitled to time off—in effective command of the semiconductor laboratory. An exciting sense of freedom possessed him. It was a chance, thought Kilby, to come up with something different.

From extended interviews with Kilby, T. R. Reid, author of a vivid history of *The Chip,* composed an inventor's fugue, which went something like this: "If Texas Instruments was going to do something . . . it probably had to involve silicon." Fair enough. . . . What could you do with silicon? Obviously transistors . . . diodes . . . already being done. But by contrast silicon did not make very good resistors and capacitors, both crucial to managing power and storing it in most circuits. It would be original, certainly, to make a silicon resistor. But it would also be absurd to use preciously purified silicon to make a device that cost a penny in carbon. Yet, thought Kilby, who had contrived various novel capacitors for Centralab, one just might be able to produce a silicon capacitor. Its performance wouldn't approach that of the standard metal and ceramic capacitor but it would do the job . . . particularly in the low power world of solid state. . . . For that matter, you *could* make a silicon resistor. . . . And come to think of it—this was the idea that would revolutionize electronics— "if you could make all the essential parts of a circuit out of one material, you could manufacture all of them, all at once, in a monolithic block of that material."

This was the integrated circuit: a group of transistors and other components interconnected on one tiny piece of semiconductor. It could be a hearing aid, a computer memory cell, a radio oscillator . . . in the end it might even be logic for avionics or a hand-held calculator (Kilby would later patent a calculator chipset, together with a tiny thermal printer for the readout). Kilby did not know what the limits would be. But his thought process captured the essence of the semiconductor revolution in America.

Because it consisted of a series of second-best solutions—inferior resistors and capacitors, for example—which ended only in a radical drop in manufacturing costs, Kilby's concept would have been unlikely to emerge from a laboratory of pure science. A breakthrough in product design, it likely would have sprung neither from a capital equipment producer nor from a semiconductor factory, both of which focus on existing products. Because it threatened the jobs of computer engineers who made their living combining electronic components into elegant configurations on circuit boards, the new device would not have been invented or accepted readily in the computer industry. A synthesis of technical and economic speculations based on materials science, circuit design, processing techniques, and a wild hunch, it was the kind of solution most doggedly obstructed by the usual divisions of labor and specialization in large companies. And it happened at Texas Instruments, in part, because most of the company was not there.

When the company returned from vacation, however, things began to go wrong. Kilby's boss, Willis Adcock, was enough intrigued by his new employee's crabbed circuit sketches to set him to work producing two prototypes. But TI was then moving toward a new system of diffusing impurities onto the wafers in hot ovens, where they could be processed like so many cookies. Because the first diffused devices were germanium "mesa" transistors, Kilby agreed to make his prototypes on a piece of germanium. It was a fatal step away from the microcosm.

The device was called the mesa because in shape it resembled a flat-topped southwestern mountain rising above the sands of a usually silicon substrate. Because the emitter and base were in the protruding part, the mesa was accessible from above and was isolated from neighboring mesas by air. Previous transistors had lead wires sticking out from their two ends below the surface of the block of silicon or germanium. Thus they could not be pushed together or made on one piece of semiconductor. Mesas could be built in groups on one chip without the wires getting in the way. The wires between the mesa tops could be added later. The mesa seemed a godsend for integrated circuits, and indeed it would play a critical role in their development at two companies. But only Jack Kilby would actually use mesas to make integrated circuits.

So it was that for his revolutionary device Kilby chose a fatefully obsolete material and a fatally obsolescent transistor design. On a sliver of germanium less than one half inch long, with spidery gold wires awkwardly soldered from one mesa top to another, Kilby contrived one of the ugliest little devices since the original point contact transistor.

Hearing of the device back in Mountain View, California, however, one engineer working at Fairchild Semiconductor quietly brought forth some similar designs he had been working on. His name was Robert Noyce, and most people in the industry regard him as the true inventor of the integrated circuit.

The son of a minister in an Iowa farm community, Noyce as a boy had created a flying machine and used it to glide down safely from the tops of barns. During his course through Grinnell College, MIT, and Philco-Ford Semiconductor, this "can-do" confidence grew into a sense of high destiny in the world.

Then he came a cropper. At Shockley Semiconductor Laboratories, also in Mountain View, where Noyce worked from 1955 to 1959, the great scientist had treated Noyce like a youth in his twenties learning from the master. In Noyce's presence, Shockley would call former colleagues at Bell to check out any novel results the young man achieved.

For example, before Esaki's Nobel Prize-winning invention of the tunnel diode was announced, Noyce presented to Shockley a detailed proposal for such a device. But Shockley was not interested. He never seemed quite to realize that the young men he had assembled in Palo Alto—for all their humble beginnings and bulging Windsor knotted ties—were in many ways superior to the polished authorities he had left behind him at Bell Labs.

Eventually the group left Shockley in favor of an offer from Fairchild of Syosset, Long Island, which was interested in establishing a semiconductor firm. Shockley is now said to see Noyce as "traitorous." But by holding the eight key men from Shockley Labs together, Noyce succeeded not only in saving Shockley's most precious legacy to the industry but also in creating the team that would bring the world of electronics massively back to Shockley's original vision of a field effect transistor (FET). Elegantly simple and easy to miniaturize, eventually the FET would be a key to the microcosm.

The eight defectors from Shockley Semiconductor were Noyce, Moore, Julius Blank, Victor Grinich, Eugene Kleiner, metallurgist Sheldon Roberts, Jay Last, an expert on photo optics from Corning Glass, and Jean Hoerni, a physicist with two doctorates. In a sense they were the founders of Silicon Valley. While rapidly expanding their numbers—hiring among others a tall, burly young engineer named Charles Sporck from a GE components factory in Schenectady, New York, and Andrew Grove—they set to work to redefine the industry.

Together the group swept past TI in integrated circuits and in the process transformed Kilby's hunch into one of the most important inventions in the history of technology. Known as the planar integrated circuit, Fairchild's concept comprised the essential device and process that dominates the industry today. In the 1960s, it gave Fairchild a lead in the new phase of the industry comparable to the vantage that TI had achieved in the 1950s manufacturing silicon transistors. Ultimately it moved the industry deep into the microcosm, and put America on the moon.

The first steps toward dominance came from the team of Moore and Hoerni. Setting out to design large diffusion furnaces that could process scores of silicon wafers at one time, they began batch-processing the very kind of transistor that TI was then making in germanium: the mesa. The precision of the diffusion method, as mastered by Gordon Moore, allowed Fairchild rapidly to make its mark in this technology.

Nevertheless, there were flaws in the mesa. For one thing, it still needed tiny gold wires hooked up to the promontory. For another, its exposed surfaces tended to attract contaminants. This problem became Fairchild's historic opportunity.

To protect the mesa from contaminants during manufacture, Hoerni and Moore began playing with the idea of depositing a layer of silicon dioxide over its surface. But running a thin silicon dioxide film up and down the mesa's steep slopes was nearly impossible, even with the new diffusion method. It would be better, Hoerni thought, to have a *plain*. Thus he stumbled into the historic idea of flattening the mesa. Then he saw he could use the flat layer of silicon dioxide for two purposes at once. Unlike the oxides of other elements, silicon dioxide could both protect devices from chemicals during fabrication and insulate them during their electrical operations.

Suddenly an entirely new and better approach to semiconductor production began to emerge in his mind. The diffusion process made it possible to create transistors not by adding layers to the top, but by diffusing the impurity "dopants" into the surface. Hoerni proposed that a single transistor be created, with a flat surface and with all its regions accessible on that surface. Such a device, which Hoerni called "planar," would solve the problem of contamination because a relatively flat topography could be protected nicely by a planar film of that excellent insulator, silicon dioxide, that could easily be grown on silicon by heating it in an oxidation furnace with pure oxygen or steam.

This concept revolutionized the semiconductor world, for it prompted Bob Noyce to think of integrating large numbers of electronic components on one flat chip of silicon. Neatly disposing of the nagging, labor-intensive problem of bonding gold wires to each electrode (as was necessary in Jack Kilby's mesa IC), Noyce proposed to interconnect the parts by aluminum lines evaporated onto the insulating oxide surface. The aluminum could be connected with each transistor through holes in the oxide.

While Kilby had invented two integrated circuits, Noyce was ready to show how to mass-produce them. And because Noyce was clearer in specifying the mode of interconnection and insulation, he was allowed to share in the IC patent—after some ten years of litigation between TI and Fairchild—despite conceiving of it and building prototypes well after Kilby.

The decision was just. The image of those mesas, with wires running between them like gold transmission lines above the intervening desert, countervailed the claim in his patent application that they could be "laid down" on the surface. They couldn't. The wires could not be laid down without an insulator and Kilby had offered none. Indeed, none would work on germanium mesas. Planar silicon dioxide insulation was what made the IC a reality and made Noyce the real inventor (with key assists from Hoerni and Moore).

Just as important as these conceptual advances in the move to the microcosm, however, was an entire chain of manufacturing techniques that Moore, Sporck, and the other Fairchild engineers perfected over the following years. Although integrated circuits operate with incomprehensible speed, they can take long months to produce. Even in 1989, with the most modern equipment, the usual fabrication process—often after many man-years of design work—usually takes some six weeks or more. During Fairchild's world-beating rush of 1959 and 1960, the process dragged out over a year.

The diffusion step, in which gaseous impurities or dopants slowly sink into the surface of the substrate, occurs at a temperature approaching the melting point of silicon (1200 degrees Celsius), and takes a period of hours. Then comes the photolithography: creating a pattern on a chip by exposing its surface—covered with a light-sensitive chemical called photoresist—to light through a glass photomask inscribed with the design.

In effect, the designs are projected onto the wafer like a slide on a screen, but with all the lenses reversed to miniaturize the image rather than magnify it. After the design is set by exposing the photoresist to light, the resist is etched away in accordance with the desired pattern. The designs are "developed," hundreds on each wafer at a time, much like a photograph of hundreds of chips.

By repeating these steps for several layers, the pattern of transistors and other IC components is created. Finally, a scheme of holes, or vias, is developed in the oxide insulator. These holes are for the metallization: the evaporated layer of aluminum, photo-etched into precise patterns, that would actually *integrate*—by interconnecting—the elements in the circuit.

Thus eliminated were all the thousands of wires—sticking out from both ends and from the top of each discrete component—that made transistor manufacture and packaging a laborious, expensive, and frequently unreliable process. At every potentially disruptive step, particularly during the etching away of unwanted metal, the Fairchild inventors used silicon dioxide to protect the device.

These general techniques embraced thousands of exacting particulars, and it is the details that are everything in semiconductor work. Suffice it to say that it took nearly three years for other companies to master the intricate interplay among the different steps. By the time others worked it all out in the mid-sixties, Fairchild had become the dominant company in the microcosm, moving fully into "the invisible sphere" that Planck had defined.

Like their intellectual precursors in physics, these callow young men could find no easy idiom to describe their achievements. Their feats defied every analogy to the world in which they—and all of us—were raised and in which our very language was formed. They had to deal in sizes that confound any metaphor of the minute, from motes to mites; in numbers, of precise trillions of drifting or diffusing electrons, that dwarf the merely astronomical; in speeds—nanoseconds (billionths of seconds)—that render agonizingly and viscidly slow the snapping of a finger. Even to begin they had to create clean rooms that flout any standard of the immaculate, any concept of sterilization, any simile of snow.

Cleanliness for these purposes is judged in particles of a diameter more than 1 micron per cubic foot. "Class 100," then necessary in a semiconductor fabrication area, measures one hundred times cleaner than the some 10,000 such particles per cubic foot in the operating room of a hospital. One of these infinitesimal particles athwart a transistor channel looms in the micrographic photos—as in the common speech of the Valley—as a "boulder." Now taken for granted in an industry moving to class 10 and below, these conditions had to be created for the first time at Fairchild.

Like TI before it, Fairchild achieved its breakthroughs with virtually no government assistance while its largest competitors—chiefly the vacuum tube companies—were receiving collectively hundreds of millions of dollars in grants. But when the government needed a way to miniaturize the circuitry for its Minuteman missiles and its space flights, it did not use micromodules or any of the other exotic technologies it had subsidized. It turned first to Fairchild rather than to its early favorites and beneficiaries. Fairchild's lack of military entanglement in the late fifties finally allowed the company to get the bulk of military and aerospace contracts in the early 1960s.

Meanwhile, at TI, Haggerty's balance between defense and civilian production went severely awry. Military contracting grew rapidly and TI lost some of its creative edge, particularly in the metal oxide silicon (MOS) technologies of Gordon Moore's parable. One day in 1966, Willis Adcock, Kilby's former boss, quit TI and walked dazzled through Fairchild, marveling at their efficiency with the planar process. The invisible sphere had shifted again.

CHAPTER

6

Flight Capital

All the rewards of the quantum economy stem from its fast-rising ratio of mind to matter. It was only when transistors became cheap enough to be chained together by the thousands and then millions in logic devices that this ratio began to soar, and mind emerged as the dominant source of value in the economy's dominant products.

The gradual mastery of silicon and the idea of integration were giant technical steps into the microcosm. But silicon transistors wired together, or even integrated in small numbers on single chips, still remained chiefly replacement parts for vacuum tubes and vacuum tube modules. Good for radios and hearing aids, they did not even suggest the possibility of a computer on a chip, or a major semiconductor memory.

The overthrow of matter in a technology of mind could occur only at levels of density many thousand times greater than Fairchild had attained. Only by linking many thousands of fast switches in a area the size of a pinhead (or scores on a pin point) could engineers make switches virtually free and begin to create mind-scale machines. Only such feats of miniaturization would yield the paradoxical bounties of size and space, speed and cost, that epitomize the microcosm.

In the 1960s, even after Fairchild had been manufacturing Noyce's bipolar microchips for several years, this achievement was far from sure. In fact, few researchers could even imagine it. The field effect transistor of Shockley's dream and the metal oxide silicon technology of Gordon Moore's parable, both crucial to the epoch of very large scale integration (VLSI), were mere concepts or cobbled experiments.

It was by no means inevitable that these vital tools could be perfected and fabricated by the millions. To achieve such goals within twenty years after the invention of the transistor and within ten years after the creation of the integrated circuit would require heroic efforts by a small group of brilliant men.

Let us then leave behind the view of businesses as functions of abstract forces, economic and technological trends, scientific discoveries, corporate plans and strategies. Instead, let us contemplate the future of the industry in the form of a dark, curly-haired teenager, lying face down in the wet furrows of a muddy field somewhere near the boundary between Hungary and Austria late in the fall of 1956. The young man is hard of hearing, but his mind echoes with the yelp of dogs, the tromp of soldiers on a nearby road, the maundered words of a hunchbacked smuggler who had taken his money in exchange for directions to the border. He cannot turn back; in Budapest, they are arresting his friends.

He shivers at the sight of an old cemetery nearby, frosted by the flare of an illumination grenade. The flare passes. He gets up and slogs forward, thrusting a cloth bag before him, then rivets himself again into the dirt to avoid a new frost of light. He has no clear idea of where he is. There are no signs to inform the international traveler, You are now entering Austria, gateway to Silicon Valley.

He knows that many have followed a similar trail through the night, sprinting a final course across a mistaken border and ending a circle of errors near where they had begun. Rounded up, searched, jailed, dispatched in trains to Russian oblivion, they would begin no exciting companies, make no breakthroughs on the electrical properties of the silicon silicon-oxide interface. Another fierce dog barks close by; a voice challenges in rough Hungarian; the young man gasps for life and breath. Could he be in Hungary still?

András Gróf, now Andrew Grove, president of Intel, is blasé about his break to the West. He says his story is commonplace in Silicon Valley. Indeed, another of Intel's founders, Les Vadasz, now vice president in charge of strategic planning, was maneuvering toward the border of Hungary unbeknownst to Grove at about the same time. Confronted by soldiers near a bridge to Austria, Vadasz hid behind a truck, awaiting a moment of distraction, then headed across the span one fraught step at a time before passing the middle and bursting into an exultant sprint for the free land beyond. Grove would not meet Vadasz until 1965 at Fairchild Semiconductor.

Perhaps half of the key contributors to Intel's success came to the company from foreign countries. Israelis, Chinese, Greeks, Italians, Japanese all played parts in launching the crucial metal oxide silicon and memory technologies that would change the world.

In that distant muddy field, Grove's prospects seemed, for a long moment, grim. The voice in the night above his stricken figure, however, turned out to have come from a bilingual Austrian. Grove had made it across the border after all. Finding his way to a refugee camp near Vienna, he devoted his days to walking the streets of the city with other mud-caked Hungarians, going from one immigration service to another looking for a "magic carpet" to the United States. Although no magic carpet appeared, the International Refugee Committee finally offered a lurching Liberty ship, crammed with other refugees, heading for the United States.

Grove spent fifteen winter days in passage on the stormy Atlantic. Half deaf, sickly, and seabegone, he arrived in New York nearly penniless and unable to understand spoken English. After failing to find a technical college that would give him a scholarship, he enlisted at New York's free public institution, City College.

Like so many immigrants before him, he found CCNY—more than the Statue of Liberty—the true gateway to America. He acquired fluency in a streetwise New York accent and a job working for the head of the Chemical Engineering Department. Eventually his success in America was certified by *The New York Times:* HUNGARIAN REFUGEE GRADUATES FIRST IN CLASS AT CCNY, leading 4,621 graduates.

Walking from subways through the sleet and slush to the campus, he also learned to dream of California sun and surf. Having married an Austrian American from New Hampshire in the summer of 1959, Grove ended up as a graduate student at Berkeley in 1960. There he immersed himself in work at the chemical engineering lab. Specializing in fluid dynamics, a branch of applied physics, Grove was able to publish four papers from his thesis and gained his Ph.D. in chemical engineering. But as he proceeded through his last year, he began to feel that he lacked the mathematics and physics background to excel in his chosen field. In his final semester at Berkeley, therefore, he tentatively and unofficially turned to the apparently easier field of solid dynamics—semiconductor electronics.

Although he had been elaborately educated to design missiles or oil refineries, Grove then sent off his résumé to all the semiconductor companies. After misasdventures with older interviewers at TI and GE, he was delighted to discover that everyone at Fairchild was in their twenties, including the imperiously brilliant head of research, the crewcut, bespectacled Gordon Moore.

A sheriff's son from San Mateo with a doctorate in chemistry from Caltech, he struck Grove as the most impressive of the some fifty semiconductor men he had met in his interviews. Moore proved his genius in short order by detecting in this young scholar of fluid dynamics the potential to become a valuable member of the Fairchild solid state laboratory.

Grove's eventual rise to the top of American industry at Intel is one of the implausible yet regularly repeated sagas of individual achievement which explain—in all cold realism—the triumphs of Silicon Valley and of the American economy since World War II. It seems melodramatic to depict the future of then nonexistent corporations hanging in the balance near the borders of Hungary in 1956. It seems romantic to see the future of a technology depending on the willingness of America to serve as a refuge for immigrants. Yet the astonishing feats of the semiconductor industry, like the earlier wartime exploits of the Manhattan Project with the atomic bomb, depended on just such accidents and open arms. The current technological miracles of miniaturization were anything but inevitable. Many other directions and duller destinies were possible in the world of postwar electronics.

The entire history of the vacuum tube illustrates the likelihood of major mistakes in the development of technology. While ever-growing billions were invested in ever more elaborate refinements of this retrograde way of managing electrons, solid state devices already were known to scores of thousands of experimenters with crystal sets. Many investigators had pointed the way toward semiconductors, exemplified by the "whiskered" galena crystals used during World War II. Yet because no one knew how to manufacture pure crystals in volume, the whisker diodes never escaped a reputation as "crude" devices, and inferior systems flourished. The best talent flocked to vacuum tubes and nuclear physics as the glamorous technologies, and government and business lavished resources on them while largely ignoring the promise of solid state.

Even within the semiconductor world itself, a few experiments by Bardeen and Brattain sufficed to cause fifteen years of infatuation with bipolar junction devices, rejecting the simpler, smaller, and cheaper field effect transistor advocated by Shockley. Through the late fifties into the early sixties, research at the major companies focused on an agenda of micromodules and silk-screened integration on ceramic substrates. TI long explored the seemingly logical but actually unpromising idea of a computer not on a chip but on a whole silicon slice or wafer (which now can hold some four hundred microcomputers). Indeed, the leading scientists and engineers and the best advised business magnates in America—from William Shockley to Thomas Watson, Jr., and even Pat Haggerty himself—favored, charted, and financed lines of development that led directly away from the massive integration of the microcosm. It was a mere handful of men, prominently among them Andrew Grove, who made it all happen, in ironic fulfillment of the technology that Shockley pioneered and then passed over in the 1950s.

Grove had attracted the attention of Bob Noyce within a week of arriving at Fairchild by writing—at the behest of a boss leaving for a convention—a compendious mathematical treatise on silicon capacitors. One of the four key methods of manipulating electrons, along with resistors, diodes, and transistors, a capacitor uses a conductor under an insulator to collect charge and then release it.

The crucial device that would store information in most semiconductor memories—and would in essence govern the "gates" of all field effect transistors—the silicon capacitor was key to the next breakthrough in the industry. Noyce sent the new employee a brief note: "I read your report on . . . capacitors. It was very nice work. [signed] Bob." With this touch of the scepter from on high, Grove knew that at last he had made it across the border into the world of semiconductor electronics.

His advantage over most of the other Fairchild personnel was his knowledge of computers. While preparing his thesis at Berkeley, he had learned how to program in the higher-level scientific language, FORTRAN. Although he knew little about semiconductor capacitors when he began his report, he quickly mastered the available materials and translated the mathematics into a FORTRAN program. Thus he was able readily to work out all the technical characteristics of the device, in crisp formulas and computations, within a few days, and gained a reputation as an expert on the emerging technology of MOS, designed to fulfill the Shockley dream of a field effect transistor.

MOS simply stands for metal, oxide, silicon (or semiconductor). The three key layers of a MOS device were *metal* wires laid down on top, an *oxide* insulating layer underneath, and a *silicon semiconductor* substrate. It resembled Noyce's bipolar integrated circuits. But by contrast, the silicon of the new devices was all of one polarity, and thus they were simpler to create, miniaturize, and operate in the microcosm. Grove's ascendancy at Fairchild and later at Intel arose in great degree from providing solutions to the key problems of this technology that had baffled so many researchers since it was proposed by Lilienfeld in 1926.

The difficulty was not in the concept, a MOSFET. Expounded lucidly by Shockley in 1952, it had been patented in one form by John T. Wallmark of RCA, clarified by M. M. Atalla at Bell in the same year, and built by several technicians at RCA in 1962. But all the devices were sadly slow, brittle, skittish, and in general hopelessly unstable.

It was primarily a materials problem with MOS itself. As Grove remembers: "You put one of these MOS devices on a curve tracer [a ubiquitous instrument that displays the electrical characteristics of a device on a CRT screen]; the characteristics would collapse and expand before your eyes." This had led Fairchild the year before, in 1965, to cut back radically on its efforts in MOS and to recommit its future to bipolar circuits; it prompted RCA to withdraw from MOS production, and deterred TI from any serious venture in the field.

The materials problem was the same essential syndrome that Bardeen a decade before had ascribed to "surface states"—the molecular loose ends left by the abrupt cutting off of the crystalline lattice at the top of a device. The problem had intimidated even Shockley and devoured many a career in its maw. "People had been fooling around for years, like struggling in quicksand. The more they struggled the deeper they got, inventing new riddles to resolve the earlier perplexities," Grove recalled. He did not intend to end his life as a researcher lost among footnotes to the Nobel Prize-winning works of Shockley and Bardeen. In 1966, he assembled a team of himself, Bruce Deal, the grower of Fairchild's purest oxides, and Edward Snow, a leading chemical engineer, and together they "went for the simplest model."

Once again it was less genius than pertinacity that carried the day. Snow happened to have written his thesis on the migration of sodium in quartz crystals. Deal was fanatical about purity; he did not accept the various hypotheses focusing on defects in the oxide itself. Daunted by the futility of earlier analyses, Grove was eager for a new approach.

After six months of frustration, going through endless tests and specifications of devices, curve tracer printouts, and chemical formulas, they discovered that the problems with surface states were far less significant than previously supposed and could be controlled by a pure oxide layer. The Grove team also uncovered the real culprit in MOS: namely, sodium contamination—salt—in the evaporators. The remedy was scourging them free of all residues from the sodium hydrochloride and related chemicals that always had been used in cleaning the machine.

The effect was virtually magical: the instability nearly vanished. Grove, Snow, and Deal won a Franklin Institute Award for their paper on the subject, and then Grove won the prestigious Ebers Award of the Institute of Electrical and Electronic Engineers (IEEE) for further researches on MOS. By 1967, he and his colleagues had cleared the way at last for the mass production of MOS field effect transistors, and Grove himself had published his classic text, entitled *The Physics and Technology of Semiconductor Devices*.

In his chapter on the Field Effect Transistor, he concluded that "the MOS transistor . . . has become potentially the second most important device next to the bipolar transistor. In fact, in many integrated circuit applications the MOS may eventually become the more important one." As it happened this cautious affirmation of MOS, like so many predictions in the industry, turned out to be a drastic understatement. Ultimately, MOS would rule the future and Andrew Grove would lead the move to MOS.

The MOSFET is one of the supremely elegant consummations in the history of technology. Harking back to the inspiration of Shockley, the founder of the field, and reaching ahead into its most farflung future, MOS triumphs through its beautiful simplicity. Rather than three or more diffusions of impurities to create the various polarities of charge, the MOSFET requires merely one to create the unipolar source and drain. Rather than needing complex isolations, the MOSFET isolates itself. In the very operation of the transistor, the field effect voltage on the gate creates an isolated channel in the substrate. Rather than space-hogging resistors, the MOSFET uses its own transistors, only slightly elongated.

To all the students of electronic engineering around the world, spraining their minds over the baroque intricacies of bipolar circuits, with their majority and minority carriers—drifting and diffusing, tunneling and avalanching, scooting and scattering among a mystifying array of "space charge regions"—the MOSFET was a benison of order. Smaller, easier to build, the MOSFET produced higher-fabrication yields and greater device densities. During the 1970s, it rushed the semiconductor industry down its most precipitous curve of experience deep into the microcosm.

Andrew Grove therefore had played a central role in bringing Fairchild to the threshold of a new era. But Fairchild would not enjoy the fruits of his work. Following the path of venture capital pioneer Peter Sprague were scores of other venture capitalists seeking to exploit the new opportunities he had shown them. Collectively, they accelerated the pace of entrepreneurial change—splits and spinoffs, startups and staff shifts—to a level that might be termed California Business Time ("What do you mean, I left Motorola quickly?" asked Gordon Campbell with sincere indignation. "I was there eight months!").

The venture capitalist focused on Fairchild: that extraordinary pool of electronic talent assembled by Noyce and Moore, but left essentially unattended, undervalued, and little understood by the executives of the company back in Syosset, New York. Fairchild leaders John Carter and Sherman Fairchild commanded the microcosm: the most important technology in the history of the human race. Noyce, Moore, Hoerni, Grove, Sporck, design genius Robert Widlar, and marketeer Jerry Sanders represented possibly the most potent management and technical team ever assembled in the history of world business. But, hey, you guys, don't forget to report back to Syosset. Don't forget who's boss. Don't give out any bonuses without clearing them through the folks at Camera and Instrument. You might upset some light-meter manager in Philadelphia.

They even made Charles Sporck, the manufacturing titan, feel like "a little kid pissing in his pants." Good work, Sherman, don't let the big lug put on airs, don't let him feel important. He only controls 80 percent of the company's growth. Widlar is leaving? Great, he never fit in with the corporate culture anyway. Sporck has gone off with Peter Sprague? There are plenty more where he came from.

"It was weird," said Grove, "they had no idea about what the company or the industry was like, nor did they seem to care. . . . Fairchild was just crumbling. If you wish, the semiconductor division management consisted of twenty significant players: eight went to National, eight went into Intel, and four of them went to Alcoholics Anonymous or something." Actually there were more than twenty and they went into startups all over the Valley; some twenty-six new semiconductor firms sprouted up between 1967 and 1970. "It got to the point," recalled one man quoted in Dirk Hanson's *The New Alchemists,* "where people were practically driving trucks over to Fairchild and loading up with employees." Eventually, Noyce and Moore, the key founders, departed to form Intel, and the jig was up. Fairchild Semiconductor simply dissolved before the Eastern management ever figured out what was going on.

When Syosset finally got the news, transmitted effectively through the collapse of its stock, Carter and Fairchild resorted to the usual panacea: money. If they could not run a semiconductor firm, perhaps they could buy one. It was an idea that would occur to scores of large corporations and ambitious governments alike over the years. But it almost never worked.

In this case, as one Fairchilder recalls, "All of a sudden, in walk eight or ten guys with great suntans," all set to take over. They represented the entire management team from Motorola's Semiconductor Division in Phoenix. In the lead was Lester Hogan, inventor of many of the crucial components for microwave systems. His followers were inevitably dubbed "Hogan's Heroes." To them, Fairchild lavishly granted all the money, flexibility, and stock options it had denied its previous employees. But it did not suffice.

Grove explains: "A company is really its middle management. If middle management is gone, nothing will put that company back together. At first, Fairchild's middle echelons were very good, and the fact that the top management left would not have mattered if the new leaders were backfilled in right away." But in this case, "the middle management left also. And after that I don't care who was going to buy the company or how much money they were going to pour into it, it wouldn't work. It's like bone cancer. Once your bone marrow is eaten up, nothing can replace that." Grove didn't try. Within days he followed Moore and Noyce to Intel.

CHAPTER

7

Intel Memories

Computers are dumb and can only do a few things. But they make up for it with speed. Any computer designer has to be a speed freak.

Since as a general rule, the more the power, the faster the switch, you can get speed by using high-powered or exotic individual components. It is an approach that works well for Cray, IBM, NEC, and other supercomputer vendors. Wire together superfast switches and you will get a superfast machine. You also may get a furnace, but liquid nitrogen or other refrigerants can shed the heat.

This rule—that more powerful, faster switches mean faster computers—might be termed the macrocosmic route to speed. It not only leads to high-performance machines; it also leads to bigger and more costly machines. It stands in the way of the microcosmic vision of widely distributed computer power—dispersing mind through the material settings of life.

The other choice for speed is to use low-powered, slow switches. You make them so small and jam them so close together that the signals get to their destinations nearly as fast as the high-powered signals. If you then can use more switches and arrange them more ingeniously for your application, you may even get a machine that for its own purposes can outperform a supercomputer. This approach

works well in the human brain and in a variety of graphics workstations, circuit simulation accelerators, laptop terminals, and other special-purpose computers.

This rule—that low-powered, slow switches mean faster computers
—is the microcosmic route to speed. Low and slow has been the
secret of success in semiconductors from the outset. Shockley substituted slow, low-powered transistors for faster, high-powered vacuum
tubes. Teal replaced fast germanium with slower silicon. Kilby and
Noyce substituted slow resistors and capacitors as well as slow transistors on integrated circuits for faster, high-powered devices on modules and printed circuit boards. Slow and low wins the race is the law
of the microcosm. Above all, slow and low was crucial to the rise of
Intel and the personal computer industry.

Intel is the most important company in the history of the microcosm. All the key components of the personal computer—the working memory, the software memory, and the microprocessor CPU—
emerged during the magical first three years of the company's existence, between 1969 and 1971. Until Intel's breakthroughs, computers were large and cost a minimum of tens of thousands of dollars.
After Intel's three-year siege, computers could be built for a few
hundred dollars. This change was the microcomputer revolution.
Intel did not win with mirrors, or by "bowling over its competitors"
by the marketing power and prestige of Noyce and Moore, as some
have said. Intel prevailed by creating an all-silicon stage for microcosmic miniaturization. Thus it launched a still ongoing transformation of the computer industry, eventually curing its addiction to the
high-power, high-speed, high-expense route to fast computers, and
showing the allure of slow and low.

At the outset, Noyce and Moore had no such transforming agenda.
They rejected Carver Mead's proposal for a rapid move into technologies for personal computing. Instead, they envisaged offering replacement parts for mainframe computers, beginning with silicon
memories. In the late 1960s even this loomed as a risky and exotic
idea.

Securely in the macrocosm, memories during that era were made
not of sand but of iron. Taking advantage of the natural binary poles
of magnetism, these devices were the epitome of the high-power,
high-speed approach. Still dominant were the ferrite core devices—
delicate looms of wire strung with tiny magnetized ringlets—invented
by Jay Forrester at MIT in the early 1950s and built with the help of

his aide, Kenneth Olsen. Forrester has since become an innovator in econometrics. In 1957, while Noyce was founding Fairchild, Olsen went on to launch Digital Equipment Corporation. But neither Forrester nor Olsen would ever offer a product more enduring and influential than their magical loom with its tiny ferrite rings or "cores."

Not only does the core survive as the storage device in some working computers; not only is its combination of permanence and accessibility in some ways still hard to beat; but its essential structure remains discernible—with the help of electron microscopy—in the awesomely smaller warp and woof of all semiconductor memories. All silicon memories are simply microcosmic translations of the core. Indeed, the idea for a semiconductor memory originated with Robert Norman, later to invent an important logic family at Fairchild, while he was working on transistor "driver" circuits for core memories. In 1955, he told an International Solid State Circuits Conference that the cores themselves were unnecessary; you could make the entire memory out of transistors. Most of the subsequent history of semiconductor memories can be seen as a drive to fulfill this vision of Norman's: a silicon core.

Resembling to the eye a loosely woven fabric strung across a tiny square frame, the ferrite core system consists of a network of hairlike wires—half strung vertically, half horizontally—with a tiny ringlet at each intersection and with a third wire threaded back and forth through all the ringlet "cores." Like the latitudinal and longitudinal lines on a map, the cross-hatched wires give a unique address, consisting of a row and column for each ring in the weave. The iron rings bear binary information by the direction in which they are magnetized, clockwise or counterclockwise, representing on or off, positive or negative, one or zero.

To change the direction of charge—that is, to *write* information into a core—an amount of current just *half* sufficient to reverse the magnetic field on the ring is sent down each of the appropriate vertical and horizontal wires. Only the ring at their junction will receive the full jolt necessary to change its field. The rest of the cores, hit only by one live wire, remain in their previous state. To *read* the information on a ring, on the other hand, an attempt is made to flip its field toward zero. If it is positive, or clockwise, the ring in reversing will transmit an electrical pulse to the third wire. As the output line, this wire takes the pulse to an amplifier, where it is magnified and sensed, and then, to terminate the complex choreography, is shipped back to its original position. If the core is at zero, or counterclockwise, no pulse travels down the output lines, and a zero is communicated.

In such core systems, both the reading and the writing processes can occur almost instantaneously at any randomly chosen address in the memory. Thus the ferrite core device is a random access, read-write memory: the sort commonly known as a RAM. Of course, useful data does not often come in individual bits. To hold data in bytes (eight bits) of information or more, the necessary number of frames—called planes—may be piled one on top of another. The eight bits in a particular byte of data (typically signifying a numeral digit or letter) may be located at the same address on eight different planes in a memory stack.

As memory advanced into silicon, determined engineers at Olsen's Digital Equipment Corporation (DEC) and elsewhere clutched their magnetic lifesavers and defied the microcosmic undertow. Moving down the learning curve through the mid-1970s, they reduced the ringlets to one one hundredth of an inch in diameter and packed millions tightly together in one storage console. As late as 1977, for example, the top of the line at DEC—the $60,000 PDP-11/70—held some four million bytes (thirty-two megabits) of core memory, using semiconductor cells only for immediate buffer and register storage within the computer's central processing unit (CPU).

One of the reasons for the survival of the ferrite core memory in a microcosmic world was its permanence, or non-volatility. Unlike most semiconductor RAMs, ferrite cores hold their information after the system is turned off; it is a *non-volatile* RAM. One of the long-term goals of the semiconductor industry is to create a large, cheap, non-volatile RAM on a silicon chip and finally relieve computers forever of the need for pumping iron.

In 1968, Noyce and Moore did not seem a prime threat to the ferrite core. Even as late as 1971, a prestigious McGraw-Hill compendium entitled *Computer Memory Technology* presented an exciting entry from Sperry-Univac on "Plated Wire: A Long Shot Is Paying Off"; a proposal from Britain's national telecommunications lab for a "high density waffle-iron memory . . . whose performance [hold your breath!] approaches that for plated wires"; and a confident offering from Litton Industries on "a post and film memory that delivers NDRD capability but avoids the problems of creep." Only two of the eleven chapters were devoted to the prospects for semiconductor memories.

Silicon storage was seen as prohibitively tricky, exorbitantly expensive, and inordinately difficult to produce, except for a few specialized uses. When Intel was founded, the largest silicon memory held only 256 bits, and Texas Instruments published a text that could envisage large-scale semiconductor memories only by employing an entire silicon *slice*—impossible because of the difficulty of producing such a large wafer without at least some defects in the silicon crystals.

The data-processing priesthood, in particular, had no reason to abandon the Forrester loom. Its economics tended to confirm the dominance of mainframe computers in their guarded sanctuaries. Like the disk and drum devices for outside mass storage, the ferrite core required elaborate and expensive support equipment that afforded large economies of scale—and low costs per bit—for large computers and tended to impose higher cost per bit on small computers. The ferrite core was a bulwark of a macrocosmic computer industry, an impossible barrier to widely distributed and portable computer power.

Nonetheless, Noyce and Moore proposed to begin not by replacing the core, but merely by expanding the short-term or working memories coupled to the CPUs of mainframes. The usual semiconductor design for this purpose was called the static RAM. Static RAMs, however, used a kind of power and space-hogging cell called a flipflop latch, because it "latched" either open or shut and thus could hold a bit of data. Requiring some five transitors for each memory cell, static RAMs could not provide memory chips with sufficient capacity to compete with cores in large memory functions.

As Andrew Grove had discovered on arriving at Fairchild years before, this space problem had led to interest in MOS capacitors as vessels for storing information. A small fraction of the size of a flipflop, a MOS capacitor is the essence of simplicity; merely a layer of conducting substance under a layer of oxide would hold a charge while a transistor switch at each cell governed access to it from the crisscrossing read-write lines. But MOS capacitors had a crippling weakness. Their charges flowed away so quickly that most researchers barred them as possible memory cells.

Encouraged by Grove's capacitor studies, however, Noyce and Moore decided to go with the flow. They would solve the leakage problem by continually replenishing the cells. Every two milliseconds (thousandths of a second) the device would have to knock off for about fifty microseconds (millionths of a second) to perform a *refresh* cycle during which all the capacitors would be recharged. A pause

that refreshes, it rendered the RAM "dynamic" rather than "static": a
DRAM, pronounced dee-ram. Like a perpetually sinking ship which
never goes down, the DRAM would be convulsed constantly with
activity—leaking, pumping, bailing, accumulating, dumping, shifting
—all in perfect time to a pulsing clock. This ostensibly cumbersome
process was enough to make the dynamic RAM a certifiable kludge,
and led many an engineer to contemplate once again the joys of
pumping iron. But it reduced the device's size, cost, and power consumption enough to give Noyce and Moore the hope of competing
with cores. It was a bizarre product, but on it they resolved to bet the
future of their new company. Noyce prepared once again to jump off
the end of the barn, this time hoping to be held aloft by infinitesimally
tiny and extraordinarily leaky buckets.

Although none of the large electronic firms chose to focus on dynamic RAMs, the success of Noyce and Moore in attracting venture
capitalists—and the willingness of these two geniuses to invest
$250,000 of their own money—did not go unnoticed in Silicon Val-
ley. In particular, another Fairchild alumnus joined two men from
IBM, where the essential DRAM cell was invented, to form a company called Advanced Memory Systems. AMS also set out to produce
RAMs, as did several other firms. Although no one at Intel seems to
remember it today, AMS actually beat them to the marketplace with
a 1,000-bit device, while Intel struggled with another of the surface
state problems that long bedeviled MOS and loomed even more
threateningly in MOS DRAMs. Because the data charge was so tiny
and temporary, it could be nullified by an equally infinitesimal field
on the oxide surface. The pressure once again fell heavily on Andrew
Grove, as operations manager in charge of engineering and manufacturing, to correct the problem.

Grove realized that the future of Intel was at stake. Although it was
producing a couple of very small static memory products, the "1103,"
as the DRAM was called, "was our first mass-produced device, so if
that hiccupped, Intel would have been a very different company. . . .
I literally was having nightmares. I would wake up in the middle of
the night, reliving some of the fights that took place during the day
between various people working for me." Intel's leaders had committed themselves to the most delicate, fussy product in the history of
the industry. They had announced it, advertised it, and introduced
prototypes. But the thing kept erasing itself with surface charges. It
would not do.

Bending over his desk until all hours of the night, Grove pored through piles of data from curve tracers and other test instruments. He proposed new experiments, and supervised changes in the structures of the devices, then proposed yet new tests. Finally the results from the fabrication area began to improve. Yields rose from 5 percent to 10 percent and more. They had surmounted the worst. It was a splendid achievement. Grove later recalled:

> The 1103 was a brand new systems approach to computer memories, and its manufacture required a brand new technology. Yet it became, over the short period of one year, a high volume production item. . . . Making the 1103 concept work at the technology level, at the device level, and at the systems level and successfully introducing it into high volume manufacturing required, if I may flirt with immodesty for a moment, a fair measure of orchestrated brilliance. Everybody from technologists to designers to reliability experts had to work to the same schedule toward a different aspect of the same goal, interfacing simultaneously at all levels. . . . This is a fairly obvious example of why structure and discipline are so necessary in our operations.

Structure and discipline were important indeed. But Intel prevailed over the ex-IBMers at Advanced Memory Systems and all other contenders chiefly because Grove's memories were deeper in the microcosm. They used gates made of silicon rather than of metal to make the transistors switch and they used silicon wires to interconnect the transistors.

Before Noyce, Moore, and Grove left Fairchild to begin Intel, an Italian immigrant in his twenties named Federico Faggin achieved one of the greatest triumphs of the slow is faster school. Building on sketchy initial findings at Bell Laboratories, Faggin put into production a system for using horribly slow polysilicon instead of fast aluminum for the controlling gate of a field effect transistor. He also invented a way to use polysilicon contacts to electrodes. Working with Thomas Klein and Les Vadasz, he thus eliminated metal entirely from transistors.

These innovations meant that on the transistor level of a computer, wires and switches could merge. Both could be made of various kinds of clean and compatible silicon. The metal layers on top of the chip could be reserved for power supplies, remote signaling, and other special purposes. For actual operation of transistors, metal would not be needed. Essentially taking the M out of MOS, these familiar initials came to signify three kinds of silicon: monocrystalline on the substrate, oxide as the insulator, and plain poly for the gates and short-range wires.

Getting rid of the metal was the single most important force propelling Intel's headlong rush into the microcosm. Not only does metal adhere to silicon less smoothly and cleanly than polysilicon does; metal also miniaturizes less well. Most important of all, though, metal melts at far lower temperatures than silicon.

The low melting point of aluminum means that no high-temperature process steps, such as the diffusion of impurities for a transistor's source and drain electrodes, can be performed after metalization. Therefore in metal gate devices the source and drain would be created and oxidized first in diffusion ovens. Then engineers would have to perform the exquisitely difficult step of aligning and laying down a metal gate precisely on the channel between these two electrodes.

The slightest misalignment would make the transistor leak. To avoid this problem, the gate had to overlap the channel on both sides, extending the metal farther than necessary and creating capacitance that more than nullified the speed advantage of aluminum. This process step alone—aligning the metal—strictly limited the possibilities of continuing to miniaturize transistors.

Polysilicon gates eliminated this alignment step entirely. Using Faggin's new method, the gate—no matter how small—would align itself!

Rather than creating the source and drain first and then trying to put the gate exactly between them, the new system permitted the silicon gate to be deposited first. The polysilicon of the gate itself would serve as a mask to define the locations of the source and drain. The diffusion would create doped regions for source and drain through holes on both sides of the gate but would not penetrate the gate itself to the channel below. The result was a perfectly self-aligned silicon gate.

Self-aligned silicon gates and contacts permitted a far smaller DRAM chip than the competition could offer. Since a silicon wafer costs about as much to process regardless of how many chips are on it, a smaller memory is cheaper to build. Because all-silicon transistors could be crammed more tightly together, the memory could operate faster. In the end, Intel triumphed because it was deeper in the microcosm.

With working memory plunged into the microcosm, Intel proceeded to put the software in as well. It indeed took two different kinds of memories mostly to replace the ferrite core. The DRAM for fast-working storage, and a new, non-volatile memory which, unlike the DRAM, kept its contents when the power went out. Even more than the DRAM, this new memory was dependent on Faggin's silicon gate process.

The name of the new invention was the erasable programmable read-only memory—the EPROM. Like so many inventions, it was almost entirely unexpected by the company. The inventor was Dov Frohman-Bentchkowsky, a young Israeli Grove had discovered at Berkeley. After inventing a promising but unorthodox kind of non-volatile memory based on metal nitride oxide, he began seeking a device better attuned to Intel's manufacturing processes.

Experimenting with silicon gates, he arrived at the idea of a floating gate. Above Faggin's silicon gate, he added yet another layer of poly-silicon, floating, as it were, in an oxide insulator. Sealed in the oxide, the floating gate would be free of any direct electrical contact. If such a floating gate could be charged efficiently, Frohman reasoned, it could provide a permanent way of storing data, because the charge would be surrounded with insulation.

The EPROM cell thus would consist of a metal contact above an oxide-enclosed floating gate, all bonded to a silicon gate transistor below. Frohman proposed that the charge be loaded by applying a large enough voltage to the metal electrode to create a temporary "avalanche" breakdown of the insulating powers of the oxide around the floating gate. Then he proposed that this charge be erased by exposing the chip to ultraviolet light.

For this concept of the Floating gate, Avalanche-injection, Metal Oxide Semiconductor (FAMOS) EPROM, Dov Frohman was to gain a place in history for himself and enhance the history of Intel. More than fifteen years later, ironically enough, the dream of metal nitride oxide still captivated the creators of non-volatile memories among some of Intel's rivals, from Hitachi to NCR. Frohman and Intel thus were a key source of both the competing concepts in non-volatility: Intel's and the competition's. But Frohman went with the floating gate, and he has never looked back. Speaking for Intel in 1983, he said, "We believe floating gate will rule the world."

By the end of 1970, Frohman managed to produce a prototype for a 2,000-bit floating gate EPROM. One of the most capacious memory chips in the business, twice as large as the 1103, the device provoked Grove and Moore to incredulity. In its distinctive package—with a quartz window on top for the ultraviolet (UV) erase—it seemed as much a kludge as the DRAM. But by November 1971, the

1702 EPROM would provide a perfect complement for Intel's even more famous invention—the single chip microprocessor—in the company's announcement of "a new era in integrated electronics." Created largely by Faggin, Intel's microprocessors would ultimately bring nearly all the world of computing deep into the microcosm.

CHAPTER
8

Intel Minds

The development of computers can be seen as a process of collapse. One component after another, once well above the surface of the microcosm, falls into the invisible sphere, and is never again seen clearly by the naked eye.

This process began when firms stopped using discrete transistors in their computer CPUs and turned to invisible devices wired together in integrated circuits. Leading the shift to ICs were Digital Equipment's PDP-8 minicomputer in 1965 and IBM 370 mainframe in 1970. But both used hundreds of ICs and combined them with electromechanical memories and peripherals.

The fall of the macrocosmic computer did not turn into a rout until that amazing November day in 1971 when Intel announced not only the DRAM for working memory and the EPROM for software storage but also a microprocessor, absorbing the entire central processing unit of a computer on one chip. Combining the three inventions, it was probably the single most important announcement in the history of microelectronics. All the world's personal computers—in 1988 some 40 million in the United States alone—owe their basic structure to Intel's triple play. Sinking the Von Neumann architecture into the microcosm, and thus extending its life, the Intel advertisement set the essential course for the entire computer industry—for better, or for worse—for the following twenty years.

After introducing the silicon gate, Federico Faggin was the key figure in making the microprocessor's sketchy concept and flawed instruction set into an operational device. He described the invention of the microprocessor:

> Because there is a certain inevitability about inventions of this sort, which have often been anticipated by futurists, the real contribution lies, first, in making the idea work. What is needed next is believing in it passionately enough to supply the emotional energy to carry on the inevitable struggle until the new idea is firmly rooted in the world and has taken on a life of its own. It is a work of intellect and love.

The intellect behind the machine's conception began to become evident in 1954 near Rochester in upstate New York. . . .

Ted Hoff, a gangly teenager just graduated from high school and on his first summer job, did not like the plan for the new signal system. It had been designed by a team of engineers at his father's employer, General Railway Signal, to warn trains moving along a side spur if another vehicle was barreling down the main track.

A shy young man, soon to enter Renssalaer Polytechnic Institute (RPI), Hoff deferred to elders; he did not speak up against the device. But it offended his sense of economy. Because the signal would have to work perfectly—be failsafe—the design team had piled on extra circuitry to assure that it would operate under all conceivable circumstances. At a certain point, though, a more complex device becomes more prone to breakdown. As Hoff already sensed, in engineering more is often less.

Hoff suggested that some amplifiers be removed and replaced with simpler diodes. This change eliminated active devices—amplifiers— that might go into oscillation and issue false signals, and substituted passive sensors that could not misfire. Thus he simplified the switch and made it safer. Nonetheless, he was surprised when his name was listed on the patent application. This railway signal switch turned out to be Marcian E. ("Ted") Hoff's first patent.

Fifteen years later at Intel, perusing a complex new calculator proposal from a Japanese firm called Busicom, Ted Hoff was reminded of the original railroad circuit kludge. Busicom's idea was a ten-chip system designed to run a family of calculators programmable for various uses through plug-in, read-only memories (ROMs). It was essentially a good idea, later adopted by TI and Hewlett Packard. But at that point Intel could not afford to devote so much of its scarce design resources to one special project.

Ready to reject the proposal out of hand, Hoff marveled at the contrasting elegance of his PDP-8 minicomputer from Digital Equipment Corporation. Designed by Gordon Bell, the PDP-8 could be considered a primitive example of what is now called a Reduced Instruction Set Computer (RISC). Bearing a rudimentary instruction set, it needed a lot of specialized software. But with its versatility in jobs from data processing to factory automation, it was working a revolution in the computer industry under the bemused nose of IBM. Busicom was asking for a more complex design merely to function as a calculator. It affronted Hoff's sense of technical economy. Bell's design told him that less could be more.

What made the PDP-8 so useful despite its small processing powers was its relatively large memory. Hoff had just come off a memory project; Intel had several silicon memories in the works. Why not use this ability for the Japanese? Why not replace some of this customized or "random" logic—and incidentally finesse a few forty-pin IC chip designs—by using some of Intel's new memory technology?

The idea of substituting memory for logic was alien to the Japanese, but it made eminent sense to Hoff, a computer buff. For example, such arithmetic tasks as multiplying and dividing can be performed by creating a permanent set of circuits to do each of these functions in response to a complex instruction. Such specific dedicated circuitry is called random, or hard-wired, logic. But such random logic is unnecessary in many cases since multiplying and dividing can be conceived as forms of multiple adding or subtracting. The random logic can be replaced by software code in memory that tells the computer's adding circuits to conduct various repetitive looping procedures which are equivalent to multiplication or division. The result is a somewhat slower and more complicated process, but it allows a simpler and potentially faster computer. Freed of special circuits for every task, the computer can become a more useful and versatile machine.

This concept in design—replacing logic with fast memory—reached a new level in the late 1980s with the introduction of a new generation of RISC processors closely coupled to fast static RAM memories. RISC was presented as a radical breakthrough by some. But it merely reflected the essential principle of all general-purpose computer design. As memory increased in speed and reliability, soft-

ware could become more complex and powerful, and central processing units could become simpler and faster. RISC was less a novelty
than a new testament to the power of Von Neumann's original vision.

With a similar substitution of memory for random logic, Von Neumann had made his historic proposal to take the software program
out of the hard-wired logic of ENIAC, the first working electronic
computer, and put it into memory. Rather than reconnecting thousands of wires, the programmer merely could punch in the new instructions as if they were new data. This concept, in fact, could be
described as the key invention that changed the ENIAC from a glorified calculator into a programmable computer.

Now, of course, a glorified calculator was what Busicom wanted.
But why not give them a programmable computer instead? Rather
than creating ten specialized chips, why not design a small set of
generalized chips which could be programmed to perform the ten
tasks: to execute all the necessary calculations and run the calculator
keyboard, display, and other functions that otherwise would require
of Intel many man-years of tedious designing? It was the essential
RISC recipe, before its time. Hoff made such a proposal and his
bosses told him to go ahead.

In the face of resistance and skepticism from the Japanese, Hoff and
his team reduced the number of chips to three. One was a read-only
memory (ROM) for storing the various mathematical programs and
operating systems; one was a RAM for working memory; and one
was an entire central processing unit (CPU) in less than two thousand
transistors on a single piece of silicon: essentially the computer on a
chip prophesied by Carver Mead a year or so before.

When, in October 1969, executives arrived from Tokyo to choose
between the two designs—Intel's and their own—they chose Intel's,
but with no enthusiasm. Hoff recalls no elation himself. "I had mixed
feelings. We had no prototype and the message we received was,
'Hope it's as good as you say.' "

Hoff's counterpart at Busicom was a young Japanese named Masatoshi Shima who also had been thinking about problems of computer
architecture. An equally formidable intellect, Shima came to the project through a series of accidents, beginning with a misbegotten effort
to launch a small rocket using gunpowder he made by hand in his
high school chemistry laboratory. As he carefully followed the formula, he claims to have had the mixture exactly right, except for some
details that he overlooked. The mixture exploded, and as he pulled
away his right hand, it seemed a bloody stump. At the local hospital

a doctor with wide experience treating combat wounds felt lucky to save the boy's thumb alone.

This ordeal taught the teen-aged Shima that "details are very important." In the future he should "pay attention to all the details." But the loss of his fingers convinced his parents—and later several key Japanese companies—that the boy should not become a chemical engineer, even though he had won his degree in chemical engineering. Thus Shima ended up at Busicom chiefly because it was run by a friend of one of his professors.

Beginning as a software programmer, Shima moved on after six months to a calculator project. In this work he reached the conclusion that he needed to learn more about computer science. Reading volumes on computer hardware and systems and on electronic logic devices, he concluded that the entire logic system of the machine should be made with integrated circuits, which struck him as a fundamental and permanent advance in the field. He also was pleased with his own progress, moving down a hierarchy of abstractions from software programming to logic chips. He had just resolved to begin an intense study of integrated circuits, when he was abruptly summoned back home by a higher authority, his parents. They felt that they had at last found a job for him worthy of his talents. That is how in 1968, as Bob Noyce and Gordon Moore made plans to launch Intel, Masatoshi Shima, age twenty-four, left electronics, apparently for good, and became a policeman in Shizuoka.

Shima's parents were very proud. Because of his education, his salary would rise rapidly and he would become a respected figure in the city force. According to his parents, his life was set. They would no longer have to ache with worry at the trials of their partly crippled son in distant cities.

No sooner did Shima report at the police headquarters, however, than the protests began. He was too young and too cocky and too disfigured. He had not attended the local university like all his colleagues. He read too many books. They badgered him to the point that within three months he decided that the honor and power and prospects of the job were not worth the pain or the separation from integrated circuitry. He returned to Busicom, where the president of the company assured him their next calculator would be based completely on ICs. As it turned out, this would not be entirely correct. . . .

The new calculator instead would be based on the world's first microprocessor, an entire CPU on a single chip, combining in essence hundreds of ICs. Shima would help build the microprocessor in the

United States where no one cared about his lost fingers and where, working for Intel, he would become one of the world's most important circuit designers. Under Faggin's guidance, Shima's obsession with detail would merge with Hoff's vision of "less is more" to transform the computer industry.

In 1970, however, when Busicom sent Shima back to the United States to review the progress of the calculator project, he was furious to discover that no work had been done since he left six months before. In his broken English he kept complaining to Faggin, newly assigned to manage the project: "I came here to check, there is nothing to check. This is just idea." Shima's calculator project had been hopelessly delayed.

Finally Shima was persuaded by Faggin that he himself had just joined the firm two days before, that Hoff had moved on to other projects, and that there was nothing to do but get to work. On that point, Shima and Faggin at last had a meeting of minds.

Working closely with Shima, the young Italian both trained him as a chip designer and brought the project back to life. Within the nine months before Shima returned to Japan, they produced working samples of what had become a four-chip set in MOS. The additional chip was a simple register to connect the calculator to its outputs. Otherwise, with several architectural improvements by Faggin, they had executed the design essentially the way Hoff had envisaged it.

As a calculator set, using four-bit numbers, the device naturally operated a four-bit "bus," or channel, to transmit information through the system. It was thus a "four-bit" microprocessor, and was named the 4004. Comprising 2,300 silicon gate transistors, some four hundred more than estimated by Hoff, the CPU was contained on a tiny chip one sixth of an inch by one eighth of an inch. Although Busicom had full ownership of the device, in exchange for a lower price the Japanese company gave Intel use of the design for non-calculator purposes. The general-purpose microprocessor was on its way.

Like many epochal inventions, the microprocessor was greeted without enthusiasm even within the company that produced it. The sales staff, blind to the capacity of such inventions to create new markets, saw it as a replacement part with a potential demand of approximately ten thousand, or one fifth of the minicomputer trade. The board of directors saw it as a systems device, requiring heavy support and software, which would distract the company from its principal markets in semiconductor components. (They would have been entirely correct in this view, except for the emergence of the microprocessor as the most important semiconductor component of all.) Gordon Moore was enthusiastic, but none of the uses he envisaged—mostly for home conveniences—had been fulfilled by 1988.

Nonetheless, the chairman of the board, Arthur Rock, eventually succumbed to the arguments of Noyce and Moore, and the company resolved to make its splash into systems. In the November 15, 1971, issue of *Electronic News* appeared the momentous announcement from the two-year-old company. A half-page advertisement proclaimed nothing less than "a new era of integrated electronics." It seemed like the usual hype in an electronic press that announces a new era every month or so. But in retrospect, Intel's presentation was a model of decorous understatement. In the continuing saga of the collapsing computer, Intel was reducing an entire CPU to one chip.

Hoff's innovation was a perfect complement for Intel's new memories. Without the microprocessor, silicon memories are merely a minor improvement for the giant computer. Without silicon memories, the microprocessor is only a further form of integrated circuit, which unless joined to a large system of cores can perform only one fixed task. It was Intel's threefold offering—with CPU, working memory, and permanent software storage all together—that created a vortex of compounding efficiencies which ended by pulling the entire world of computing into the microcosmic domain.

The reports from the scene suggest that Intel at this moment was in some degree blinded by its own success. But a glimpse of the future from the center of the sphere offers a better guide than any number of bulky volumes of market research in the shadowy domain of the old establishment. Not only did Intel learn much from the process of developing and producing these devices, but the firm also learned vital lessons from its customers.

Intel knew that for some reason their 4004 had sold far better than anticipated. It was essentially a calculator chip, with all the limitations of four-bit addressing and calculator architecture. But its reception prompted Intel to pay new attention to a project it had initiated late in 1969, at the behest of Victor Poor of Datapoint, to develop a CPU for the Texas firm.

Although Datapoint envisaged a design with sixty separate packages of hard-wired, random logic ICs, Hoff, fresh from finishing the instruction set of the 4004, felt that the new device too should be put on one chip. Seeking a second source, Datapoint commissioned a design also from Texas Instruments. In a little known historic anomaly, it was TI's team, under Gary Boone, that finished ahead. TI produced the first eight-bit microprocessor and won the industry's first microprocessor patent. But Datapoint could not make it work.

Using the silicon gates that had given Intel such an advantage in DRAM dimensions, Faggin's eight-bit creation was fully operational and 60 percent smaller than TI's metal gate design. Thus Intel's chip was much cheaper to produce and package. But by the time the chip was ready for use, Datapoint was no longer interested in it and gladly yielded the rights back to Intel.

The difference in capacity between the two Intel microprocessors —the 4004 and 8008, one four-bit, one eight-bit—ostensibly suggests a factor of 2. But the eight-bit device could tap directly some 64,000 memory addresses, while the four-bit was limited essentially to 256. The eight-bit is thus 250 times as powerful in the crucial area of direct memory-addressing capacity. While a machine taking data in four-bit "nibbles" is suitable for dealing with the four-bit codes of binary-coded decimal arithmetic for a calculator, an eight-bit machine is appropriate for the eight-bit language of the American Standard Code for Information Interchange (ASCII) used in nearly all word processing and data handling.

After announcing the 8008 in April 1972, Intel disbanded its design team, started no new microprocessor project, and apparently planned to return its focus to the continuing battles in memory. But the world would not allow it. Not only did the 4004 begin selling heavily for scores of unexpected uses, from automating avocado sprinklers to playing 21; the 8008 created a still greater rush of demand in various computer applications. With engineers around the world writing for samples, the new era announced by Intel startled the company itself with the force of its global impact.

It was not until April 1974, however, that Intel finally mastered the microprocessor. In that month was launched the product which— with a short detour via Exxon—finally would establish Intel and its microprocessors as the dominant force in the electronics of the decade. The new device was the 8080, and its key designer under Faggin was none other than the young Japanese programmer from Busicom, Masatoshi Shima, whom Faggin had summoned back from Japan to work for Intel.

Using Intel's newest process technology and a new instruction set by Stanley Mazor, Hoff, and Faggin, the 8080 was ten times as fast and capacious as the 8008. Within the next five years the world would see hundreds of 8080 microprocessor applications, including scores of computer terminals faster and more powerful than the Datapoint specifications. Indeed, by the end of the 1980s, many of Intel's microprocessor chips alone constituted systems more sophisticated than the majority of mainframe computers at the time that Hoff set to work on the design of the 4004.

The microprocessor that managed to establish the Intel instruction set of the 8080 most firmly in the computer industry was also designed by Federico Faggin and Masatoshi Shima. But this time Shima was working under Faggin in an unprecedented setting: a spinoff from Intel, improbably financed by Exxon. This firm was called Zilog, and Shima, who executed most of the design himself, began by paying Intel (and himself) the flattery of imitation. He essentially reproduced, on his Z-80 chip, the Intel 8080 instruction set. But he added onto the same chip various controller functions that had been relegated to peripherals in the earlier design.

Because the CP/M operating system designed by Digital Research, Inc., and then used in most business software was written for the Z-80, it became one of the most popular of all microprocessors for early personal computers. Because of Faggin's shrewd choice of features to integrate, the Z-80 was also used in thousands of other products. Finally sold for a few dollars apiece and used for a wide range of toys and appliances both in Japan and elsewhere in Asia as well as in the United States, the Z-80 became the leading microprocessor in the world in unit sales in the late 1980s (though Intel's advanced devices led by far in revenues and profits).

Although Intel was enraged by the raid on its microprocessor team, in the end Exxon could be seen as financing an independent design and marketing test for Intel's microprocessor series. Intel's entire subsequent microprocessor strategy pursued the path of alternating advances in power (the 8080) followed by advances in integration (the Z-80 bringing peripherals on board) conceived by Faggin and Shima at Zilog. After designing the Z-8000—a chip with a 32-bit internal address "bus" and relatively small success—Shima felt cramped in the arms of Exxon and returned to Intel as head of its design group in Japan.

Meanwhile, Motorola's microprocessor began to challenge Intel's for the largest computer markets. Its 6800 family achieved great popularity in itself and also earned the tribute of imitation by ex-Motorola designers Charles Peddle and William Mensch at a small firm called MOS Technology. When Commodore bought the firm, it used Peddle and Mensch's microprocessor, the 6502, in most of its best-selling personal computers. Since Steven Wozniak happened to have some extra 6502s in his garage when he was putting together the first Apple in 1976, the Motorola based 6502 architecture became the most widely used in early personal computers.

Then, in 1979, Motorola launched what was the fastest and most potent microprocessor of all, the 68000, with 16-bit input-output and a 32-bit internal architecture. Potentially capable of addressing two to the thirty-two memory locations (4 billion bytes or four giga-bytes), it was adopted for the Apple Macintosh and nearly all high-powered computer workstations. Because the 68000 had essentially no limits in direct memory access, it seemed likely to eclipse the Intel 16-bit 8086. The 8086 could only work in 64-kilobyte (64K) seg-ments and with a total of 640K of applications program memory was too small for ambitious software—such as complex graphics—and a perplexing obstacle for applications designers to overcome. Mean-while, Mensch, working as a consultant, personally contrived a 16-bit version of the 6502, and sold it to Apple for the Apple //GS. In memory addressability and processing speed, every one of these chips was technically superior to the Intel 8086.

Nonetheless, amid all this apparently powerful competition, Intel taught the industry a crucial lesson. In the computer business, and in the related rivalry in microprocessors, the crucial market share is the share of total software written for your machine—share of the minds of software developers. In terms of the learning curve, the crucial experience is the accumulated experience of software writers with your computer's instruction set and operating system. Software standards are more important than hardware power. Even a substantial edge in processing power will fail the Drucker test of a tenfold increase. But an array of specialized software packages, written for a particular mi-croprocessor, are infinitely more valuable than MIPS on benchmark tests unusable for the needed application.

Intel's supremacy in microprocessors came from putting the soft-ware needs of the customer ahead of the technical visions of hardware designers. Through a series of shrewd hardware kludges, Intel re-mained supreme in its chosen field. It cleverly contrived the 8088, a chip that still drives computer scientists to frenzies of scorn. Combin-ing a 16-bit chip with an 8-bit bus and retaining the awful Intel segmented memory, the 8088 seemed to botch its communications and data storage, destroying most of the 16-bit advantages in speed and memory. But the device didn't botch the memory and communi-cations that counted: Intel remembered the scores of thousands of customers for its previous processors and communicated with the thousands of software writers who had invested many years in the Intel architecture.

Partly because this misshapen chip allowed usage of all the software and peripherals designed for the 8080 and Z-80, IBM adopted the 8088 for its first personal computer. Microsoft wrote the MS-DOS operating system for the 8088 and scores of thousands of software designers somehow learned to live with 64K segments. Including the 8086, the 186, 188, 286, 386, and 486, all adopted for some purposes by IBM and the major Japanese computer firms, the Intel family of microprocessors—kludges all—seemed assured of dominance well into the 1990s.

By launching the computer-on-a-chip, Intel gave America an enduring advantage in this key product in information technology—an edge no less significant because it was achieved largely by immigrants from Hungary, Italy, Israel, and Japan. Intel's three innovations of 1971—plus the silicon gate process that made them the smallest, fastest, and best-selling devices in the industry—nearly twenty years later remain in newer versions the most powerful force in electronics.

Intel's success is so complete that it constitutes the industry's greatest problem. Perhaps even more than IBM, Intel has become overwhelmingly the establishment in personal computing. It is the most powerful bastion of the Von Neumann architecture that its microprocessors perpetuated. It is the heart of the American computer industry. The huge challenge to its leadership is to manage a transition away from the architectures of its own dominance, before it is too late. This challenge will require innovative skills as powerful as Intel commanded in its early years. But it will also demand entrepreneurial skills. The microprocessor saga itself offers a clue.

Intel's revolution was not only unforeseen by the market but unwanted even by its first customer, Busicom. Hoff's impulse to give Busicom more than it wanted was a classic entrepreneurial expression of Say's Law: supply creates its own demand. Similarly, almost no one seemed to want the personal computer. It was widely regarded as an impracticable fantasy until it became a widespread reality.

Intel and its American rivals now face the challenge of giving the computer industry more than it thinks it needs—a machine that anticipates the increasing convergence of computer and consumer electronics. If anything is clear in electronics, it is that the television display, VCR, compact disk, video game, training tool, telephone, and other home appliances are on the way to becoming digital computer systems, full of microprocessors and DRAMs, and interconnected by fiber-optic cables. As the computer collapses into the microcosm, the Intel revolution of 1972 is rapidly moving out of the office and into the world.

The question is whether Intel has the vision and courage to follow its technology into the world and become an electronics giant, or whether it will remain in its profitable niche of IBM standard computer hardware. The coming era will require a return to the entrepreneurial spirit of the industry's infancy, when the marketeers of semiconductors overthrew matter in business structure and strategy, and no one was afraid of the Japanese.

CHAPTER

9

The Curve of Declining Entropy

Firms that win by the curve of mind often abandon it when they establish themselves in the world of matter. They fight to preserve the value of their material investments in plant and equipment that embody the ideas and experience of their early years of success. They begin to exalt expertise and old knowledge, rights and reputation, over the constant learning and experience of innovative capitalism. They get fat.

A fat cat drifting off the curve, however, is a sitting duck for new nations and companies getting on it. The curve of mind thus tends to favor outsiders over establishments of all kinds. At the capitalist ball, the blood is seldom blue or the money rarely seasoned. Microcosmic technologies are no exception. Capitalism's most lavish display, the microcosm, is no respecter of persons.

The United States did not enter the microcosm through the portals of the Ivy League, with Brooks Brothers suits, gentleman Cs, and warbling society wives. Few people who think they are in already can summon the energies to break in. From immigrants and outcasts, street toughs and science wonks, nerds and boffins, the bearded and the beer-bellied, the tacky and uptight, and sometimes weird, the born again and born yesterday, with Adam's apples bobbing, psyches throbbing, and acne galore, the fraternity of the pizza breakfast, the Ferrari dream, the silicon truth, the midnight modem, and the seventy-hour week, from dirt farms and redneck shanties, trailer parks and Levittowns, in a rainbow parade of all colors and wavelengths, of the hyperneat and the sty high, the crewcut and khaki, the pony-tailed and punk, accented from Britain and Madras, from Israel and Malaya, from Paris and Parris Island, from Iowa and Havana, from Brooklyn and Boise and Belgrade and Vienna and Vietnam, from the coarse fanaticism and desperation, ambition and hunger, genius and sweat of the outsider, the downtrodden, the banished, and the bullied come most of the progress in the world and in Silicon Valley.

Among media pundits, these people often fail to pass social muster. In 1987, for example, a visitor from *The New York Times* discovered the Valley to be full of unsophisticated rubes and obsessives. Prowling various computer and chip firms with her nose in the air sniffing for rot, she found it in rich "computer wizards" with the "Silicon Blues," "rigid" goons and "workaholics" who have "missed the meaning of life," who lack "harmony" and holistic feeling and languid afternoons in museums of modern art, and think Voltaire was the inventor of the volt.

Compared to the dedication and art of the average Silicon Valley engineer or entrepreneur, however, the ideal of balanced hedonism, cultural refinement, and mellow moralism exalted in the media is corrupt and feckless. Forty-hour weeks never wrote a Computer-Aided-Engineering software package, or solved a problem on the silicon-oxynitride interface, or launched a new ion implanter, or got a Midwestern conglomerate run by failsafe-financiers to take a chance on a new transistor. The fanatical energy and commitment of outsiders drove American electronics to the top and saved the world economy.

Little Wally was one of the outsiders who made it all happen. But he almost didn't make it at all.

"I was your sandy-haired, blue-eyed college kid who thought because he could bench-press his own weight he was going to make the world safe for democracy. I was going to kick ass." Now he was in the trunk of a car bleeding fast and virtually dead. His skull was cracked, his nose squashed, his ribs broken. He was gagging on blood and couldn't see. It was not at all the way he wanted to be remembered. But the priest leaned over him gravely and recited his last rites.

Like many fatherless boys of the Chicago slums, Wally was a fighter. A 95-pound kid struggling for respect and attention from his peers, his life was changed by an ad in a sports magazine. It was the same ad that every tormented youth of that era read and clutched to his skinny chest: the ridiculous and uncannily targeted story of the weakling on the beach. The bully kicks sand in his face and steals his girl. He goes back to his room and lifts weights or applies "dynamic tension" until he can avenge his humiliation . . . And sometimes he scrutinizes more closely the sand.

After a couple years of weightlifting, he had nearly doubled his weight and the street fights began going his way. Then he went off to college in Urbana. Returning on vacation, he saw one of his friends attacked on the street by a local bully. It was time for the moment of heroic revenge in the dynamic tension ad. Wally stepped into the fight. And he was doing fine, well ahead on points, until his friends abandoned him and the bully's friends pinned back Wally's arms and beat him into the ground and stomped on him and carved his face with a can opener.

It took all the best efforts of the Little Company of Mary Hospital on 95th Street to put him back together again. Thirty years later, his nose remains crooked and he still has a depression on his forehead "where they took the bone out to kind of fix my nose." It must have been a harrowing experience of betrayal and trauma. But Wally does not today include this story in an interview transcript on the trials of his life. The brush with death in the car trunk happened after he got to college. The worst had already happened earlier, when Little Wally had to face the public in "inappropriate attire."

Things had been looking bad for some time. His mother, one Catheline Finn, was fifteen when he was born, and his father was already fading from the scene. Wally's life continued that way, ghetto pickings, for most of his childhood. Two of his uncles died in prison of accumulated petty offenses. His mother lost interest or lost track, as she produced a series of further kids with another man. He was finally saved by his paternal grandfather, an electrical engineer, who brought him in from the streets, corrected his grammar, and pushed him to excel in school.

In retrospect, the older man gave Wally his chance. But to the boy, full of emotional hungers, the grandfather then seemed a mean and tyrannical figure. Nothing the grandfather could do could restore the lost maternal love of his childhood or save young Wally from the painful trials and humiliations of a rejected boy foraging for his manhood on the streets of the city.

The worst of all was the dinner with the mayor. Little Wally was smart, no doubt of that. He had won the history contest at Earl Elementary School on the South Side and gone down to Mayor Daley's banquet to get his prize. On his arrival, he realized he was the only kid without a suit. There they were, the spruce youths from Oak Park and New Trier and Beverly Wood, all looking like miniature high finance, and Little Wally was in a sweatshirt and torn jeans. When he came back to school the next Monday, the principal told him the teachers had been embarrassed by his appearance. Did perhaps his family need some sort of welfare assistance? It was mortifying: a key difference between the ghetto youth of then and now.

But some things don't change. From despair sprang fantasy. Little Wally would grow up to be a movie star. When he was fifteen, his grandparents took him for a summer visit to California and the fantasy sprouted fronds of green. "I looked there at the palm trees, the green grass, the hibiscus, and I thought, Jesus, this is really it. I'd have a house above Sunset Boulevard and I'd have a Cadillac convertible, and I'll be a big guy." He visited Santa Monica and saw the beautiful girls and he knew they would lavish on him for a lifetime the attentions his mother withheld. His Chicago peers would gnash their teeth and pull their hair, seeing at last how hugely they had misjudged his nobility and worth.

Back in Chicago, getting up at 4:00 A.M. to deliver the milk before school, he would think about California. In the evening, serving as an usher in the local movie theater, he would dream about Hollywood. On the weekends, cleaning up the broken bottles, the trash and the vomit at a bar, carrying out the garbage cans, unloading endless crates of beer for the coming night, his mind would wander to the Santa Monica Beach. In the summer, lugging cement bags upstairs for a construction gang, he fantasized about great wealth (maybe even a hundred dollars a week in the money of the time).

In January 1954, he graduated from high school as valedictorian, won a George W. Pullman scholarship, and changed his name from Wally to Jerry. Since his grandfather was Walter Jeremiah Sanders the first, and his father was Wally Junior, Jerry had suffered for years as Little Wally. Going off to college, he at last could leave the name behind. But you should not forget that seething and struggling young man. Little Wally would never be hard to find in the life of Jerry Sanders, entrepreneur.

Accepted by MIT and Carnegie Institute, he resolved on the University of Illinois in Urbana because it was cheaper. Eventually he followed his grandfather in the study of electrical engineering. "I never had a real driving desire to be an engineer, but I always had a driving desire to be a success." He graduated with honors in 1958, but his parents didn't come to the ceremony. "They were 135 miles away, and God, that was really an inconvenience."

After graduation, he went to work in Los Angeles for Douglas Aircraft at $500 a month. Still dreaming of Hollywood, Jerry thought it was very fateful and significant that at Douglas he should immediately be assigned to the exact same desk just vacated by Ty Hardin, who had supplanted Clint Walker on *Cheyenne*. Sanders began spending his time off strutting down the Santa Monica sands. Finally, one of his co-workers told him that if he wanted to go into the movies he should actually talk to an agent, rather than depend on his animal magnetism alone to draw them to him inexorably down Muscle Beach. It was an idea. Wally's friend knew a friend of Billy Wilder, who knew of a woman, who knew her way around the business.

Sanders went to see the woman, a worldly middle-aged Hollywood hand, and they hit it off immediately. But she had bad news. She told him he could never be a leading man with a face like his. "You got a good body and you've got kind of a rugged nice look about you, but look," she said, reaching for a sheaf of photographs, "I got a guy here who can't get arrested in this town. He's gorgeous and he's suave and he can't get any work." His name was Efrem Zimbalist, Jr. Wally looked at the pictures and looked at himself in the mirror and touched his potato nose and the dent in his forehead and painfully felt the dissolution of his dream.

The woman must have seen a shadow pass his face or water well in his eye. Jerry was a very emotional young man. "Why do you want to be a movie star anyway?" she asked with a note of concern.

There was something about the woman's air of cynical worldliness that evoked Jerry's trust. At last he blurted it out, "Bottom line, I guess, is women . . . adoration, adulation."

Wally would never forget her response, offered in a tone of Delphic truth.

"That just takes money," she said. "You want to have women, you produce the pictures, they'll act in them. Actors aren't worth a shit!" His world tipped giddily as he contemplated a new theory of life. Actors were replacement parts. He did not merely have to imitate success; he could actually be it. He went back to Douglas Aircraft and threw himself into his first design job.

Sanders turned out to have a keen feel for technology. He began by designing an electronic pitch-trim compensator: a device for controlling the elevator on the airplane tail to maintain steady vertical pitch during flight. He was then given the job of creating the air-conditioning control system for the DC-8, using Zener avalanche diodes from Motorola to control the potential at the required 28-volt level (Zeners exploit the steady voltage of diodes in the breakdown mode).

In his dealings with the semiconductor company, he made an important discovery: Motorola's salesman knew little about the technology and he knew a lot; yet the salesman got to drive a company car and take clients out to lunch at restaurants in Santa Monica that served fancy steaks and salad with Roquefort cheese dressing. That was enough to get him out of aerospace and into sales at Motorola Semiconductor in Phoenix.

When Jerry joined the firm in 1960, Motorola was at the beginning of a ten-year upsurge under C. Lester Hogan, who would take it from $3 million to over $200 million in sales and would make it the world's second-ranked semiconductor firm behind Texas Instruments and first in profits. But after a few months it made the terrible mistake of assigning Sanders to a selling job in Chicago.

By 1961, Sanders was acquiring a deeper sense of what was important to him in his profession and it wasn't selling diodes in Chicago. He might have come to Douglas to be near the beach and to Motorola for the Roquefort cheese dressing in Santa Monica. But he moved from Motorola to Fairchild Semiconductor because of Robert Noyce's integrated circuits and Tom Bay's understanding of them as the marketing chief. He found Noyce "simply awesome," and Bay, at once formidable and friendly, unique among marketing bosses. They filled the key gap in his life: they were male role models who could channel his raw energy and restless intelligence into the pursuit of something real. "In those days I had never met a marketing guy besides myself who knew products." At Fairchild, everyone knew products.

Furthermore, he was surprised to discover, they would even send him to Southern California. Giving up on his fantasy life to pursue the microcosm, he suddenly found himself parking a Cadillac convertible outside his new offices on 1191 Sunset Boulevard. His job was selling chips to the aerospace market and he quickly made his mark by getting a major order from Hughes Aerospace. Soon he showed himself as a rising star in the company and within fifteen months he was promoted to manage the consumer market for the entire country.

His mandate was to take $50 chips consigned for small defects to the military reject bins and sell them for one dollar to the entertainment market. He made some gains in this effort. But his big break did not come until in 1963 FCC chairman Newton Minow decreed that all future television sets would contain UHF tuners. He quickly saw that the UHF oscillator needed in every one of these tuners in every television set in America represented a huge opportunity for his company. Fairchild already had a chip, the 1211 transistor, that alone among all semiconductor devices could do the job.

Then came the bad news. RCA announced a "cool tube," a new miniaturized vacuum tube called the nuvistor that also could do the job. A major breakthrough, the nuvistor was a rugged miniature ceramic device with low power, high reliability, and high frequency. RCA set its price at $1.05. With the market for vacuum tubes reaching new highs, the tube people were not about to give up yet. The nuvistor was their most impressive feat of miniaturization and high-frequency performance. They knew that if they could not win against the transistor for this kind of application, tubes would ultimately be doomed, and the lucrative replacement business would disappear.

At the time, Sanders was selling the 1211 transistor in small orders for $150 apiece to the military. Although he knew he could bring the price drastically down for such a large market as the UHF tuner, the nuvistor seemed unbeatable at $1.05. Crestfallen and demoralized at the prospect of seeing this vast market slip away, he returned to Fairchild headquarters to see what could be done. Perhaps the company could contrive some other chip to do the job.

Instead, Fairchild surprised him by authorizing sale of the chip, at lower guaranteed performance than the military version, for $5. Once again he did the rounds. "I visited RCA and I visited Magnavox, GE, Zenith, and Motorola, and all these guys said, 'Yeah, it's all well and good, Jerry, and you've got a dynamite idea there, but the nuvistor does the job and it's $1.05.' " The transistor was better in that it did not require a deep socket and fragile wire connections; it could be soldered directly on a printed circuit board and it never wore out or disconnected with heat and use.

At the time, however, American television vendors saw maintenance as inevitable in their product. Replacement parts were seen as a key source of cash flow and an advantage against the Japanese. When the Japanese used solid state devices to create televisions that never broke down, the American companies lost most of the market. But in 1963 compactness and durability were not arguments that could usually prevail against a $4 difference in price.

Then the Zenith engineers informed him of a further problem. This one the entire Fairchild sales force assumed was terminal: a deviation from smooth operation at the highest frequencies. Sanders's sales-people were already uneasy about the amount of time they had devoted to the project and were giving him a long list of reasons why they couldn't sell the transistor. Now came this technical clincher, dubbed by Zenith "high-frequency suckout."

Watch out how you talk in front of Jeremiah Sanders. "High-frequency what!" The Zenith man explained that the transistor deviated slightly from linearity at the top of the UHF band. Did it still function as a tuner, Sanders wanted to know. Well, yes. It was an unnoticeable non-linearity. But it was another reason why Sanders would not be able to sell the 1211 for UHF tuners unless Fairchild could drop the price to $1.05. They had already dropped it from $150 to $5 and now he wanted it at $1.05. His organization of sales engineers told Sanders to forget it.

Sanders, however, already had an intuitive grasp of the remarkable dynamics of industry pricing. The greater the volume, the better and more reliable the product, the lower the price. This was a market of some 10 million TV sets a year compared to a few hundred radar stations. What could be fairly sold for $150 for a small number of military applications, could be sold for $5 and maybe less for a huge consumer market. He returned to Mountain View and made his case to Tom Bay. Bay arranged a meeting that Sunday afternoon in Bob Noyce's house in the Los Altos hills.

It was a revelation in the life of the young marketeer. There he was, an agitated kid, drinking beer on Sunday afternoon with the inventor of the integrated circuit and Noyce was listening attentively. While Sanders was almost embarassed to reveal the necessary price, Noyce took it in stride. After Sanders explained the nuvistor and what it did, Noyce smiled and said he guessed they would have to knock that one out now. "I guess you're just going to have to quote $1.05," he said.

Imagining the immense market opening before him, Sanders was euphoric. But he still could hardly believe his luck. The chip then cost $50 or more apiece to make. How could they produce it for a dollar? Charles Sporck, the manufacturing man, was not even at the meeting. How could Noyce be so sure? But Noyce met the next week with Sporck, and Sporck and Julius Blank, a Fairchild founder, left soon after for Hong Kong to set up a factory to assemble the chips in cheap plastic packages. It would be Fairchild's first product in plastic. It would be Fairchild's first product manufactured outside the United

States. "We were going to make the chips in a factory we hadn't yet built, using a process we hadn't yet developed, but," as Sanders recalls, "bottom line: we were out there the next week quoting $1.05.

"I went back into Zenith and they gave me the whole suckout story again before I could even make my case. Then I said: '$1.05.' And everyone was very quiet for a while and then they said, 'We can use it.'" Even RCA, the creator of the nuvistor, finally gave up and incorporated the chip. Replacement of heavily shielded tubes in the UHF tuner, when they wore out, turned out to be more trouble than it was worth. Noyce and Sanders had finally tracked down the new high-tech vacuum tube into its own lair and destroyed it.

"We were selling the future," Sanders said. Not only did the 1211 not have a socket, it was smaller and had an essentially unlimited lifespan. And with the price at $1.05, Sanders could say that there was no telling how low it might eventually go. Two years later, in 1965, the price in fact dropped to 50 cents.

Sporck and Blank soon had the plant under way and for many years it stood right next to the Ki-Tac Airport in Hong Kong, with a large flying "F" on the roof. Within a few years, the Hong Kong factory was producing the chip for less than 15 cents and 15 cents eventually became the price. It had dropped by a factor of 1,000. Sanders was in the microcosm at last and it was even better than Hollywood. "When you look at an automobile, you know there is a certain amount of iron in it and there is a limitation on how cheap it can be." But in a chip the bulk of the cost is in a plastic package 300 mils wide. With the chip, the price is in your mind.

From 1963 through 1965, Fairchild won 90 percent of the UHF tuner market in the United States, perfected its plastic packaging, and launched its overseas assembly. The cheaper the transistor became, the larger the market grew, and the more money Fairchild made on the part. As the price dropped, the margin remained near 30 percent, and in 1965 this one item selling for 50 cents apiece represented more than 10 percent of the company's profits. And the nuvistor was forgotten.

All this happened mostly because one sales manager would not give up. He could descry, in the teeth of a $150 price tag and a $100 fabrication cost, a highly profitable 50 cent product. This is the mentality that built the U.S. semiconductor industry and that today is making the Japanese the titans of consumer electronics. At that very time back in the sixties, for example, RCA sold videotape recorders for several thousand dollars for industrial uses. But only the Japanese saw in that large and complex device the essence of an elegant system that could be sold by the millions for $300.

Selling below cost is the crux of all enterprise. It regularly transforms expensive and cumbersome luxuries into elegant mass products. It has been the genius of American industry since the era when Rockefeller and Carnegie radically reduced the prices of oil and steel. Above all, it was the crucial principle of the semiconductor industry in its ascent to the pinnacles of American business. In the saga of the 1211 transistor, Jerry Sanders had shown an intuitive grasp of the key entrepreneurial dimension of the assault on the microcosm: Learning must rise and matter decline as a share of the value of products.

Today, however, while the Japanese dominate market after market by obsessive use of the curve, many American companies and business experts believe that the strategy is somehow obsolete or reprehensible. Even Jerry Sanders himself has taken to charging the Japanese with "selling below cost." Many analysts associate the learning curve with Texas Instruments and other American semiconductor firms that have sometimes stumbled in recent years. The fashion at many business schools is now to say that the experience curve strategy has failed.

As a conscious and principled strategy of entrepreneurship, TI did in fact originate the learning curve. But nothing that happened in Texas or in Silicon Valley in any way undermines the experience curve as an indispensable principle in the entry into the microcosm.

Many semiconductor companies had noticed the radical drop in costs that attends increased volume in chips, but they imagined that it was a peculiarity of their own industry. Often the learning curve was confused with the wafer yield curve (the increase in the proportion of working chips as volumes rise), or with the density curve in integrated circuits, as in (Gordon) Moore's law, which predicts a doubling of memory capacity every year.

In reality, all these special insights pointed to a broader principle. Following the inspiration of Patrick Haggerty, TI collaborated with Bruce Henderson of the Boston Consulting Group and with his protégé William Bain of Bain and Company to subsume all these curves into the more powerful general theory of the experience curve. Applicable in all industries, this principle asserts the famous drop in costs of between 20 and 30 percent with every doubling of accumulated volume. The thousandfold drop in the price of Fairchild's 1211 transistor, for example, corresponded to a more than four-thousand-fold rise in its accumulated unit sales.

The semiconductor people, however, are correct in believing that the curve works with special power in their own industry. Graphed against a given period of time, the curve tends to be especially steep for information technologies. Because the ratio of intellectual value added to material costs starts high for information products, volume increases often come with great speed. When matter plays so small a part in production, there is less material resistance to increased volume. In chip production, for instance, volume is increased not chiefly by processing more silicon but by placing more circuits on a given area, which is almost entirely a function of accumulated knowledge or experience.

One of the early summaries of the general theory of the experience curve appeared in the Texas Instruments report to its stockholders in April 1973 as the company prepared to enter the consumer fray selling Jack Kilby's calculators against the Japanese.

The keys to world competitiveness, said the report, were market share and market growth. TI should focus its resources to gain an expanding share of those markets that were growing fastest. In a fanatical pursuit of market share and market growth, the company would have to build capacity before demand for the product was sure; it would have to lower prices without assurance that new markets would emerge; it would have to enlarge volumes to increase learning and lower costs, without knowing competitors' plans; it would try to share its semiconductor experience among other products.

Above all, the company would have to stress quality, in the knowledge that the cheapest goods are normally the best goods, with the largest markets and the largest volumes. Volume without quality is not production; it is merely waste. By the same token, however, quality without volume is not production; it is research and prototyping. Design quality must focus on constant improvement of production in volume.

Obviously a high-risk approach, it would provoke all the familiar charges of "dumping" goods and "selling below cost" and "predatory pricing" that ring in the ears of all industry leaders on the experience curve. But as TI president Fred Bucy pointed out at the time, "It is absolutely mandatory to compete successfully with the Japanese." TI used the strategy to blast into the calculator market in the early 1970s and hold the leading share for ten years of explosive market growth.

Even though TI's advantage in experience applied only to chips—

which in turn comprised a steadily diminishing portion of the cost—
TI charged on from calculators into digital watches. Partly a piece of
jewelry, this product posed more difficult problems. But TI nonetheless became the driving force in the market, pushing the price below
$10 and outperforming all established American watch firms for sev-
eral years. It outlasted four leading semiconductor companies, including Intel and National, and drove mighty Gillette out of the fray
before finally giving way to the Swiss and Japanese in 1981.

This saga is often seen as illustrating flaws in the strategy. But the
strategy does not assure enduring dominance to all the firms that
follow it. It merely shows how winning companies will achieve their
victories and that the companies with the largest share of whatever
market—large or small—will tend to have the highest long-term profits. Ever since that seminal stint at TI by the Boston and Bain consultants, they have been working out, applying, and proving ever more
widely the strategic implications of these microcosmic insights. In the
process, Henderson first, and then Bain still more successfully, have
created major companies themselves: converting their own growing
experience into an industry of knowledge and their knowledge into
further experience at companies around the world.

Patrick Haggerty had been correct; upholding an essential insight
of Say, the microcosm was volume-driven. By launching the company
down the learning curve in the production of plastic packaged devices
in volume, the UHF tuner had prepared Fairchild—and its successor
Intel—to exploit the huge markets for integrated circuits, memories,
and microprocessors. In the microcosm, you had to be willing to lose
money on every unit and make it up by volume. Sanders's coup made
the company ready for the ferocious world rivalry that would emerge
over the next decade in integrated circuits, memories, and microprocessors in cheap plastic packages.

The learning curve is the most fundamental fact of microcosmic business strategy. It registers the overthrow of matter in an economy. At
the beginning of any production process, uncertainty is high. In information or communications theory, such uncertainty is defined as
"entropy." The measure of this entropy is how much information,
measured in "bits," is required to resolve the uncertainty. Technically,
there is no connection at all between this informational entropy and
the entropic change of energy from concentrated and useful forms,
such as oil, into useless gases and other wastes. But in fact these two

forms of entropy—informational and thermodynamic—do join in a rough way in all economic processes.

In any manufacturing activity, both forms of entropy play a role. Energy is being used up and uncertainty is being reduced. Analysts of production processes usually focus on material entropy: the irretrievable consumption of useful energy: the *increase* in physical entropy. But more important is the reduction of informational entropy: the replacement of uncertainty with knowledge as total production experience rises.

In both forms, the pace of entropy replacement defines the curve. At the outset of any production process—whether making cars or transistors—uncertainty is high. Workers do not fully understand their jobs. No one knows exactly how well any of the machinery will function. A high level of uncertainty in a process means that its managers must supervise it closely, keep high levels of supplies for emergencies, maintain high manufacturing tolerances or margins for error. Without a substantial body of statistical experience created by high production volumes over time, for example, managers cannot even tell whether a defect signals a serious problem recurring in one of ten cases or a trivial one occurring once in a million cases. Quality control requires statistics over an adequate number of units.

Early in the life of a product, uncertainty afflicts every part of the process. An unstable process means energy use per unit will be at its height. Both fuels and materials are wasted. High informational entropy in the process also produces high physical entropy.

The benefits of the learning curve largely reflect the replacement of uncertainty with knowledge. The result can be a production process using less materials, less fuel, less reworks, narrower tolerances, and less supervision, overcoming entropy of all forms with information. This curve, in all its implications, is the fundamental law of economic growth and progress.

The successes of the microchip industry, however, would come not only from more rapid moves to high volumes, but also from more rapid introduction of new products, down new curves of cost. The microcosm impels what Carver Mead calls a compound learning curve. The major innovations launched every decade or so—the transistor, the integrated circuit, the microprocessor—rekindle the curve and produce a compound rate of interactive advance much greater than the usual 30 percent rate of gain enjoyed by a single commodity when accumulated output doubles.

The situation for any one company with many products inter-twined on compound curves of learning is far more complex than a single curve would suggest. Among the best at managing the com-pound curves of the microcosm would turn out to be the Japanese, learning their lessons from Bucy, Noyce, and Sanders.

CHAPTER
10

Japan's Microcosm

As Peter Drucker said, "What one man can do, another can do again." Distilling discoveries of science, a set of technologies, and a philosophy of enterprise, the microcosm is far too big for any one country. Even its products are mostly made of ideas—waves that suffuse the mindscape of the world.

Whether captured for a while on magnetic disks, or flickering in a dynamic RAM, or vibrating electromagnetically around the globe, these ideas tend to flee confinement. They might originate secretly in an intimate interplay at a computer's cathode ray tube (CRT) screen. But once they are widely shared, the ideas become the knowledge of others and there is no way they can be stuffed back into the tube—no way the general shape of the ideas belongs to anyone anymore as defensible property.

The microcosm is at large and mostly free, in every sense of the word. It is public in a way no widget, ingot, automobile, restaurant meal, barrel of oil, or other good can be. Unlike material products that are used up as they are used, microcosmic knowledge grows as it spreads.

Many Americans imagine that quantum technology was owned by the United States and then filched or copied by Japan. But if the microcosm is too big for one country, it is no country's property and it affords no separate destinies. Americans and Japanese alike can possess its fruits only as long as their governments and people alike observe its universal laws and disciplines.

Not only did Japanese companies and individuals make major and original contributions to the journey into the microcosm, they did it in roughly the same way as the Americans did. The myth of MITI—Japan's Ministry of International Trade and Industry—is as dubious as it is durable. According to the myth, a prescient group of bureaucrats at MITI planned, designed, guided, and funded Japan's efforts in chips and computers, using and coordinating the resources of Japan's huge business organizations along the way.

This vision is about as true as Japan's counterclaim that still larger Pentagon subsidies explain the success of Silicon Valley. Yet the MITI myth wins by acclamation around the world. It arises in great tapestries of Marxist history and in narrow liberal surveys of technological rivalry. It provides convenient grievances in trade litigation against dumping, anti-trust offenses, and non-tariff barriers; it animates endless congressional hearings on "industrial policy"; it enthralls business conferences and annual meetings mobilizing for government aid. It is the great alibi of American capitalists as they struggle with sometimes superior rivals in the East, and it is the inspiration of European socialists as they summon a new five-year plan for information technologies. It is a wonderful boon for MITI itself. Its officials wax warm and expansive to the American intellectuals and politicians who come to learn of the marvels achieved in Japan simply by giving power to intellectuals and politicians.

So the academic consensus ascribes the Japanese ascent to a government plan. In the late 1970s, the far-seeing chieftains of Japan Inc. resolved to transform their country into an information economy. Seeing the microchip as the heart of computer and communications technology, MITI in 1977 launched a $300 million program of governmental guidance, subsidies, and collaborative research and development. Central to the effort was a coordinated drive for the import, adaptation, and near theft of U.S. capital equipment. By the early 1980s, these measures bore fruit in domination of the DRAM, the keystone of the computer age.

In explaining economic achievement, academics are inclined to exalt government bodies dominated by intellectuals. In fact, however, the MITI plan was largely a stream of clichés about the inevitability of an information economy, and the government funds were dwarfed by Pentagon monies in the United States.

By the early 1970s, the Japanese had already moved close to the U.S. in many semiconductor technologies, including DRAMs. Under Dr. Sueo Hattori, NEC began producing 1K DRAMs at its new Kyushu factory in 1971, just months after Intel. The Japanese were as good as the Americans in fabricating semiconductor products as early as the 1950s. What stopped them from exporting to the U.S. was the huge U.S. lead in innovation, embodied in patents won by Kilby, Noyce, Hoerni, and Bell Labs that, together with a 4 percent tariff, took 15 percent off the top of any Japanese export revenues. This obstacle diminished during the 1970s, when the Japanese began to acquire countervailing patents and licenses.

Operating out of a tiny independent laboratory, Shumpei Yamazaki, Japan's "patent man," invented low pressure chemical vapor deposition (LPCVD), crucial to semiconductor production, and later adopted by Applied Materials and others. With no government support, Yamazaki also made several advances in nonvolatile memory and MOS processing. Matami Yasufucu of Fujitsu bought Yamazaki's patents and used them to help break into the U.S. market. Hitachi already had patents on a reorientation of the silicon slice necessary for moving beyond two-inch wafers. NEC had an agreement with Fairchild and Toshiba with RCA. As soon as the Japanese could circumvent or surmount the patent hurdle, their enduring strengths in manufacturing would move to the fore.

In the critical area of capital equipment, the Japanese made key innovations themselves in vacuum technology and robotics. Nonetheless, the Japanese relied chiefly on U.S. gear. More important than MITI, however, in bringing American capital equipment to Japan was the mighty ministry of TEL, an institution never visited by professors and scarcely known by any. Its story begins on a cold February day in Chicago in 1959.

The Japanese youth, thin and dignified, with a shy, toothy smile, was in town to sell resistors, capacitors, and tubes to Motorola. Compared to American devices, he knew that his stuff, mostly from Fujitsu, was "junk," but he also knew his job was to sell it. His firm, Nisho Iwai Trading Company—and his country—desperately needed foreign exchange to buy the crucial American technologies that would animate Japanese productivity. He would try to offer a better price than the Americans. Yet more likely he would have to struggle even to make himself understood in his functional but hesitant and heavily accented English.

Leaving the airport bus, he could scarcely make his way down the pavement in the face of the blasting gusts of wind, full of icy grit from the streets of the city, to the door of the YMCA where he would spend the night. He paid the man at the front desk $4 and went up to his cubicle. He saw a shower room. Perhaps a hot shower would warm him up. He took off his clothes, stepped out onto the frigid concrete, and turned on the water. None came. He tried all the other outlets. Nothing. He rushed back to the bed with its thin blankets and lay there shivering. Welcome to Chicago, U.S.A.

He thought of his father's life back in Kyushu, the remote and barren southern island of Japan. His father had run a prosperous business before the war, but had lost everything after it. He had tried to scrape out a living by farming, but Kyushu was scarcely arable and was far from big city markets. Now the old man was ill and despairing of his life. The youth wondered what he was doing in America trying to sell inferior products in a land of plenty when he was needed at home.

The next morning he headed out to the Motorola purchasing department in the Chicago suburbs. By the time he reached the plant, taking an endless series of buses, there was already a line of sleekly dressed American salesmen in the office ahead of him. He checked in and sat down to wait. The names were read out. General Electric, Sprague Electric, Transitron, and then, sure enough, Nisho Iwai America. He showed his line of gear to the purchasing agent. To the Japanese youth's surprise, the man got briskly down to business, ignoring his lack of credentials and his problems with English. Fast in fast out, was the Japanese salesman's fear. But the Motorola agent made a substantial order on the spot.

The young salesman shook his head in amazement as he left. It could not have happened that way in Japan. All large users would require an introduction and a long proof of performance. It occurred to him that "the United States was a successful nation because it was open to the unknown." It fostered new business because large businesses would gladly purchase goods from obscure vendors if the price was right. From the Motorola plant he proceeded on to a Magnavox TV facility in Tennessee and the same thing happened again. He got an order for some 200,000 vacuum tubes.

Wiry of build and nearly six feet tall, this Japanese youth was named Tom Kubo. His home base was a dingy fourth-floor apartment off

110th street on the Upper West Side of Manhattan. He was paying $60 a month rent out of his monthly earnings of $350. He remembers that IBM was offering $500 per month to workers in the ghetto near his apartment. As an overseas representative of a major company, he was worth 30 percent less than line labor in America. He was at the bottom of the heap.

Nonetheless, he was excited to be in the United States, captivated by the magical giants who bestrode this mythical land. He recalls Roger Maris hitting home runs at a heroic pace, John Kennedy stirring the country with his campaign for President, and visitors from Japan all thronging off to see *West Side Story* and *My Fair Lady,* then booming on Broadway. He remembers scores of movies he attended to learn the language and the culture. But Kubo was not in New York as a sightseer.

Beyond his early assignments selling Fujitsu components and Kobe Kogyo transistor radios, his mandate was extended to find American products to sell in Japan. This meant taking visitors from Nisho Iwai and its client companies to Western Electric in Allentown, Pennsylvania, and RCA in Somerville, New Jersey. Kubo also worked as an interpreter for many Japanese notables, including Leo Esaki of Sony on his trips to Bell and RCA after inventing the tunnel diode in 1957, and Endo of Tokyo Broadcasting System.

The key to Kubo's success and the inspiration of his subsequent activities, however, came less from his access to world-renowned corporations than from his discovery of several smaller firms. Although semiconductors have become a mass-manufactured commodity, the equipment used to create the chips has usually originated in small establishments ruled by one or two high-technology craftsmen. Even if these key men sometimes work within a large company structure, whether Bell Labs or IBM, they usually operate nearly alone until the final stages of development.

In 1960, long before many Americans in the industry had heard of Fred Kulicke and Al Soffa, Kubo visited their cramped machine shop in Horsham, Pennsylvania. In a tiny office dominated by a rumpled bed, he met the two engineers and found they were developing a wire bonder: a machine that could swiftly and reliably attach the tiny gold wires from a transistor die to the leads on its package.

Later he visited an almost equally small firm in a former textile mill in Westboro outside of Boston, where a young man named Ken Olsen was making printed circuit boards for voltage meters. Olsen grandly called his shop Digital Equipment Corporation. At Airborne Instruments Laboratory in New York, Kubo encountered planar transistors for the first time and listened to a formidable engineer tell him of the immense future of this technology from a new firm called Fairchild. Kubo sensed a coming explosion of possibilities in the industry; the technology seemed like a "great surprise package," a *piñata* of magical treats, about to burst upon an unsuspecting world.

Encouraged by Kubo's early successes, Nisho Iwai in 1960 dispatched two other representatives to investigate the possibilities in America. One was a short, smooth-faced young man named Tom Kamo, and the other was a stocky engineer, like Kubo a few years older, named Tom Kodaka. They also went forth to the small laboratories and facilities, usually dominated by one or two men, who were designing and assembling new capital equipment for the emerging semiconductor industry. Among their finds were Honeywell, building a new diffusion furnace, and VEECO, a small firm in Long Island making vacuum gear for deposition of metal conductors on a chip.

When Kulicke & Soffa began producing its bonder, Kamo and Kodaka closed the deal while Kubo dealt with Nisho Iwai's customers in New York. At first the two American engineers in Horsham were dubious of the three Japanese Toms, all still in their twenties. K&S made Kamo "sweat it out" for a week in a sweltering three-dollar hotel room in Philadelphia before they would even allow him to visit. But of approximately $1 million in bonders sold in 1960, one third ended up going to Japan. These unexpected orders generated most of the profits that allowed K&S to expand rapidly through the early 1960s. Kubo kept sending his visitors from Japanese electronics firms off to Horsham to see the breakthrough in automatic packaging. Kubo also alerted his colleagues to DEC and other computer firms.

In 1962, Kubo received a cable from Nisho Iwai headquarters, saying his father was critically ill in Kyushu. The company offered to pay half the cost of flying him back. Kubo delayed. Then, a few days later, a telex told him his father had died. The news of the death of this man, who had once seemed so powerful and then collapsed into such impotence and misery, shook the young man to new reflection on his own life. He felt a newly fierce resolve to escape the downward spiral that had caught his father and other men of the older generation.

At the same time an event in business heightened his sense of personal possibilities. For some months he had been watching friends at Sony engaged in an apparently forlorn effort to sell "Tummy TVs" in America. Transistorized boxes with five-inch screens, they demonstrated world leading mastery of high-frequency transistors. But the set cost more than the nineteen-inch consoles of the day. Many Japanese and American observers laughed at this small firm that operated out of a shoddy little office at 584 Broadway and presumed to sell all its products under its own name. But then Kubo saw the transistor TV become a large success. Somehow it confirmed to him what Kennedy was saying. "America might be a big country," Kubo thought, "but it was still a new frontier."

One day at a business equipment show at the New York Coliseum, the plot thickened. Kubo met a disheveled man in his sixties, wearing a wrinkled raincoat, hovering around the Japan Victor Corporation booth. He grilled Kubo about Japan, about the JVC paper tape-recording process, about the possibilities for importing Japanese products into the United States, about the prospects of starting a new import-export electronics firm. No direct proposals had been made but the idea of a new company stuck in Kubo's mind. He brought it up with the other two Toms and they too were interested.

Returning to Japan later in 1962, Kubo visited his friend at Tokyo Broadcasting, Mr. Endo, to ask for advice. Kubo said he had an American investor who was interested in financing a new import-export company to sell American semiconductor production gear in Japan and Japanese electronics in the United States. The chief problem was MITI. Even for a tiny joint venture involving just a few thousand dollars of initial capital, MITI required a special license. Defecting from an established trading company with close links to MITI, Kubo could see no way to secure permission. Nisho executives told him he would end up a taxi driver. "Not so bad," Kubo thought. With the Tokyo Olympics coming up, taxi drivers could earn some 50 million yen a year ($20,000) compared to the 30 million yen ($12,000) he made as chief of Nisho America's electronics division.

Endo, however, was then approaching retirement age at Tokyo Broadcasting and offered to help. He induced TBS to put up 40 percent of the money, or a total of some $6,400, to finance his *amak-aduri,* or post-retirement career. The man in the wrinkled trenchcoat, whose name turned out to be Seymour Ziff, came through with another 40 percent, and the remaining 20 percent was scraped together by the three Toms. Kamo managed to get a $1,000 investment from his mother. The total initial outlay was $16,000. With a distinguished firm—Tokyo Broadcasting—now owning a large share of the company, Endo chairing it, and Japanese citizens with a controlling interest, MITI at last was willing to approve the deal.

Although lacking any facilities at all at the time, the firm would be called Tokyo Electron Laboratories. The name conveyed that this was no mere trading company but a value added trader. For stationery, always a prime asset of a company with nothing else, Kubo reverse-engineered the imperial logo of the other laboratories he knew best: Bell. But business was slow. Kubo remembers stewing away in his windowless Tokyo office borrowed from TBS while the rest of the world watched the Olympics on TV and taxi drivers reveled in extra income.

Back in the United States, the two Toms failed to lure any firms away from Nisho Iwai. But Fred Kulicke did give them introductions at Fairchild, which was the market leader in IC testers, and at Thermco Products, the new leader in diffusion furnaces. In early 1965, Thermco became the first equipment firm to sign up with TEL.

To sell Thermco's product in Japan, Kodaka pulled off a headhunting coup that no search in Japan by an American company could have achieved. He hired a young man named Ryozo Sato, on the ascendant at Hitachi, who had set a company record of 330 hours of overtime in one month: that is more than 82 hours a week beyond Japan's usual 45. Sato came to the United States for six months' training at Thermco and greatly impressed Thermco founders Karl Lang and John Krickl.

The next step was to sign up Fairchild, which at the time was under contract with Kamamatsu, another trading company that, like Nisho Iwai and unlike TEL, provided little or no service after sales. On a trip to Japan, Bob Noyce and Fairchild president Rich Hodgson visited Kubo at the shabby company offices in Tokyo. Impressed with the concepts and enthusiasms of the Toms, they agreed to sign up with TEL. The next year, Fairchild agreed to give TEL rights to sell the company's ICs to Japanese computer companies. Then the world's leading company in both integrated circuits and IC testers, Fairchild, gave TEL the foundations for its emergence as a leading participant in Japan's semiconductor industry and TEL gave Fairchild's tester division monopoly profits in the Japanese market.

As early as 1968, TEL's efforts in the United States had given it a broad line of production equipment with dominant market share in Japan. Japanese customers could use TEL for purchase and service of Thermco diffusion furnaces for doping and oxidizing the wafer, Electroglas contact printers for inscribing the designs on them, VEECO evaporators for laying down the metal layer, Electroglas probe stations for testing chips still in wafer form, and Fairchild testers and VEECO leak detectors for final evaluation of the packaged product. Dispersed among many distributors, TEL's competitors were inferior in service and comprehensiveness. Nonetheless, in the early 1970s, the product line began collapsing.

Hitachi had resolved to teach the firm a lesson for luring away the tireless Sato to sell and maintain the Thermco ovens. In two years, Hitachi's subsidiary, Kokusai, produced a more reliable diffusion oven, sold it for a 20 percent lower price, and began routing Thermco from the market. Meanwhile, the other key part of the TEL line, the Electroglas printers and probers, began breaking down under the stress of prolonged use. Complaints poured into Tom Kodaka's office.

These products were the heart of TEL. To repel the Hitachi threat, Kodaka proposed to Thermco that they set up a joint venture with TEL in Japan under Sato's command to manufacture furnaces specifically for the Japanese market. With TEL and Thermco each holding 47.5 percent of the stock and Sato holding 5 percent, the new firm emerged in 1968. While Sato worked hours comparable to his award-winning stint at Hitachi, the joint venture regained all the ground lost to Kokusai and achieved a 70 percent share of the market.

The problems with Electroglas, on the other hand, seemed so severe that Tom Kodaka decided to drop the company as an embarrassment to TEL. But an aggressive young TEL executive named Larry Yoshida thought he could find a solution. Yoshida went off to Silicon Valley and contacted five engineers he had met in Menlo Park who had designed a workable contact printer. Their firm was called Cobilt, and Yoshida managed to introduce its machines to Fujitsu, NEC, and Hitachi as rapidly as they were launched in the United States.

Steadily improved by TEL, Cobilt contact printers were to produce most of the 16K and 64K DRAMs with which the Japanese first broke massively into the U.S. market. By indices of cost and performance, the Cobilt machines may well have outperformed Perkin-Elmer's projection systems that almost completely displaced Cobilt in the United States. Keeping Cobilt alive in Japan long after it had expired in America, TEL helped finance future Cobilt projects.

Meanwhile, Yoshida proposed to Electroglas president Jerry Lucas that it create the world's first automated probe testers. Lucas agreed that it was a splendid idea. Manual probe testing was a costly process that required large numbers of women to look through microscopes for long hours, marking the bad chips in red. Yet unlike assembly and test—laborious backend functions often farmed out to low-wage Third World factories—probe testing is essential to every wafer fabrication line.

Electroglas in 1977, however, was being partly purchased by a major TEL competitor. Under the leadership of industry pioneer Ken Levy, Cobilt hired engineers from Electroglas who one year later introduced the first fully automated prober at an Institute of Electrical and Electronic Engineers (IEEE) show in New York. But under financial pressure, Cobilt had to drop the product. Yoshida persuaded TEL to buy the Japanese rights at a cost exceeding TEL's total profits in 1976.

Initiated in Japan, developed in the United States, and built by TEL in Japan, Cobilt's automated prober eventually gave TEL 70 percent of the Japanese market. It provided another critical workstation on the production lines that gave Japan its entry into DRAM markets. Cobilt president Ken Levy went on to head KLA, a world leading producer of chip inspection gear also marketed for many years in Japan by TEL.

For all its contributions, however, TEL still lacked major products at the most ambitious end of the market—million-dollar wafer steppers, E-beam writers, and ion implanters. Success came to TEL in a tangled web. A Varian scientist, Dr. Lewis Hall, had invented high vacuum ion pump technology, which could produce a clean vacuum suitable for many chipmaking processes. NEC had purchased the rights to it among other Varian products and formed a joint venture to sell the devices.

As later with Fairchild's chips and Intel's microprocessors, however, the NEC management quickly stopped selling the imported items and began imitating them. In 1978, Larry Yoshida persuaded Varian chief Larry Hansen to leave NEC and launch a new joint venture with TEL. By the end of the year, TEL-Varian had emerged as an important force in Japanese semiconductors, eventually winning a 60 percent share of the Japanese market in ion implanters.

Together with an array of other equipment, Varian products on the advanced edge of process technology made TEL Japan's leading vendor of semiconductor production gear, the fasting-growing company on the Japanese stock exchange, and the winner of *Nihon Keizai Shinbun*'s award as Japan's best-run company in 1983. By 1986, it was the leading producer of semiconductor capital equipment in the world, slipping ahead of Perkin-Elmer with the rise in the value of the yen, and was selling an array of other American technology, including

Kurzweil's speech recognition system and Sun Microsystems workstations. In 1988, Varian bought rights to sell the TEL line in the United States.

During the entire ascent of the Japanese semiconductor industry the key capital equipment came from Tokyo Electron Laboratories. With no help from MITI, TEL—an independent company competing against the giant electronics *keiretsus*—had become one of the most impressive of Japan's many formidable companies. And as Kubo notes ironically, much of TEL's equipment went to the barren island of Kyushu, once farmed by his father, but now rich as "Silicon Island."

No one will deny that Japan's triumph in semiconductors depended on American inventions. But many analysts rush on to a further theory that the Japanese remained far behind the United States until the mid-1970s and caught up only through a massive government program of industrial targeting of American inventions by MITI.

Perhaps the leading expert on the subject is Makoto Kikuchi, a twenty-six-year veteran of MITI laboratories, now director of the Sony Research Center. The creator of the first transistor made in Japan, he readily acknowledges the key role of American successes in fueling the advances in his own country: "Replicating someone else's experiment, no matter how much painful effort it might take, is nothing compared with the sweat and anguish of the men who first made the discovery."

Kikuchi explains: "No matter how many failures I had, I knew that somewhere in the world people had already succeeded in making a transistor. The first discoverers . . . had to continue their work, their long succession of failures, face-to-face with the despairing possibility that in the end they might never succeed. . . . As I fought my own battle with the transistor, I felt this lesson in my very bones." Working at MITI's labs, Kikuchi was deeply grateful for the technological targets offered by American inventors.

A man who has lived with MITI and Japanese electronics for most of a lifetime, Kikuchi might be expected to relish the high repute of his former agency. But Kikuchi rejects the notion, constantly encountered in his travels, that MITI was crucial in guiding, shaping, and even financing the Japanese semiconductor industry. As he says, "Ignorant delight harbors untold danger." He acknowledges that the agency serves as a strong and intelligent advocate in the chambers of government for policies favorable to technical enterprise. It brings Japanese researchers together in fruitful collaborations (which like most such initiatives tend to produce less than they promise). But as Kikuchi insists: "The driving force behind the growth of Japanese industry has been the human motivation and eagerness of industry itself."

He explains: "Whether or not a country can pull together and wield its collective strength, at least as far as I can ascertain from Japan's case, depends more upon the initiative and fire of the people themselves than on the government." He fears that the Japanese government itself will fall for the myth of MITI and lose touch with the real sources of the nation's success.

The microcosmic pursuit has always been global. The United States has led only because the United States has been most open to global forces, immigrants, and ideas. There are no separate destinies in the microcosm. There is just the difference between being free and in tune with the deeper disciplines and possibilities of the time, and being entangled and stultified by the materialist superstitions of a grim past of nationalist bondage and poverty.

Throughout the era since World War II, the United States and Japan have alternated as favored sites for progress in the quantum economy. The United States began to give way to Japan in the late 1970s, when American firms first slipped from the center of the sphere.

CHAPTER

11

The CMOS Slip

What is this? Approaching the podium, white-haired and with a gilded crown. He is draped in frangipani leis, royal in a radiant velveteen robe, open to show a grizzled chest and a French bathing thong above tapering sun-rouged legs. Fun day on the fast track. Is it Rocky Five, the gray panther, out to defend American honor? Or merely Hugh Hefner, emerging ruddy from a pool of decadence? Or a new contender for Iacocca's laurels in the pugnosed papacy of American business? Or is it the Pope himself, escaping temptation in the tropical wilderness, smiling broadly in triumph through an honest but battered face?

He spreads his arms wide and gives a beatific smile. The crowd roars. In full cartoon regality, he stands there in splendor, breathing in adulation, envy, admiration, awe; breathing out confidence, optimism, hope, hokum, triumph, travesty. It's Walter Jeremiah Sanders III, at work in Waikiki, preparing to address the 1984 worldwide sales meeting of Advanced Micro Devices, the firm that he launched after leaving Fairchild in 1970. In less than fifteen years, AMD had become one of the world's leading chip firms.

The throng waits in tense anticipation. Perhaps Sanders is going to bless them with thousand-dollar bills. Perhaps he will dump a galore of gold watches. Or redump some Japanese DRAMs with a scowl. Perhaps he will spiel a sparkling speech. You cannot be sure. He embodies a dream, a drama, a challenge and a prize, a gamble and a gauntlet for his troops, nearly all grown men, in the fight of their lives, piling up points in the middle rounds, but well aware of the battles still to come. Like them, Sanders began in sales, and soared to this. In America, or at least in AMD—for Sanders is worried about America—anything is possible.

Later it will be gold-plated watches, hurled by the hundreds across the stage of the convention center next door. Gold watches, the reward offered to the retiring champions of conventional American business, gold watches a symbol of trash to the men of the microcosm, gold watches pure garbage to these paragons of the picosecond in the flush of youth and soaring stock options and book-to-bill. Gold watches go with an arthritic groan. But pick one up if you want, Sanders sneers. Go ahead if you dare. The crowd laughs uproariously at the very idea. The watches will serve as tips for the hotel help.

"Who's the girl next to Jerry?" someone whispers. She is Hollywood gorgeous in every way—lithe, pneumatic, sinuous, fluorescent —the living blond image of Mattel's Barbie and the incarnation of every salesman's motel dream—draped in pink, reeking sex and success and catered charm, propped up on high heels, gazing out at the crowd through blank blue eyes. "Oh, she's just Jerry's weekday girl. You should see his real girlfriend. From Vassar. She visits on weekends."

They are proud of their Jerry. But these are troops. There will be none of their women on this campus—the Hilton Hawaiian Village —until the sales force has made it through a weeklong gauntlet of technical seminars and even endured a speech by an economist. In 1983, as they swept into the biggest boom in the history of the industry, the world's most expensive economist, Alan Greenspan, told them of the impending great depression. In 1984, just before their market went down the tubes, a supply sider celebrated the Roaring Eighties. After the speeches and seminars, some of the force would fly in their women for a brief fling on the beach. But most would be charging off, heads full of technology specs, to sell their chips to a suddenly glutted world: What *VLSI Update* called the "highly volatile REM market (as in random excess memories)."

On this night of welcome, Sanders gives a brief speech, weaves back through the crowd, and leaves with his prize. But early the next morning he will be jogging along the beach, working on the aging body, the crinkled skin, the loneliness, all the exquisite traps in the trappings of wealth, power, hedonism, hard work, relentless demands. If you look closely you will see peering out through his light blue eyes, with a note of hurt and surprise, somewhat lost in the endless shuffle of success and celebrity, the often-worried face of Little Wally of the Chicago South Side.

He has a right to be worried. He may not know it but he is heading for a crash. As Little Wally sees it, he is fighting alone, an American independent, wearing the "power colors" of red, white, and blue, against "the sovereign power" of the Japanese government. Five huge Japanese competitors would like to beat him down and he will be lucky to escape intact in the trunk of a conglomerate. Will he make it? Will he even survive as an independent company? Or will he have to install the German flag of Siemens above the three triangles of AMD? He will soon have plenty on his mind.

For fiscal 1986, ending in March, AMD's revenues will plummet by one third, to $575 million, and its earnings will drop by nearly 100 percent, to a penny a share. For calendar 1985, AMD will be displaced from its position among the world's top ten semiconductor firms by Mitsubishi, a Japanese maker of ships and steel and cars and VCRs and also, incidentally, a bank. What are you going to do when the bully kicks sand in your face, asks Little Wally. Pump iron? Run for the Senate? Possibly. Sanders is a natural politician. But first you run to the Senate, House, Department of Commerce, the Trade Representative, the courts, to whoever will listen in Washington, to the insistent complaint that the Japanese do not trade fair. . . .

One year earlier, Andrew Grove of Intel—the once bewildered, half-deafened immigrant landing amidst the garbage clutter of New York with scarcely a word of English—returned to the city on a marketing trip. He stayed at New York's premier hotel, the Helmsley Palace, and dined with a reverent writer at the Helmsley Trianon. In this regal renovation of the mansions of Villard, the restaurant opens in cavernous red velvet at the top of a broad sweep of stairs under glittering chandeliers. Visible through the arched and spangled court-yard is St. Patrick's Cathedral in Gothic splendor across Madison Avenue.

For five years, Grove had served as president and chief operating officer and was soon to become CEO of what journalists have recently discovered to be one of America's fastest-growing and most admired companies. The journalists are wrong. For nearly two decades, Andrew Grove had imparted the nervous energy and bristling intelligence that made Intel quite simply the most important single company in the world.

Throughout the 1970s and early 1980s, Intel was decisively more important than IBM, which missed all the critical new computer markets, from the minicomputer to the PC, and entered the domain of small machines on a crest of Intel's creativity and Apple's ghastly mistakes. In 1981, IBM staked its future in the largest new computer markets on the microprocessor architectures of Grove's firm, and in 1982 invested some $250 million in a purchase of nearly 20 percent of Intel's shares. Many observers thought Intel had become dangerously dependent on the leviathan, but the IBM PC and its followers were even more dependent on the inventions of Intel.

The true measure of Intel's contribution is not its link to IBM or its compounded annual growth rate of 40 percent or its decade of average margins over 20 percent. More significant is that more than half of the revenues of the world semiconductor industry—from Silicon Valley to Japan—derive from parts and processes that Intel invented, pioneered, launched, or perfected: the DRAM, the SRAM, the EPROM, the E-square, the microprocessor, the silicon gate process, among many others. Three quarters of all the world's computers dance their data through central processing mazes conceived and designed at Intel.

If this is the information age, as some assert, Grove's company invented most of the vessels and machines by which the information is summoned, shaped, and transmitted. If Intel is recognized as the real continuation of Fairchild, as it is, the company's claims of preeminence—and Grove's reasons for pride—become still stronger. With Grove providing the governing text—*The Physics and Technology of Semiconductor Devices*—Intel led the world into the microcosm. Now he was offering a major text on management as well, *High Output Management*.

Still, among the fine wines and rich sauces at the Trianon, fears beset him: of vulture capitalists preying on his company's best engineers, of uncomprehending American politicians failing to enforce the rules of fair trade, and above all, the specter of Japan. Japanese cartels conspiring against America's manufacturing base, Japanese firms stealing Intel's key innovations, the Japanese government subsidizing an array of ruthless rivals, bent on monopoly, the Japanese refusing to trade fair. Andy Grove was still running across a field in the night beset by the sounds of marching troops and more and more uncertain that he would ever get to the promised land.

In the mid-1980s, both Intel and AMD suffered through a wrenching crash in their markets. But soon Intel returned to its accustomed role as the fastest-growing large firm in the industry, leaping up the list of top ten world IC producers even in the face of the overvalued yen. AMD, meanwhile, was not absorbed by Siemens or anyone else. Instead, AMD took over Monolithic Memories and surged past the billion mark in sales even without including the MMI contribution. Both Intel and AMD launched formidable new 32-bit microprocessors as company spearheads. Yet both Sanders and Grove continued unabated their tale of woe.

Sanders gave speeches with titles such as "Can U.S. High Technology Survive?" He continued to lament the conspiratorial strategies promoted by the Japanese nation against his company and country. Grove similarly issued a long series of complaints against the sinister designs of Japan. Both continued to cite dubious statistics from the Semiconductor Industry Association that compare U.S. and Japanese shares of semiconductor markets, leaving out IBM and other captive producers from the U.S. market and including all the chip-producing Japanese conglomerates in the Japanese market. Even the $4.5 billion in Japanese losses in the industry's recession of 1985 and 1986—perhaps twice the American losses—emerged in the visions of the two leaders as evidence of a ruthless plan.

These complaints from the industry's two most articulate voices echoed through many other firms in Silicon Valley and through the national media, and reechoed in Washington, D.C., where bureaucrats smoothly extended their lists of the needy to include steel, textiles, farmers, semiconductors. Powerfully but unintentionally, Sanders and Grove spread the idea that America's semiconductor firms—still producing nearly half of world output and ahead in most leading edge markets—were losers. And if they could not do without major government aid, who possibly could?

Perhaps Grove's most shrewd observation, though, came in an address to the Rosen Semiconductor Forum in 1985 on the subject of cancer: "The cancer I'm talking about is related to the fact that we are all getting increasingly ensnared by self-doubt. . . . Perhaps the Japanese success . . . has reached the point where it's really gotten to us all; we can't think straight any more."

These are brilliant men in the midst of a global struggle. But Grove is correct: They are no longer thinking straight. What they say diverges starkly from the real condition of their companies and their industry.

Nonetheless, the anxieties of these two leaders are not mere hypochondria alone. They reflect two sharp and powerful changes under way in their industry—two key prophecies of Carver Mead unfolding in a tidal convergence.

One was the new balance of power stemming from a long-term shift in chip designs away from the semiconductor industry and toward the computer and systems firms. Although benefiting the Japanese firms with their thousands of systems products, this development chiefly stemmed from scores of entrepreneurial American companies. A long-term siege of transformation, it is the story of the breakthroughs of Carver Mead and his students: Mead's Way.

The other shift was the continuing drive down into the microcosm of ever denser and cheaper devices. In propounding Moore's law in the late 1970s, Gordon Moore had predicted a doubling of chip densities every year, until the mid-1980s, when the graph swoons. But the Japanese did not swoon. Although some U.S. firms still pioneered in this domain, the Japanese were increasingly taking the lead in manufacture of the densest chips. This development was rightly unsettling to Grove and Sanders; but, at least in their public alarms, it was not fully comprehended by them.

What Grove and Sanders and other industry spokesmen sensed was that the large Silicon Valley firms were slipping from the center of the sphere. The key to a continuing drive into the microcosm was a technical breakthrough as important as the shift to silicon gates that orbited Intel in its early years. Because the shift was less abrupt, it was less fully understood. But it impelled a change in the fortunes of firms and nations as significant as Faggin's invention.

The new microcosmic breakthrough was called "Complementary MOS." Like the silicon gate, it emerged amid the turmoil of the mid-1960s at Fairchild, but was exploited elsewhere. Frank Wanlass delivered the first paper at the International Solid State Circuits Conference in 1963 and acquired the patent for Fairchild. RCA built the first devices in 1964 and introduced the first family of products in 1968. But unlike every other major process advance into the microcosm, CMOS became a forte of the Japanese, and they exploited its potential most aggressively. It was this technological discontinuity that played the largest role in the discomfiture of Grove and Sanders and the ascendancy of the Japanese in key semiconductor markets.

CMOS, pronounced "seamoss," is the ultimate triumph of slow and low. CMOS was not just slightly slow. As its leading protagonist T. J. Rodgers of Cypress recalls, it was "super slow," perhaps a thousand times slower than its higher-powered fast competitors. It did use less than one tenth of the power of the prevailing NMOS and bipolar processes. But Intel and AMD assumed that NMOS power usage could be reduced over time. Throughout the industry, engineers were impressed by a paper from IBM in 1974 that showed NMOS with an advantage in cost and speed down to submicron geometries.

Rather than using one kind of transistor, whether negative or positive (NMOS or PMOS), throughout a device, CMOS uses both kinds throughout. But it ties them both to a single input in each cell. A negative-input voltage that would turn on a P channel would choke off an N channel—and vice versa for a positive voltage. Therefore the combination of N and P devices in each cell means that one transistor is always off. Thus power can flow through the system only in tiny spurts as the devices switch, giving CMOS its uniquely low-power consumption. Like a canoe paddled on only one side at a time, a CMOS device both requires less power and causes less disturbance (heat) than rival technologies. But just as you can run a lot more canoes on a lake than motorboats, you can pack more CMOS devices on a chip. In essence, by switching voltages or fields more than current or power, CMOS moved the industry deeper into the microcosm.

As Ralph Gomory, IBM's research director, has said: "In computing we don't do any real work. We just transform one pattern into another." The impossible dream of computing is a heatless, perpetual processor of patterns. With infinitesimal power usage, CMOS was nearly the ultimate information device—the perfect slow and low process—coming closer than any previous technology to the frictionless interplay of ideas.

Yet very few Americans saw this unique microcosmic appeal. Grove virtually ignored CMOS in his text. Intel resorted to CMOS only for its unhappy venture into digital watches, which like all battery-powered devices require CMOS. When the firm skulked out of the watch business in 1979, it disbanded all CMOS production. Robert Noyce declared that NMOS would suffice for another decade. With Sanders's disdain for consumer markets, AMD never pursued CMOS at all until forced by competitive pressures in the mid-1980s. Entirely committed to the fast and high route to speed, IBM not only ignored CMOS for a decade but also tried to get by with metal rather than silicon gates on its transistors.

The problem was not chiefly a failure of technical insight on the part of the Americans. In the 1970s, Carver Mead, David Hodges of

Berkeley, and James Meindl, then of Stanford, perhaps the industry's leading scholars, all pointed to the large benefits of CMOS. Even at Intel and AMD, some engineers could see the ultimate need to switch to this low-power technology. As T. J. Rodgers puts it, their models showed that at high densities and tiny geometries, "both bipolar and NMOS started to fry."

Even these far-seeing engineers, however, regarded CMOS as chiefly a *technical* imperative at submicron geometries. Many of them got the message when Hewlett Packard led the entire industry by launching a 32-bit microprocessor in 1982 with 400,000 transistors on one chip. This amazing device failed chiefly because its closely packed NMOS circuits required an elaborate cooling system.

U.S. firms were quick enough in recognizing the attractions of CMOS as a high-density technology. What Americans generally failed to see was the centrality of CMOS in fulfilling the larger destiny of quantum technology. They failed to see that CMOS was indispensable to the immediate next step in collapsing the computer. And they failed to see that collapsing the computer was not merely a future technical opportunity. It was a marketing imperative.

The Japanese concentrated resources on CMOS not because they comprehended the technology better than Intel, AMD, and the other American firms, but because they understood the consumer market. They used CMOS to collapse the computer not only into watches but into calculators, not only into toys but into cameras and televisions and scores of other consumer products. Then, in 1983, they used it to collapse full-function computers to laptop size. The Japanese understood better than most American semiconductor technicians that eventually most information technologies, from minicomputers to oscilloscopes, from telephones to televisions, would be portable. Anything truly portable would have to be low power, and low power meant CMOS.

The collapsing full-function computer would not merely mean CMOS microprocessors or controllers. It would also mean CMOS static RAMs in large volumes. When American engineers examined the first Hitachi CMOS static RAM in 1978, they focused on its high complexity and cost compared to NMOS devices of the same speed and density. They pointed to its relatively slow performance compared to NMOS or bipolar devices of similar cost. They assumed that low power could not compensate for its large expense. But for portability, low power was all that mattered.

As it happened, it was an American firm—Tandy Radio Shack—

that with Japanese help in 1981 conceived the first successful laptop computer. But when Tandy proposed the Tandy 100, only the Japanese were ready to build it or its key components. They provided a CMOS version of Intel's 8085 microprocessor made by Oki, CMOS static RAMs made by Hitachi and Toshiba, and liquid crystal displays (LCDs), a low-power screen technology. Invented by George Heilmeir of TI, LCDs were exploited chiefly by the Japanese in their consumer products such as watches, small calculators, and even finally color TVs. The key microcosmic force on the marketing level, the collapsing computer dictated CMOS not as a long-term technical need to be phased in for advanced parts but as an immediate mandate of the marketplace.

Having seen the marketing need, the Japanese were set to benefit from the law of the microcosm: low and slow is ultimately faster. Putting cooler and slower transistors closer together led to more efficient and even speedier products. Moving aggressively down the learning curve in CMOS, Toshiba, Hitachi, NEC, Fujitsu, Oki, Sharp, and other Japanese firms soon were making CMOS microprocessors, DRAMs, gate arrays, and other devices, not just for portable computers but for nearly all relevant applications. Very quickly it turned out that most of the fastest static RAMs would be made with low and slow CMOS.

While Intel and AMD fumbled with various CMOS approaches, the Japanese were taking Intel and AMD designs and converting them into CMOS for low-powered applications. Thus the Japanese could make money selling Intel microprocessors and memories for good prices long after the chips came out.

In time CMOS chips, slow and low though they were, would be used in some of the fastest minisupercomputers in the industry and compete in speed with bipolar devices while still using one tenth of the power. Toshiba managed to use its edge in CMOS to dominate production of one-megabit DRAMs outside of IBM. Most important of all, the Japanese edge in slow and low allowed them to break massively into the U.S. computer market for the first time.

Although the United States continued to design the best laptops, from Tandy's 200 to Zenith's spearhead Supersport line, the Japanese CMOS edge allowed them to produce most of the components of the collapsing computer. In the late 1980s, the Japanese still held less than 10 percent of the personal computer market in the States. But Japanese CMOS producers held some 35 percent of America's laptop business. Including Zenith machines manufactured in Japan with mostly Japanese CMOS components, the Japanese were making more than three quarters of all laptops sold in the United States in the late 1980s.

Fortunately for the U.S. industry, however, the highest performance end of the CMOS market stayed mostly in America. In the early 1980s, entrepreneurial startups had entered the breach to exploit the speed advantages of slow and low. George Huang left HP and John Carey joined him later to form Integrated Device Technologies. AMD's process prodigy, T. J. Rodgers, left to form Cypress. With the help of a large investment from Seymour Cray, Tom Longo left Fairchild to start Performance Semiconductor. Performance created fast CMOS logic and memory chips that could interface with Cray's superfast gallium arsenide logic.

Scores of other companies emerged to use CMOS in specialized designs. Perhaps the most remarkable single product was Xicor's CMOS one-megabit electrically erasable programmable read-only memory (E-square PROM), a non-volatile device which uses a unique quantum tunneling process to load a floating polysilicon gate. Within five years the collective revenues of the new CMOS firms excelled Intel's and AMD's CMOS revenues put together and gave the United States a large base of advanced CMOS production skills and technology.

Cypress and IDT were exceptionally galling to Sanders. Both of them pursued an explicit business plan of converting all AMD's best-selling bipolar and NMOS designs to CMOS. Cypress's largest selling products were CMOS versions of AMD programmable logic devices. Like the Japanese, the U.S. CMOS entrepreneurs could capture the benefits of AMD's design and marketing skills by making better low-power versions and there was nothing AMD could do about it.

By the late 1980s, Intel, AMD, Motorola, Micron, IBM, and other U.S. companies finally achieved proficiency in CMOS and had largely closed the gap with the Japanese. But the agonies of the early 1980s, ascribed to Japanese dumping, government subsidies, and unfair trade, in fact were attributable mostly to Japanese prowess in the mass production of CMOS gained from their success in consumer products. Because consumer products were mostly based on various ways of collapsing computers and putting them in portable devices, from tummy TVs to laptop computers, the market orientation of the Japanese took them deeper into the microcosm. Pursuing slow and low for watches, the Japanese soon found themselves making some of the fastest chips as well.

Grove and Sanders had correctly sensed that in some key ways the leading Silicon Valley firms were slipping from the center of the sphere and the Japanese were entering it. But they still did not quite understand what was happening to them. The American executives might have found a clue in the history of Intel itself, the last firm to lead so decisively into the microcosm of processing. Rather than recognizing the impact of Intel's silicon gates, competing firms such as Advanced Memory Systems and Mostek ascribed Intel's gains to the marketing and technical prestige of Noyce and Moore.

"We beat them with the technology, but we were just bowled over by Intel's marketing clout," said Tom Lloyd of AMS. L. J. Sevin of Mostek made a comparable claim. Today, American firms are groping for similar alibis for their failure to outperform the Japanese. But like Intel in the early 1970s, the Japanese simply managed for several years, in one crucial way, to move faster through the silicon gates of the microcosm.

CMOS was by no means the only Japanese strength. It does not explain, for example, the earlier Japanese successes in NMOS DRAMs. But Japanese prowess in mass manufacturing of components for consumer products contributed to this earlier triumph also. In the mid-1980s, meanwhile, yet another "threat" from Japan was looming on the horizon. Let us visit it with another of the great American semiconductor entrepreneurs, Charles Sporck of National.

CHAPTER

12

Mountain of Memories

Charles Sporck, president of National Semiconductor Corporation, had been warned of the new threat on the horizon. After all, he and his company, long a leading American producer of microchip memories, were to be its victim. In a larger sense, he felt, the target was America, its culture, its industrial base, its economic future. With little aid or understanding from the U.S. government, he knew he would have to contrive a strategic defense and he knew it would not be easy. As the helicopter approached Tateyama, with Sporck crouched inside, all six feet five of his lanky frame, surrounded by spruce Japanese businessmen and engineers, he seemed in a sense a captive, visiting the site of his own impending demise in the company of his own executioners.

Earlier visitors from National had described Tateyama emerging from the mists as a gigantic scar, carved by bulldozers out of the top of a mountain. As the company helicopter dropped for a landing, they said, the scar opened into a vast flattened space that chilled them chiefly by its size and its message of a major new manufacturing facility to come. It would be yet another huge addition to the ever-expanding mobilization of money, men, and machines for the mass production of microchips.

Now, just a few months later, Sporck himself hovered in another helicopter above the same site. Tateyama stretched out below as 100,000 square feet of roof for an immense building beside a sparkling lake. Sporck thought it must be a shell; after all, in the United States it took nearly a year to build a new semiconductor facility. But brought inside for a tour, he saw array on array of glistening new equipment, ready for operation, and interconnected by automated channels and robotic arms, sealed behind thousand-foot interior walls of glass. The size of the effort and the funds available for it chilled the grizzled veteran of Silicon Valley.

Erected in just four months, the factory complex was directly aimed at the market for what was once National's leading product—and, indeed, the prime product of the entire world semiconductor industry —namely, the dynamic random access memory (DRAM). In the United States, entrepreneurial firms such as Sporck's played the key role in DRAM development and marketing. First mass-produced in the early 1970s by then tiny Intel, technically perfected in later generations of the product by Mostek, then a new startup near Dallas, and manufactured most efficiently by Texas Instruments and National Semiconductor, the DRAM was the pride of American entrepreneurs like Charles Sporck.

By the mid-1970s, Sporck had brought National's manufacturing skills to a new pitch in the output of DRAMs. Ever since, DRAMs had been a prime product of National Semiconductor. In 1982, National proudly opened a huge plant near Salt Lake City in which just some seven hundred workers would be able to produce four million devices a month. With state-of-the-art equipment and three modules of 25,000 square feet apiece, the facility cost virtually as much as Tateyama. From Utah, the devices would be sent to Asia for the labor-intensive work of plastic packaging. The plant represented a renewed commitment by National Semiconductor to the bitterly competitive international rivalry in DRAMs.

In 1985, however, with Japanese giants taking some 85 percent of the world merchant market (exclusive of DRAMs made for in-house use by companies such as IBM), smaller American firms seemed to be fading from the field. Intel, virtual inventor of the DRAM, withdrew from the commodity memory business. After losing several hundred million dollars on the product, Mostek also abandoned DRAMs and was bought up by the French nationalized firm, Thomson, now merged with the Italian firm SGS (centered in Phoenix). The most aggressive of all the U.S. entrants, Micron Technology of Boise,

Idaho, created the densest and most ingenious DRAM designs in the industry. Then Micron too foundered, when prices collapsed and yields disappointed. Texas Instruments was the only U.S. company outside IBM that kept pace with Japanese mass-production technology. But it had long produced all its leading-edge DRAMs at its plant in Miho, Japan. An era of American enterprise in semiconductors was coming to a close.

Process skills and technologies acquired in making DRAMs by the hundreds of millions can lower costs for the entire product line. With large volume, the effects of process changes can be statistically tested over large numbers of chips; with niche products it is often hard to tell what is going on. Thus leadership in DRAM manufacture can often give a company low costs and reliable processes for making more specialized chips. In the eyes of many experts, therefore, the Japanese superiority in DRAMs implied eventual domination of the entire industry. It was widely believed that companies like Sporck's, with total revenues under $2 billion, would either retreat to niches making chips with sales too small for the giants, or they would leave the semiconductor business entirely.

Charles Sporck was one of the experts who found this logic difficult to answer. Now bearing down on the world with a Samurai mustache and a passionate belief in the need to beat the Japanese, he began pounding tables in Washington as well as Silicon Valley. He called to the U.S. government for compensatory aid in making new investments and in breaking into the Japanese market. Since American firms held 55 percent of European chip business, twice the share of the Japanese, Sporck reasoned that on a flat playing field the U.S. companies would do far better in Japan than 11 percent (which the Semiconductor Industry Association, by treating Japanese captives as open markets, claimed was the U.S. share).

The problem with DRAMs, however, was more complex than Sporck was suggesting. National Semiconductor and other U.S. merchant semiconductor firms were losing share not only to the Japanese but also to U.S. conglomerates. In fact, if the "captive production" of IBM and AT&T was included in the totals, Japan's share of the American market in DRAMs would fall from an overwhelming 85 or 90 percent to around 65 percent in 1988. Many experts concluded that the future of products like commodity random access memories belonged to large systems companies that could use much of their own product, reap technological benefits applicable to other lines, and write off nominal losses in DRAM sales against profits in computers or telephones or consumer goods.

IBM might not make profits from selling DRAMs but it gained skills and reliable in-house supplies that enhanced the profits of the entire company. The Silicon Valley firms chiefly devoted to producing chips rather than the systems that used them would be permanently at a disadvantage against all the conglomerates, U.S. or Japanese, Korean or German, that both used chips and had other products to tide them over during the inevitable slumps in semiconductor markets.

Sporck's leading competitor in 1986, for example, was probably Hitachi, often the world's largest producer of commodity DRAMs and one of the world's leading users as well. Hitachi also produces videocassette recorders, computers, televisions, washing machines, printers, elevators, and power turbines. Sporck tried to compensate in some degree by importing Hitachi computers, by manufacturing point-of-sale terminals for supermarkets, and by producing chips for the U.S. military. But these products still comprised less than a quarter of National's revenues. National's line was still heavily based on commodity semiconductors under pressure from the Japanese electronics giants.

Now Sporck looked down on Tateyama, a new and essentially different kind of threat from Japan, bearing a different portent for the industry. The man who chopped off the top of this hill and erected this massive factory came from none of the established Japanese electronics firms; he makes no major systems products to soak up inventory in bad times; he received no government aid or encouragement for his entry into the industry. His name is Takami Takahashi and in twenty-five years he had made his company, called Minebea, the world's largest maker of miniature precision ball bearings, producing some 35 million a month.

These bearings, used in virtually all instruments with closely calibrated movement—from disk drives and videocassette recorders to strategic missiles—are ground to a size of one millimeter in diameter and to a tolerance of one tenth of a micron. Made in highly automated cleanrooms where a speck of dust a half-micron wide looms like a boulder, these ball bearings play a critical role in an array of the world's highest technology products. Like Sporck, Takahashi stressed manufacturing discipline and technology and built his company to over $1 billion in sales. Then he suffered a life-threatening disease and began to contemplate the meaning of his career.

One of Japan's most flamboyant entrepreneurs, providing components for some of the world's most celebrated companies, Takahashi had seen high honor and acclaim going to the makers of automobiles and microchips. He had no interest in making cars. But chips . . . what were chips anyway, he asked himself, but a miniature component of the sort in which he specialized. "A chip," he suggested, "is merely a miniature ball bearing—flattened out, with a picture on it."

In addition, it seemed to Takahashi that the chip industry was going awry. The DRAM, like a ball bearing, had become the epitome of a mass-produced component. Yet he saw that the business was increasingly dominated by firms like Hitachi, Toshiba, and IBM, which specialized in making complex systems such as computers and telecommunications switches. To him, this division of labor seemed appropriate only for the many chips that themselves had become complex systems. But DRAMs, however advanced a product they had been in the past, had now become a commodity produced by the billions. Nonetheless, compared to ball bearings that sold for under 50 cents apiece, DRAMs often brought what seemed to Takahashi the princely sum of several dollars.

Yet even at these prices the manufacturers claimed that it was often hard to make a profit—that they made DRAMs less for their profits than for their use as technology drivers to improve manufacturing efficiencies for other chips. If manufacturing efficiency was the key, Takahashi thought, why should DRAMs be made by companies that specialized in designing and producing *systems*? They wanted manufacturing efficiencies; he would show them manufacturing efficiencies. He would launch a new company, NMB Semiconductor, and build Tateyama, a totally automated production facility that ultimately might produce as many DRAMs and other memories a month—some 35 million—as his other factories produced ball bearings.

Many experts said it could not be done. DRAMs represent the most demanding feat of mass production in all world commerce. None of the complex procedures is easy to automate. Automation itself, moreover, is no final solution to the problems of dust and contamination. Machines collect and shed particles and toxic wastes nearly as much as people do. Chip experts derided the view that these ten-layered and multiply patterned electronic devices, requiring hundreds of process steps, resembled ball bearings in any significant way.

Takahashi knew all that. But experts had derided almost every decision he had made throughout his career. "Successful people," he says, "surprise the world by doing things that ordinary logical people think are stupid." The experts told him he could not compete in America with New Hampshire Ball Bearing. He ended up buying it. The experts and bankers had told him not to build his biggest ball-bearing plants in Singapore and Thailand. Those plants are now the world's most productive. The experts told him not to buy two major facilities in the United States, full of obsolescent equipment and manned by high-priced workers. But those facilities now dominate the American market for precision ball bearings. Now the experts told him he couldn't make DRAMs. He knew he could. "If I listened to logical people," he says, "I would never have succeeded."

The fact that many prestigious companies, from ITT and Siemens to Sony and Sanyo, had at various times tried and failed to become major factors in DRAMs did not daunt Takahashi. They were *systems* companies. The chip manufacturers might not know it, but DRAMs were now more like ball bearings than like systems. When the makers of a product do not understand that it has become a commodity, "that is the time I like my chances," says Takahashi.

Takahashi resolved to banish humans from the cleanrooms by building a uniquely robotized plant. He would contract with the world's chip designers to design the product and he would buy capital equipment from outside suppliers. His business was putting machines together to make a production line, and at managing this business he knew he was among the best in the world.

To do it, he attracted leading facilities designers and builders from many of the other major Japanese companies, including Sanyo and Sony. To recruit them, he employed the method used in the late 1960s by Japan's Henry Ford, Soichiro Honda, then the world's leading producer of motorcycles. To enter the automobile industry at the top, from an industry widely believed to be far less demanding, Honda had intrigued and challenged the nation's mechanical designers by declaring his plan to build the world's fastest racing car and win at Le Mans. Takahashi announced his determination to build the world's supreme factory. He would create an independent company to build a plant fully automated to produce vast quantities of what was widely regarded as the world's most challenging product. The engineers came running to his service.

The result was Tateyama. A two-story structure, it houses all the cleanroom fabrication lines and control systems on the top floor and the supply systems for distilled water, special gases, and production chemicals in the basement. With all the remote control and monitoring equipment, supply lines, and other peripherals outside the processing rooms, no persons enter the fabrication areas. Automated systems move wafer containers along channels of air running the length of the ceiling above the line. Robotic arms move the containers into each processing station, from wafer cleaning through ion implanting and photolithography to bonding and packaging. If any part of the line malfunctions, it is automatically lowered to the basement for maintenance.

Each month four groups of forty employees work fifteen shifts of twelve hours each. They collectively produce nearly 6 million chips (some 70 million a year). Each employee thus produces 40,000 chips a month. Since each chip contains at least 600,000 transistors, each worker produces some 21 billion transistors a month, or 250 billion a year. In 1960, ten years after the invention of the transistor, workers were proud to produce a dozen a day; today, each worker averages nearly 1.5 billion on a twelve-hour shift. Transistors then cost $5 apiece; now, much faster and more effective transistors go for a ten thousandth of a cent.

American companies also produce billions of transistors for ten thousandths of a cent apiece and the productivity of their workers has risen by a factor of well over a million since the 1950s. But their levels of overall productivity may be as low as a tenth of Tateyama's. Including the thousands of workers devoted to assembly and packaging operations in Indonesia and Thailand, for example, Sporck has thousands of workers doing the job of forty at NMB.

Charles Sporck can demonstrate conclusively that it is more economical to send the chips to Bandung and Malacca for backend processing (test and assembly) than to purchase automated gear and do it at home. He can show that with the American cost of capital, a plant like Tateyama would be a white elephant. The Japanese semiconductor experts agree. They told the press that Tateyama could not possibly pay off—the company could not write off the equipment over the four years of its probable useful life and still make a profit. Takahashi smiles, and talks of logical men.

In entering the DRAM business from a field considered far less exacting, Takahashi is not unique. He followed in the footsteps of an American entrepreneur named J. R. Simplot, an Idaho billionaire in his mid-seventies who made his fortune mass-producing fried potatoes for McDonald's. Together with a small group of other Idaho magnates, Simplot propelled Micron Technology, then a small consulting firm in Boise, into the midst of the global DRAM fray.

Unlike NMB, Micron first depended more on its skills in memory design than on its manufacturing prowess. Under the leadership of the Parkinson twins—Ward, a former Mostek chip designer, and Joe, a former New York lawyer—Micron began by assembling the industry's leading group of memory designers and layout artists. In each generation of DRAMs, from 64K to one megabit, Micron developed the smallest and thus probably the most cheaply producible chip in the industry. In 1983, the firm attempted a major design innovation at the 256K level that worked technically but failed to pay off. As the industry slumped in the mid-1980s, Micron suffered two years of severe losses. In 1985, Jerry Sanders predicted that it would never again have a profitable quarter. But Simplot defied the experts and continued to support the firm. In the late 1980s, Micron, which like NMB already did nearly all its test and assembly at home, began focusing on process innovation. It reduced the number of mask steps needed for DRAM production to about half the number needed by its Japanese rivals. At the end of 1988, with profits of $98 million on revenues over $300 million, Micron was one of the few companies in the industry with net profits on its total DRAM production during the decade. In 1989, Micron moved into new facilities that tripled its capacity.

With initial capital investment ten times Micron's, however, NMB could move far faster than the Idaho firm in defying the experts. In 1988, with annual revenues of $200 million from a $200 million investment, NMB Semiconductor was already two thirds the size of Micron. By 1989, NMB was profitably producing semiconductors at a run rate near $350 million, still with only forty workers on a shift and all the chip designs purchased from outside.

Whatever happens, Tateyama's meteoric ascent demonstrates that DRAMs are a commodity component—a flat bearing with a picture on it. If you looked closely, some said, you could descry Takahashi's smiling features amid the memory cells.

Some writers have used software as an analogy for the knowledge in a human mind and hardware as an analogy for the brain itself. In software, some conclude, the information is all. But in computer hardware, they assert, manufacturing and materials are decisive.

Computer hardware, however, no less than software, consists chiefly of ideas. At the heart of every computer and at the heart of virtually every piece of Tateyama's equipment was a central processing unit (CPU) usually embodied in one microchip or a set of microchips. These chips are hugely complex; each one may bear a pattern of circuitry as intricate as a street map of the entire state of California. But for all their complexity, chips also resemble the state of California in being made essentially of sand. They use rare materials such as gold and titanium in such infinitesimal amounts that total materials costs comprise less than 2 percent of the costs of chip fabrication. Labor costs are under 5 percent.

Like a book or a disk or other information product, therefore, a microchip costs between 80 cents and $2 to manufacture in volume and it sells for between 50 cents and several thousand dollars depending on the value of the information it contains. The CPU of most advanced personal computers of the late 1980s was Intel's 386 microprocessor. Like its chief competitor in CPUs, the 68020, from Motorola—which is notably used in Apple Macintosh computers—the 386 uses "words" 32 bits long and in most respects can outperform the CPUs of million-dollar mainframes of early in the decade. In volume it may cost under $2 to manufacture. Yet in 1988, the fast versions still listed for nearly 150 times that amount: close to $300.

This exemplary piece of hardware—the computer's CPU—therefore follows the microcosmic rule: it is priced according to the value of its information, which can be hundreds of times greater than the cost of its manufacture.

DRAMs, on the other hand, are the exception that proves the rule. They are the industry's supreme example of the dominance of capital and manufacturing. The densest of all microchips, capable of reading or writing between 1 and 16 million memory bits, DRAMs are by far the most difficult of all chips to fabricate. From the time the first prototypes are made, DRAM companies often go months before any usable devices are produced and years before high yields are achieved in the production process.

For all their advanced technology, however, most memory chips do not embody any unique design or conceptual content. In Japan, the United States, and even Korea, DRAM producers from Boise to Tateyama have made heroic strides in fabricating their fiendishly exacting devices. So, unlike all the other devices that contain valuable information, memory chips in full volume production may even *sell* for between 80 cents and $2.

In the late 1980s, systems firms believed that they would have to produce their own DRAMs to remain competitive in the increasingly digital realms of electronics, from computers to video games. In 1989, complex new pay telephones, printers, fax machines, and television sets all contained at least five times more DRAM memory than the actual desktop computers of 1985. Does anyone truly believe that all these products will fall into the maw of the large DRAM producers? That systems firms will be able to use their DRAM skills to monopolize the entire world of electronics?

Tateyama offers a different portent. It is one of the few firms on earth solely dedicated to reaping profits out of commodity memories. It demonstrates that DRAMs are mostly a commodity product, and will remain a commodity, appropriately manufactured by specialists in commodity components. These producers will not thereby dominate the rest of the microchip business or even the most profitable parts of the rapidly growing specialty memory chip business. Manufacturing prowess is necessary and beneficial in microchips, but it is knowledge that brings the profits. The chips that are most lucrative to produce and use are complex and innovative designs. Embodying new ideas and applications, such chips comprise the bulk of the cost of new computer hardware. Thus, in the end design-rich Intel became the chief beneficiary of Tateyama by taking over all of NMB's production in an agreement in February, 1990.

Computer hardware thus is another form of information technology like books, films, and disks. The value resides in the ideas rather than in their material embodiment. The chip design is itself a software program. Even the design of the computer's plastic chassis and keyboard may well have begun as a software program. Like a book, a spreadsheet financial package, even a film on a videocassette, a microchip design is conceived and developed on a computer screen and takes form in a storage device that costs between 80 cents and $2 to manufacture. The current dominance of such products in the world economy signifies the end of the industrial era and the onset of the age of the microcosm.

Nonetheless, the rise of complex information products in no way belittles Tateyama or any other DRAM production line. A semiconductor fab itself is one of the world's supreme information systems. To upholders of the materialist superstition it may appear to be a gigantic physical machine. But it is everywhere sustained by ideas and software, enhanced by new learning and experience, and driven by the will and ingenuity of its workers and entrepreneurs. Tateyama and Micron are brilliant feats in the architecture of ideas.

The materialist superstition often afflicts these vessels of ideas. In the late 1980s, nations reached for the DRAM as the physical key to the microcosm. They imagined that they might buy its means of production and monopolize it, and they feared that others might.

They built giant factories to catch and control it. But whether in Kyushu, Miho, Einhoven, Essex Junction, or Boise, the factories themselves—obsolescent nearly the day they were built—were the material masks, not the precious essence of the microcosm. The microcosm was a worldwide mindscape of men and ideas, and just when you thought you had it under control some crazy maker of ball bearings or potato products would rise up and build a Micron Technology or a Tateyama.

Their learning curve creed would be concept for concept identical to the creed of Henry Ford, and of Patrick Haggerty, the builder of Texas Instruments. It sums up the achievement of Charles Sporck, when he brought low prices, highest quality, and efficient manufacturing to Silicon Valley. Charles Sporck and other Americans had already brought down the price of a million bits of memory (a megabit) from $25,000 in 1975 to $100 in 1980 and to $10 in 1985. But shortly after Sporck returned from his visit to Tateyama, he announced that he was withdrawing from the production of DRAMs and devoting his Utah facility to other products.

Sporck joined with several other companies in suing the Japanese for dumping chips—below cost—in America, as if dumping chips wasn't always an essential strategy for building markets for commodity memory. The only reason Sporck had not "dumped" DRAMs during the 1985 industry slump was that he probably only had a few to dump. Like many other American producers, he seemed to be giving up. *Fortune, Business Week,* and other publications wrote requiems to the American semiconductor industry.

Many academic analysts hastened to predict that the demise of American leadership in semiconductors would lead to a general loss of American competitiveness around the world in the array of products—from cars and computers to telecommunications and TVs—increasingly dominated by digital computer systems, features, and automatic controls. If America could not compete in the automated production of high-technology goods—and could not compete in the labor-intensive output of other products—where could America compete? In leveraged buyouts, McDonald's stands, and showrooms for Toyotas, some said, as we move to a "service economy." Was Lee Iacocca, TV star and political operator, huffing and puffing his macho Americanism while quietly marketing cars for Mitsubishi, to be the new paradigm of U.S. business? Or was it Charles Sporck, gruffly denouncing Japan while negotiating with Oki for chip designs and marketing mainframe computers from Hitachi?

Was this the end? Not with a bang but with protectionist whimpers would the heroic leaders of American technology give way to the new titans from Asia? Was this industry to be the microcosm of America in decline? That was the way it looked to many observers in the late 1980s. But the American industry was not quite through yet. In the world of new technology, things are rarely as they seem from the top.

To Sporck, Sanders, and Grove, the industry seemed to be maturing. Value added seemed to be shifting from innovation to capital-intensive manufacturing. But in fact the anxieties of the 1980s derived from a maturing of those last great innovations—memories and microprocessors—and the need for a new innovation of design. That innovation was already on its way and it would launch the industry—and America—into a new siege of microcosmic surprises.

PART

The Transformation of Capital

III

CHAPTER
13

Mead's Theory

The evolution of the industry since the late 1960s seemed to fulfill all the prophecies of Carver Mead. After spurning his advice to enter the field of small computers, Noyce and Moore of Intel had launched all the key elements for collapsing the inner works of a computer: the DRAM for storing data in use, the microprocessor for calculating with the data, and the EPROM for holding permanent systems software. As the computer became smaller, it became far cheaper and more widespread. As Mead predicted, it was embedded in tools, instruments, toys, prosthetic devices, typewriters, telephones, doors, magazines, greeting cards, weapons, automobiles, pots, ovens, televisions, and pencils.

Intel and other companies in the United States and Japan were also rapidly fulfilling Mead's predictions of VLSI. Summing up the huge new densities possible in semiconductors, VLSI meant very large scale integration, with chips bearing first hundreds of thousands and then millions of transistors with features measured in angstroms. Before VLSI, systems of circuitry were built mostly by assembling groups of relatively simple chips on printed circuit boards. The dominant category of these relatively small logic ICs was a bipolar set called transistor transistor logic, TTL for short. It was often called glue logic because of its common use to interconnect larger ICs. By placing more circuitry on each chip, VLSI cut the number of chips in a system, economized on connections between chips, increased speed, and reduced power and cost.

The early Intel microprocessors were among the first important VLSI products. Primitive VLSI enabled Ted Hoff to reduce potentially thousands of logic devices to three microchips. But Mead was predicting chips not with mere thousands but with millions of transistors. Emerging from several companies, the megabit memory fulfilled this Mead vision. Four- and 16-megabit DRAMs followed, and 64-megabit devices were scheduled and targeted by many companies and national industrial policies. The movement down into the microcosm went as fast as Mead expected and brought all the prophesied benefits in cost and performance.

And yet the heart of Mead's prophecy for VLSI was more than a quantitative advance, more than bigger memories and larger microprocessors. He envisaged VLSI as a force of liberation, allowing chip designers to create new architectures rather than merely bigger and more efficient versions of existing concepts, such as DRAMs and Von Neumann central processing units (CPU). Mead was predicting a cornucopia of single chip systems that operated far more resourcefully than any of the chips of the past.

To master the complexities of these VLSI systems, however, would require radically new computerized design tools and new computer architectures. On a single chip, a conventional Von Neumann architecture would be dominated by memory and the communications between memory and CPU. To fulfill the new promise of VLSI, Mead envisaged new schemes of parallel logic and new forms of artificial intelligence that would open the computer more fully to the world and the world more fully to the computer. He foresaw computers spreading their reach into the home and pervading the domains of consumer electronics. He saw the world newly impregnated with this new technology of mind.

During the mid-1980s, however, the leading American companies seemed to have missed the meaning of VLSI technology. Rather than a vessel of single chip systems, they saw the new technology as chiefly a way to manufacture denser memories and microprocessors. Fearful of the manufacturing leadership of Japan, many Americans predicted the end of the line for any U.S. semiconductor firm without a leviathan sponsor or a Pentagon niche.

To Mead, the portents were radically different—not a period of

American decline but an epoch of fulfillment for the technology. He developed a theory of the microcosm that both predicted and embodied a complete and adequate response to the challenge of the Japanese, the impasse in computing, and the perplexity of the American semiconductor firms. Far from succumbing to the forces of a maturing technology increasingly dominated by large manufacturing conglomerates, Americans were perfectly situated to respond to the deeper message of VLSI and bring forth an age of rising entrepreneurial innovation and progress, riding the crest of a worldwide wave of creation.

Mead's theory began with an analogy to early computers. In 1975, at the beginning of the personal computer era, hapless hackers had to program their machines in machine language, the ones and zeroes that computers can use. Toggling rows of bit switches on the front panel of their mail-order Altair 8800s, they had to follow a Mickey Mouse maze even to add one number to another.

First they had to put some fourteen instructions into specific addresses in memory. Each instruction required seven instruction switches (one for each bit) and a deposit switch, for a total of 112 toggles. Then the two numbers had to be put in binary one-zero form and loaded into the memory addresses specified in the program, each also toggled in binary code, for a total of several hundred more toggles for large numbers. Then, by sending the computer to the address of the first addend, if he could recall and find it, and toggling "run" (ten or so more toggles), the lucky hacker might get Intel's miraculous 8080 microprocessor to go to work and get him his answer. That is, provided all the several hundred toggles were exactly correct, the computer would get the answer. For the hacker to get it would require a series of further toggles, beginning with the binary address of the register holding the sum.

Using computers this way required complete concentration, utterly methodical diligence, and remarkable memory. It required the user in a sense to become a computer himself. Closely resembling the regimen of the users of ENIAC (using 6,000 switches and patch cords), this trivial pursuit on the Altair gave the hacker at once a sprained brain and a keen sense of computer history. It is safe to say that if personal computers had remained like the Altair, no personal computer revolution could have occurred and relatively few microprocessors would have been sold for use as CPUs.

In the late 1970s, Apple, Radio Shack, Imsai, and other firms had solved the toggle switch problem by providing their machines with the BASIC higher-level language on read-only memory. Using plain English words such as DATA, PRINT, LIST, STOP, RUN, higher-level languages allowed people to ignore all the bits and nits. You no longer had to be an obsessive hacker to use a computer. Adding numbers, for example, you could simply command the machine to "add" the specified conventional numerals. The computer, functioning as a calculator, would do the rest.

What makes higher-level languages work is usually a compiler. A compiler is essentially a translator: it automatically converts the statements of the higher-level language into all the ones and zeroes that the Altair hacker and ENIAC engineer previously had to figure out and toggle in. For complex and specialized programs, the compiler works in several phases. It translates the special-purpose instructions into a general-purpose language and checks for consistency—"Are letters or numbers being added?"; "Is there an action instruction for each set of data?" Then the general-purpose language may be translated into various intermediate languages on the way to conversion into binary ones and zeroes. Finally, the code may be compressed for efficiency and speed.

Higher-level languages, compilers, and software packages had transformed the computer from an elite instrument or bizarre toy into a democratic tool. Carver Mead was determined to achieve a similar transformation of the process of designing chips and computers.

In the 1970s, to design a chip, you had to be like Masatoshi Shima, the author of most of the circuitry in Intel's 8080 microprocessor in the Altair. You had to be a near-obsessive master of all the details of the logic and layout, and you needed a photographic memory. Just as very few people were willing or able to twist their minds into nits and bits to use the Altair, even fewer were capable of the intellectual contortions of chip design. Mead's goal was to democratize chip design the way higher-level languages, compilers, and other software packages had democratized the personal computer. He did not want Shima's 8080 and a few others to be the last microprocessors. Using chips and software to automate the production of chips and software, he hoped to launch a new spiral of creativity in electronics.

During his regular trips to Intel in the early years, Mead had watched the design rigamarole with amazed frustration. With voluminous hand-drawn logic diagrams, followed by detailed circuit schematics, electrical checks, timing checks, design rule checks, and intricate hand drawings of the actual layout, all in endless iterations, virtually no one could understand the entire process. No one really knew what was going on.

Mead felt that Intel was being saved from the effects of this folly by Shima. A logic designer taught the intricacies of circuitry by Federico Faggin, Shima could actually contrive logic without constant simulation; he could conceive of the connections between the Boolean gates in his head; then he could remember the sizes and shapes of thousands of transistors and how they related to one another.

When Shima was asked about a feature on a chip, he would thumb through the endless pages of mylar, twisting his hair in frustration, and then bang his finger down on a tiny inscrutable point in the layout. The entire design was in fact in his head. But he combined a virtually photographic recall with an utterly methodical mind; if everyone was like Shima, the world might not need computers. In an ordinary world without computer design tools, however, only Shima and a few others could design efficient chips.

As executed by Shima and others, the actual design process was bad enough. But then, to create accurate patterns to be photographed for photomasks, intelligent and attractive women with pads on their knees would crawl over huge sheets of mylar coated with a pink plastic called rubylith. Using special scalpels called exacto knives, they would cut lines on the rubylith corresponding to transparent areas on the photomask. Then others would would follow with tweezers to remove the plastic. As a symbol of the high technology of the day, it all smacked more of low farce.

More important, Mead could see that this approach would thwart his prophecy of hugely denser circuitry. As scores of thousands and then millions more switches were placed on each chip, the design problem—like so much else in the microcosm—would grow exponentially. Using full-sized women with tweezers, the creation of a single design in the 1990s would take up much of the Bay Area, and have most of its female population crawling across mylar in knee pads.

Mead thought chip design was too important to be restricted only to a few engineers with photographic memories. His target was Shima. Methodical, encyclopedic, logical, and prodigious in memory, Shima was the exemplary expert to incorporate into a computer design program that operated like a software compiler. Mead wanted to make Shima obsolete.

This was no modest goal. During most of the 1970s, Shima's approaches dominated the industry, not only with his design techniques but also with his microprocessor designs. Shima, Faggin, and kindred circuit gurus were determining not only how to design chips but what chips to design. Their answer was microprocessors. Mead thought that this answer was sadly inadequate. Central to Mead's theory of the industry was his critique of the microprocessor, the industry's greatest pride.

Microprocessors were indeed a marvelous new exploitation of VLSI. By putting entire programmable systems on chips, they vastly reduced the cost and extended the reach of computing power. By relegating creativity to the software end, they avoided afflicting the semiconductor companies with a need to figure out how to use their chips. As the VLSI era advanced, newer microprocessors made possible the fulfillment of most of Mead's prophesies of computing power on every desk.

Mead, however, soon came to see the microprocessor as more a problem than a solution. Sometimes he would say in exasperation— or in his drive to make a point—that "the microprocessor set back the technology ten years." His listeners would gasp and he would pass it off as a joke. But if the truth be known, Mead was serious.

Many years later, in 1985, Gordon Moore of Intel made a similar point. "The problems of the past were not necessarily solved," he said in a speech, "they were simply circumvented." With microprocessors and memories, "the semiconductor industry developed a different set of markets to keep itself busy, postponing the solution of its previous problems." These problems—of design and product definition—were the focus of Mead's distress.

He understood as well as anyone in America the importance of microprocessors. Nonetheless, as Mead saw it, the first microprocessor—and most of its followers—were hopeless kludges, with endless brambles of random logic, primitive instruction sets, and awkward timing schemes, with the interconnections sprawling in a hopeless tangle across the chip. "If Ted Hoff had been my student," Mead said, "I would have had to flunk him." Instead, Mead honored the first microprocessor with a full-page microphotograph in his textbook as an example of atrocious wire-dominated design.

Mead believed that the industry was still avoiding the clear message of its technical medium. The designers were once again using VLSI, just as they used the transistor and the integrated circuit, mostly to reduce the costs of executing old ideas. Hoff actually was boasting that he followed the scheme of Digital Equipment Corporation's PDP-8, a computer advanced for its time but ghastly in its waste of printed circuit board space on wires and glue logic. Yet the VLSI copy made by Hoff was inferior in every functional respect to the original computer. It was as if VLSI was an obstacle to the designer rather than an open sesame for radically improved processors. The semiconductor designers seemed to have no concept of the potential of the microcosm.

Avoiding through software, memories, and microprocessors the need to design new chips for new applications, Mead believed, would prove a Pyrrhic victory. Neither the microprocessor and memory nor conventional design methods would be able to meet the demands for the fast, specialized systems of the 1990s. From three-dimensional graphics to continuous speech recognition, from computer simulation to interactive games, from high-definition televisions to fast modems, the new functions would require totally new designs and architectures. The industry, Mead declared, was headed for a serious economic and technical impasse.

With a better design system, however, Mead believed it would be possible in a matter of weeks to create scores of far better microprocessors for every specific application. Rather than taking years of effort from Federico Faggin, Masatoshi Shima, William Mensch, and other geniuses, single chip systems could be generated from programs that embodied much of Shima's knowledge.

Well before the invention of the microprocessor, Mead had begun working on a simplified method of chip design usually termed hierarchical. As usual when learning a new subject, he started by teaching it in a class. In 1968, Richard Pashley, later to become a major figure at Intel, urged Mead to give the class and assembled thirty students to take it.

Mead developed a new metaphor for chip design. At the time the design was executed by several specialists: the systems architect, the logic expert, the electrical circuit designer, and finally the layout draftsman. The only person with a clear vision of the overall project was the systems architect, but he had no role in the actual design of the chip. Each of the others was a master of the voluminous intricacies of his specialized duties in the design: call him a short fat man. Mead set out to create what he called a "tall thin man," who could combine the lofty perspective of the systems architect with an effective mastery of the lower and more detailed levels of the design, down to the endless technicalities of the layout itself.

The only way such a "tall thin man" could operate was by including the lower-level details in the upper-level specifications. Like the software compiler with its several levels of operation, this approach was multi-layered. It meant "nesting" the layout rules within a scheme for the electrical circuitry, which in turn would be nested in a logic design or functional specification. The tall thin man would design the chip as an orderly assemblage of functional blocks rather than as a mass of randomly placed transistors or even polygons of layout.

A top-down approach, it was the obvious method for designing any system. But in the real world of the semiconductor industry, the system top was defined by different people than the silicon topography. Thus top-down methods were only rarely applied in a systematic way. Short fat specialists ruled.

Although the hierarchical approach was clearly the right thing to do, even the best chip designers such as Shima resisted it because it meant sacrificing the efficiency of "doing things right": putting each transistor in exactly the right place, occupying precisely the minimal space. Like the original software virtuosos in the machine language of the computer, the chip designers denied with fervor that their special genius could ever be captured in a higher-level language. They would *never* give in. In a sense they were right. Most software programs do still entail the execution of certain critical or complex code in bit-level language. Similarly, most chips at the state of the art require the design of certain special features by hand. But just as the high-level software language compilers improved to the extent that virtually no programs are now written any longer in machine language, so no chips in the future—Mead insisted—would be designed in polygons.

There was a further problem with the top-down method. It deprived the chip designer of control over the design. Hierarchical layouts were recognizable by their simple regular structures, consisting of functional blocks, interconnected by buses or other orderly wiring schemes. They were intelligible to the boss.

Optimal hand-designed layout, on the other hand, often looked a lot like rectangular pans of spaghetti. Metal communication lines among the various elements would sprawl over the silicon. Just as some 90 percent of the space on printed circuit boards had come to be taken up by interconnections, so the chips now replacing entire boards would be brambled with metal. All this spaghetti might take up less space than the blocks of a hierarchical system, but only a few designers could comprehend them. The boss probably could not.

Although Mead's goal was to teach the larger possibilities of VLSI, he could see that prevailing head-in-the-silicon design techniques would reduce his University Lab to the very simplest chips. Even if a student could find the years of time to devote to a project, no mere student could gain a Shima's encyclopedic lore and mastery of systems and silicon.

By using Mead's hierarchical techniques, his students found they could design complex chips far more rapidly than previously was thought possible. But the real payoff was computerization. The process of top-down design of a chip correlated perfectly with the well-known hierarchical technique usually employed in writing and compiling software.

Mead resolved to develop a package for designing a chip on a computer and to use it to create a new kind of device. The kind of chip he decided to build is known as the programmable logic array (PLA). A combinatorial logic device (i.e., with no memory), it takes a limited number of conditional inputs (such as the presence of cars approaching an intersection) and produces a limited number of outputs (such as red, yellow, and green lights).

Made of two simple arrays of switches—a "plane" of Boolean AND gates leading into a plane of OR gates, often with a feedback loop connecting them—a PLA is well adapted to perform the functions of what is known in computer science as a finite state machine. Like the traffic light controller, the PLA finite state machine executes many of the endless traffic functions in a computer. It channels data on and off of buses to and from various memory locations. In fact, a good definition of a computer is a combination of datapaths—the arithmetic logic units (ALUs), controllers, and other logic through which data flows—orchestrated by finite state machine traffic supervisors, often implemented on PLAs.

Determining the relationships between the inputs (sensors) and outputs (control signals or traffic lights) is merely the pattern of wiring between the logic elements in the two planes of the PLA. Orderly connections between two simple blocks, these wires can be defined in the last metalization layer of wafer fabrication. PLAs thus could be fabricated and sold as standard products and then customized for a particular purpose in a final step.

The precursor of a series of programmable logic devices widely used in the industry—later including the PAL and the gate array—Mead's design was seminal. Monolithic Memories (now part of AMD) would become one of the Valley's leading semiconductor firms by introducing and developing an advanced form of PLA called the PAL, used first in Data General's Eclipse minicomputer and immortalized in Tracy Kidder's *Soul of a New Machine*. But Mead was not the inventor of the PLA. Unbeknownst to Mead, it had been created a year earlier in 1968 at Texas Instruments by a design virtuoso named Richard Probsting. (Probsting would later stun the DRAM world by multiplexing Mostek's 4K device into a standard 16-pin package, then believed to be impossible). What made Mead's PLA unique was his design technique. He created a computer program to design, simulate, lay out, and effectively test his chip.

Mead saw that all previous design was performed in a fundamentally illogical way. From broad architecture to detailed layout, any systems design must in some sense be hierarchical. As the designers move down the hierarchy, complexity grows at an exponential rate. The systems architect may merely describe a few functions—ROM, ALU, and buses—while even in the mid-1970s, the logic designer specified many thousands of Boolean gates (NAND and NOR structures). The circuit designer in turn would produce a schematic or "netlist" including scores of thousands of transistors, and the layout draftsman would limn in millions of polygons and parameters. This multi-phased process was necessary every time a new PLA or other chip was designed, even if it differed only in minor functional ways from a previous model.

After the design was completed, moreover, with much intellectual effort invested at every level, the designer essentially flattened it. He reduced it to the transistor level—with its many thousands of switches —and simulated it in this, its most complex form. He actually ran a tracking device over every polygon and re-created the circuit schematic from it. Then he compared it with the original, transistor by transistor, and to the logic diagram, bit by bit.

Mead decided to reverse this process. Rather than using a program to extract the circuit design from the layout for testing purposes, he would create a program to go the other direction: given a circuit design, it could produce a correct layout. Then he went further, creating a program that could produce both a circuit design and a layout from a definition of what the final chip should do. Given the specified truth table combinations of inputs and outputs for the PLA, it would design the entire machine, top down. It took him many months to create the program but only a week to actually design the chip. In a primitive form, though he did not name it or define it in full, Mead in 1971 had created a PLA compiler.

The implications were radical. With the timing, logical, and electrical rules incorporated in the program, the design would be correct by construction—rather than correct by the actual simulation of all the electrical, logical, and timing operations of its gates. Extended to other generic chips, the Mead approach might reduce the scores of man-years in semiconductor design to a few man-weeks. Even though a very limited example of computerized chip design—the PLA is a simple and orderly structure—Mead's invention was comparable in importance to the previous great breakthroughs also first cobbled together in primitive form: the transistor, the integrated circuit, and the microprocessor.

Mead sent a paper describing the spectacular double concept of his PLA and compiler to the appropriate committee at the International Solid State Circuits Conference for presentation at the next meeting. He fantasized the electrifying impact it would make on the crowds of assembled designers and the subsequent influence it would exert on the industry. He waited. Eventually he received a crisp letter of rejection.

The letter pointed out that there were no new types of fabrication technology or new circuit families included in the paper. "It appears merely to describe a new pattern of previously known devices." Mead had merely described a revolutionary mode of design of a revolutionary new function. But the semiconductor industry was uninterested in a new organization of what to do with a chip. It took the functions of chips for granted; it was concerned chiefly with clever ways to perform those functions more efficiently. It was still in the business of creating replacement parts.

The irony was that although the computer industry would eventually adopt the PLA, computer scientists were no more interested in Mead's breakthrough than the semiconductor designers. Computer science was an abstract discipline that eschewed concern with the actual physical implementation of architectures. Most creativity was focused on the software end and most software was custom-written for huge mainframe computers.

Computer architects regarded chip design as an occult art that they could never master. They preferred to just send in general instructions to the semiconductor companies. Or, still better, they wanted to continue buying simple TTL logic chips and standard memories off the shelf and put them together themselves on arrays of printed circuit boards. That was the language they understood.

The rejection letter thus precisely defined the problem of both industries at the dawn of the era of VLSI. The semiconductor people were chiefly filling in sockets in previously designed systems. The computer people were still responding to the semiconductor in the way that Carver Mead at the age of fifteen responded to the transistor. It was a replacement device. "It didn't do anything that you couldn't do with vacuum tubes." Until the mid-1970s, this approach meant that the computer companies hesitated to use custom components much at all. Mead's concept was ignored.

Once an eminent professorial presence in the industry, Mead during this period became known as a "tiresome" and "compulsive" gadfly. The difference between a prophet and a gadfly is that the gadfly's message is unacceptable. Mead's message was totally unacceptable; it meant the end of the line for the semiconductor industry, then happily creating better components for the computer industry, and it meant the end of the line for the computer industry, then wiring together hundreds of components on printed circuit boards.

Instead of two industries—one designing and producing components and the other producing computer systems from them—there would be only one industry, creating new designs. Embracing all products from computers and instruments to telephones and televisions, it would be an electronic systems industry. If the semiconductor industry was to survive, it would either have to become a systems industry itself, creating computers and television sets. Or it would have to become a foundry for manufacturing chips designed by the systems architects in the systems companies—the only ones who really understood what the chips had to do.

In making his case, Mead often used an analogy from book publishing. Imagine a world where only the owner of a printing press could author a book. Few books would be published, and most of them would be lined notebooks or standard texts—Bibles and dictionaries, say—with a sure market and the approval of the printers.

This was the situation in electronics. Only the owners of semiconductor factories, costing many millions of dollars, could author a chip design. No matter what great ideas for chips might be popping into the minds of computer designers or other systems inventors, virtually nothing could get built unless the semiconductor executives running the fabrication lines could envisage a market for many millions of chips. Jerry Sanders asserted the secret of success in semiconductors: "Leverage your designs, to maximize sales out of each design." Of necessity, such an approach restricted the range of creativity in the industry.

The predictable result was that relatively few different chips were created, and nearly all of them were standard designs with demonstrated markets. A commodity memory chip, for example, was like a lined notebook; its pages were essentially blank, leaving the specifications to the customer. Even the microprocessor needed writers of software to fill in the functional content.

For a while, all this suited the U.S. semiconductor industry. But it slowed creativity in electronics, in much the way a world of Bibles, dictionaries, and an occasional encyclopedia would tend to stifle literary creativity. During this period, the people in systems companies —the chip consumers, from IBM on down—knew the right chips to design but either did not know how to design them or doubted they could be designed. Conversely, the silicon sages in semiconductor firms—from TI on down—had little idea which special chips were right to design. So they stuck to general-purpose devices with big markets and a premium on manufacturing. In effect the creative designers of U.S. technology were divorced from the production engineers, and into the chasm fell the United States' competitive advantage.

As uniformity was imposed on chip designs, the Japanese steadily improved their ability to mass-manufacture the key standard products —memories and microprocessors—that the American firms regarded as their birthright. Soon the Japanese began taking these markets away and the U.S. firms began closing factories and opening law offices in Washington. It was a grim denouement for an industry too slow to change.

A gadfly harassing his friends in the semiconductor business, however, Mead came to see that he was focusing his message on the wrong people. The problem was not in the chip industry but in the computer industry. As long as systems designers thought all computers should be built essentially the same and customized with software, they would see no need for greater powers of chip design. He would have to become a prophet again, propounding new computer architectures.

As Mead had discovered in his work in computer science, the difficulty in the field began with its very conceptual fundamentals: MIT's inventor of information science Claude Shannon and his switching theory and John von Neumann's stored program computer structure. Developed for the phone company, switching theory was a brilliant system for economizing on the use of expensive and unreliable relays, vacuum tubes, and other switches. Mead's experience in the microcosm, however, had told him that VLSI switches would very soon be nearly free and totally reliable. The problem would come in the designs and interconnections.

It was solder that shook loose; it was wires that suffered metal migration from the heat of heavy use; it was wires that eroded the signal and ramped up temperature with their capacitance. The era when wires were cheap and dependable, switches were dear and capricious, and designs were simple and clear-cut had given way to an era when wires would be expensive—entailing time, torment, heat, and chip real estate—switches would be nearly free, and switching theory would be obsolete.

In Mead's view, the Von Neumann machine was out of date because it was an expression of switching theory. Economizing on vacuum tubes and extravagant in wiring, one CPU does all computation and the burden falls on the process of sending and sequencing information to and from far-flung memories. Data mark time endlessly in registers, stacks, buffers, and queues waiting for access to the lordly central computing unit, originally assumed to be an assemblage of hot and bothered tubes.

Mead likened the system to a hypothetical General Motors Corporation where only one person works and all the rest stand in lines to give him instructions. "In a standard computer system," he pointed out, "you can have millions of instruction words and one little processor doing one instruction at a time. It's absolutely crazy."

The problem continued late into the 1980s in the most celebrated of all machines, the supercomputers of Seymour Cray. Look at their front consoles, all cool pastel, look at their performance in gigaflops, leading the world, look at their multi-million-dollar prices, and the often-soaring prices of Cray's shares, and one could imagine oneself in the presence of high technology. But all that curvaceous plastic conceals a ghastly secret. Behind the Cray is the mat. It consists of hundreds of thousands of twisted wires in riotous coils and loops, interwoven like the uncombed hair of a thousand harried witches. It is far worse than the spaghetti of bad VLSI. An interplanetary visitor inspecting the mat behind the Cray would hardly be surprised to see a thousand hamsters on treadmills rolling out the data.

A Von Neumann CPU is less a datapath or processor than a datapinch or bottleneck. It all made sense when it was invented. In those days, wires, tubes, resistors, cores, and circuit boards were separate and made of different materials, and wire was the cheapest of them all. By all means add a few thousand more wires. But in VLSI the components could be made of silicon and could be crammed together on one chip, and wire was the most expensive in every way. VLSI could house processors galore but could be readily choked and strangled by layers of metalization. In fact, far into the 1980s the best production lines in the world were having trouble getting top yields from devices with more than one or two layers of metal.

To put a Cray One on a chip would be impossible not because of the switches but because of the mat, which would translate into layer upon layer of aluminum running rampant across the silicon. To use a structure in which one small CPU used a lot of wire and the bulk of the chip went mostly unused, and most of the time was spent toing and froing, stepping and fetching, scurrying back and forth between memory and mind, represented a ridiculous waste of VLSI resources.

Mead saw that the solution would be massively parallel computer architectures, using scores and even thousands of processors working together on single chips. Massively parallel systems, he said, were inherently application specific. Most particular functions would require a new parallel computer system. It would be impossible to create general-purpose parallel processors adaptable to each computing task. Thus the adoption of parallel architectures would necessarily require the use of fully computerized chip design to create the many scores of thousands of new designs that would be needed. Displacing the general-purpose computers that had long dominated the industry, computer entrepreneurs faced a challenge as far-reaching as the change in chips.

To transform the most deep-seated practices of both industries, Mead saw that he would have to develop a new mathematical model of computer efficiency that could replace the existing approaches. In order to reform the field of computer architecture, he would have to return once again to the microcosm of physics and listen to the technology.

He would return to the inspiration of that tragic early prophet of the physics of information, Ludwig Boltzmann. And he would find a new entrepreneur of ideas to propagate his vision of a new era—Lynn Conway.

CHAPTER
14

From Boltzmann to Conway

What we call solid, the scientist sees as dark. Just one century ago, the microcosm of inner space was almost totally opaque. Most scientists denied there was anything there. The world consisted of solid materials and even the existence of atoms was intensely disputed.

In the subsequent century, many scientists and engineers have shone light into this microscopic pitch. Carver Mead is one of them. But Mead himself declares foremost among the prophets a nineteenth-century Viennese scientist named Ludwig Boltzmann. A pioneering physicist with a huge beard, he died in 1906 at the dawn of the modern era. To Mead, the story of Ludwig Boltzmann still rings with the poignancy of a memory of personal triumph and personal pain.

Boltzmann's great insight was that when in the dark, you make your way by probability. By probability, the blind Helen Keller makes her way across a room. By probability, a scientist could glimpse into the dark of the microcosm.

Boltzmann did not invent probability theory. But together with the great Scottish physicist James Clerk Maxwell, he changed it from a means of approximation to an explanatory principle. He showed that whether in the form of a wire or a gas, matter would behave in entirely different ways as a result of the workings of the law of probability.

Thus he changed the relationship between information and matter. In the past, matter came first, and by probability laws the scientist could estimate its condition. After Boltzmann, probability law was seen as operating on a deeper level actually to set the condition of matter: determining the location and consistency of various objects.

Boltzmann's bold reinterpretation of the relationship between information and matter came in a series of papers written in the 1870s that radically reconstrued the Second Law of Thermodynamics, the entropy law. Entropy is the irreversible tendency of the total energy in the universe to move toward equilibrium—the tendency of both heat and cold to trend toward a lukewarm norm, of available energy to become unavailable and inert.

Until Boltzmann, thermodynamics had no use for the concept of atoms. Thus it was a bulwark of the pre-atomic view of matter. Boltzmann, however, showed that entropy grows not because of some strangely isolated law of nature but because of a heavy bias of probability in the behavior of atoms.

The Boltzmann theorem asserted that the probability of any state increases as its energy decreases. Low-energy states are thus more probable. The flow of heat from a hot body to a cooler one—to create two lukewarm bodies—can be seen as creating a more probable overall arrangement of molecules (a random mixture of fast-and slow-moving particles) from a less probable one (one uniform group of atoms dervishing and another one inert).

According to Boltzmann, any particular collision of two atoms in a box is perfectly reversible, with no gain of entropy. But a large number of collisions over time move the system irreversibly toward equilibrium as a result of statistical probability. In other words, the behavior of collective phenomena could not be simply deduced from the behavior of single particles. A collection of reversible events could produce an irreversible result.

In analyzing entropy, Boltzmann was addressing a central concept of all practical physics and engineering, and an idea influential even in economics and philosophy. Usable energy is energy in disequilibrium: water at the top of a cliff, steam in a steam engine rushing through a channel to a cooler cylinder, a high-pressure area juxtaposed to a low-pressure one, a positive voltage connected to a negative one.

For example, the flow of electrical current down a voltage slope

from a positive pole to a negative one is governed by the same Boltzmann formulas that define the flow of hot molecules toward a region
of cold. The drift and diffusion of electricity is a Boltzmann effect.
Overall, Boltzmann's statistical mechanics lays the foundations for
explaining how the properties of atoms affect the properties of visible
matter, such as viscosity and conductivity. In fact, Boltzmann's insights are so central to understanding electronic phenomena that
Carver Mead's model microcosm for teaching information technology
is called "Boltzo."

During his lifetime, however, Boltzmann aroused bitter scorn and
denunciation from physicists, chemists, and mathematicians all over
the continent. Positivists such as the physicist Ernst Mach and the
chemist Friedrich Wilhelm Ostwald denounced his "metaphysical"
belief in the reality of atoms that no one had ever seen. Classical
physicists passionately rejected his new relativistic, indirect, and probabilistic form of thermodynamics. Finally Jules-Henri Poincaré, a
leading mathematician of the day, wielded an elegant and authoritative calculus to prove that Boltzmann's theory was wrong. A prestigious figure in the French intelligentsia and first cousin of a famous
French statesman, Poincaré declared that a finite atomic system such
as Boltzmann's would eventually return to its original condition, thus
reversing direction against entropy.

Poincaré's argument was mathematically consummate, and seemed
to force rejection of Boltzman's kinetic theory of atoms. But Poincaré
made a crucial error. He assumed that the contained molecules of gas
which he and Boltzmann were describing could bounce perfectly off
the walls of the container and return to their original state through
these orderly reflections. But as Carver Mead fervently asserts today:
"The walls themselves would be made of atoms and each one would
vibrate unpredictably. Boltzmann knew that there is no way to keep
anything decoupled from the rest of the universe with all the billions
and billions of degrees of freedom in it. There's no way to keep things
from distributing their coherent energy into all those degrees of freedom. Order is lost. There's no way to get it back. So it won't ever
come back."

Nonetheless, Poincaré's elegant math prevailed over Boltzmann's
practical findings. For some thirty years, Boltzmann struggled to get
his ideas across. But he failed. He had the word, but he could not find
a way to gain its acceptance in the world. For long decades, the
establishment held firm.

So in the year 1906, Poincaré became president of the French

Académie des Sciences and Boltzmann committed suicide. As Mead debatably puts it, "Boltzmann died because of Poincaré." At least, as Boltzmann's friends attest, this pioneer of the modern era killed himself in an apparent fit of despair, deepened by the widespread official resistance to his views.

He died, however, at the very historic moment when all over Europe physicists were preparing to vindicate the Boltzmann vision. He died just before the findings of Max Planck, largely derived from Boltzmann's probability concepts, finally gained widespread acceptance. He died several months after an obscure twenty-one-year-old student in Geneva named Albert Einstein used his theories in proving the existence of the atom and demonstrating the particle nature of light. In retrospect, Boltzmann can be seen as a near-tragic protagonist in the greatest intellectual drama of the twentieth century: the overthrow of matter.

Boltzmann's ultimate answer to the prestigious critics who called him a loony metaphysician was the constant. The "k" or "kT" that relates temperature to electron energy in all states of matter, Boltzmann's constant is even more common than Planck's "h" in the guidebooks of solid state physics. kT is where you begin: the electronic threshold of activity caused merely by the temperature of the environment.

More important to Carver Mead, though, in his own battle with the computer establishment, was the deeper significance of Boltzmann's constant. Because it relates energy to probability and probability to order, Boltzmann's law paved the way to a fully integrated theory of computation, combining physics with information.

As early as 1968, Mead had decided that the emergence of VLSI, with millions of transistors on a chip, would mandate a new way to measure the efficiency of computers. Mead's answer was a broader application of Boltzmann's concepts, unifying chip and computer design through an elegant merger of computer science, thermodynamics, quantum mechanics, and the information or switching theory of Claude Shannon.

Overshadowing computer science for fifty years has been Alan Turing's pre–World War II demonstration that any digital computer can be reduced to a very simple mechanism of a movable segmented tape and instructions. Computer scientists still usually treat computers as abstract Turing machines governed by logic and information theory alone. But Mead saw that computers are also physical systems and finally succeed or fail at the physical level. By depicting the most basic laws of physics in terms of probability and disorder, which are forms of information, Boltzmann provided a way of joining the two perspectives on computation.

Boltzmann's entropy registered uncertainty as to the state of a physical system. Shannon's informational entropy registered uncertainty as to the state of a message source. Boltzmann's concept was measured in joules (an index of work) per degree of heat. Shannon's concept was measured by the number of bits needed to resolve the uncertainty of a message source. In a sense the two concepts are opposites. At the heat death of the universe, when all matter is at a (dead certain) uniform temperature and no work or message is possible, Shannon's entropy is zero and Boltzmann's entropy is maximized.

The two concepts thus are treacherous in their similarity and many writers have gone astray in linking them. But Mead applied them shrewdly and fruitfully to computing. In Shannon's terms, computing represents a kind of game of Twenty Questions—perhaps even beginning "Animal, vegetable, or mineral?"—in which each answer cuts down the number of possible outcomes. The entropy of the data is measured by the number of answer or decision bits needed to achieve the one correct solution, the unique point of lowest entropy. Mead designates this measure of necessary logical steps to a unique solution as the logical entropy of the system.

The problem is that most computing theory stops with logical entropy. Mead, however, introduces the concept of spacial entropy. Just as logical entropy represents the degree to which the data is in the wrong form, spacial entropy represents the degree to which the data is in the wrong place. Just as logical entropy is measured by the number of logical events needed to get the correct answer—to put the data in the right form—so spacial entropy is measured by the number of transmission events required to put the data in the right place.

Combining logical and spacial entropy constitutes a major step forward from conventional switching theory, which focuses exclusively on logical entropy. The inclusion of spacial entropy, for example, would penalize a Von Neumann architecture for its heavy reliance on communication events. But to gauge the total efficiency of a computing system also requires consideration of the physical entropy. The physical entropy relates to the time and the power consumed by the system or the amount of work needed to resolve the logical and spacial entropies.

Both logical and communication events entail switching. Each switching action expends a certain amount of time and a certain amount of energy which can be measured in terms of the Boltzmann constant, kT, expressing the impact of temperature alone on the device. Communication events, however, not only take time and switching energy. They also require movement along wires, which requires a charge that can overcome the resistance and capacitance of the wires and can compensate for the heat emission or entropy increase of this process. Boltzmann's constant thus provides a unit of measure that can be related to both spacial and logical entropy in the computation.

Boltzmann's constant also applies to the entropy of the entire system. The time expended and the heat shed in each of the logical and communications events steadily accumulates and measures the total work needed to run the machine. For example, Josephson Junctions and other cryogenic superconducting elements radically reduce switching energy and delays by drastically reducing the temperature in kT. But pending new advances in applying superconductivity at higher temperatures, these systems also demand heavy power usage and entropy for refrigeration (if only by being launched into outer space).

By using Boltzmann's concepts as a unifying idiom, Mead in the early 1970s developed a new law of the microcosm: a precise mathematical index of the efficiency of any information system. The Mead index favors low and slow VLSI systems in CMOS, scaled down to geometries limited chiefly by tunneling thicknesses, that maximize the use of switches, minimize the use of wires, and economize on the use of energy and shedding of heat. Such systems dictate a high degree of concurrency in the architectures, using many processors in parallel on each chip rather than one processor communicating in complex ways with a large distant array of memory addresses. Mead's concepts dictate the avoidance of wire-intensive universal clocking schemes and often favor asynchronous systems without a ruling clock. Distributed machines are favored over centralized networks. Low-power analog systems also thrive.

Some fifteen years ago, eons in the time frame of microelectronics, when switching theory, Von Neumann architectures, and 10-micron geometries ruled the industry, Mead prophesied the precise lineaments of the current era. In the mid- and late 1980s, venture capital firms, computer magazines, VLSI conferences, and university computer science departments all reverberate urgently with the Mead idiom of parallel processing, multiprocessors, asynchronous logic, hypercube parallel interconnects, custom designs, computer-aided engineering, silicon compilation, and new analog architectures.

At a time of intense competition, government targeting, and ingenious innovation abroad, this wave of new creativity is thrusting the United States ever farther into the world lead in computer technology. It is reviving the U.S. microchip industry. But for more than a decade, Mead's message was mostly ignored. His feats of intellectual synthesis thrilled his students at Caltech, but alienated most semiconductor executives as they counted up their profits in microprocessors and TTL bipolar logic chips.

The word may be critical in promoting revolution. Ideas are indispensable to enterprise. But as Ludwig Boltzmann learned, they cannot prevail without organization and entrepreneurial initiative. The Mead theory has prevailed, not a minute too soon for the fate of U.S. high technology, through an unprecedented feat of intellectual enterprise launched when Carver Mead met Ivan Sutherland, who had tired of life in Salt Lake City with Evans & Sutherland, the leading computer graphics firm. Mead asked Sutherland, one of the two charismatic brothers who have played a key role in computer innovation for the last decade, to become chairman of Caltech's Computer Science Department.

Where Mead is an inventor and visionary, Sutherland is a galvanic leader and exponent. He mobilized industry and the Computer Science Department and ultimately launched a transformation of both. And perhaps most important of all, he introduced Carver Mead to a young woman named Lynn Conway, who was working for his brother William ("Bert") Sutherland in the laboratories of Xerox's Palo Alto Research Center, better known as Xerox PARC.

Starting at MIT, Lynn Conway dropped out after two years of fine academic performance and ultimately got a master's degree in computer science from Columbia University. She worked on advanced computer architectures for IBM during the 1960s, before designing a small business system for Memorex just before its withdrawal from computers. In 1973 she joined Xerox PARC under Bert Sutherland, the more amiable of the brothers. At Xerox she worked on special-purpose architectures for image processing and became aware of "a gap between the sorts of systems we could visualize and what we could actually get into hardware in a timely way."

Call it the Mead gap. It meant that she, like the rest of the world's systems designers, usually had to forgo the advantages of VLSI in favor of TTL parts bought off the shelf and plugged into printed circuit boards. But as she discovered on her image-processing project, even though the TTL solution cost just $7,000 and took a few weeks to achieve, it was barely adequate in performance. Moreover, each further wire-wrapped board, crammed with TTL, cost some $500. In most cases, the idea, like Conway's graphics processor, would never be tried.

In those days, however, a VLSI solution was hardly better. It began with TTL breadboards for testing and prototypes, each costing the same $7,000 and taking six weeks. Then it took another year and a half and some $150,000 to get it into first silicon. The VLSI chip might be incomparably faster, more reliable, and cheaper to manufacture than the TTL board. But unless it was a foolproof, foolhardy, or top-priority project for a major company, it was never made at all.

In the Mead gap moldered thousands upon thousands of possibly brilliant ideas. In the Mead gap moldered the prime resource of American technical industry—its supreme creativity and resourcefulness in design. But the Mead gap was a concept almost entirely alien to even the most enlightened figures in the semiconductor industry. They lived not in a world of excessive designs but in a world barren of uses for VLSI.

Consider, by contrast, the Moore gap. In a speech in 1979 to the Institute of Electrical and Electronic Engineers, Gordon Moore famously declared that "besides products containing memory devices, it isn't clear how future VLSI can be used in electronic products." He pointed out that in the past a gap arose between the possible density of VLSI and the actual chips produced, because "it was difficult at the time to identify any semiconductor products whose complexity came close to the potential limit. The gap did not develop because there was a lack of effort . . . but rather because of a problem in product definition."

The two gaps were really the same. They came down to a great gulf between the people with the ideas in systems companies and the people with the fabs in semiconductor companies. Conway and Mead were both trying to close the gap. In late 1976 and early 1977, she attended a class on VLSI design given by Mead in the PARC facility and began an epochal collaboration with him. Mead had already developed his integrated theory of solid state computer architecture and design technology. "At the same time I decided to expand my knowledge from computer architecture to silicon," as Conway puts it, "I met Carver, who was coming upward from a knowledge of ICs into computers." After the class, in which under Mead's guidance she created a chip design in a few weeks, Conway decided to do something about the gap. Under the sponsorship of the Sutherland brothers, Xerox and Caltech put together the Silicon Structures Project to address the problem of VLSI design.

For the next two years, Conway coordinated her efforts under Sutherland at PARC with Mead's ongoing work at Caltech. But she was frustrated with the pace of progress. There was no shortage of innovative design ideas; computerized design tools had advanced dramatically since Mead's first efforts several years before. Yet the industry as a whole continued in the old rut. As Conway put it later, the problem was "How can you take methods that are new, methods that are not in common use and therefore perhaps considered *unsound methods,* and turn them into *sound methods*?" [Conway's italics].

She saw the challenge in the terms described in Thomas Kuhn's popular book *The Structure of Scientific Revolutions*. It was the problem that took Boltzmann to his grave. It was the problem of innovation depicted by economist Joseph Schumpeter in his essays on entrepreneurship: new systems lay waste to the systems of the past. Creativity is a solution for the creator and the new ventures he launches. But it wreaks dissolution—"creative destruction," in Schumpeter's words—for the defenders of old methods. In fact, no matter how persuasive the advocates of change, it is very rare that an entrenched establishment will reform its ways. Establishments die or retire or fall in revolution; they only rarely transform themselves.

Lynn Conway, however, invented a new way to wreak an industrial revolution. In a multi-faceted campaign that far excelled any project of MITI in its fast and far-reaching effects, she followed Mead's example and excelled it in capturing the next generation of workers and entrepreneurs where they are most impressionable: in the classroom. She began by trying to persuade Mead to write a book. In part, she failed. Like many men with a gift for creative expression in numerous fields, Mead finds writing a painful process. He persuaded Conway to serve as co-author and constant catalyst of the book, and Bert Sutherland gave her time off to complete the work. It became the first step in Conway's historic campaign to spread the gospel and the technology of VLSI design.

"In electronics," she recalled, "a new wave comes through in bits and pieces . . . I was very aware of the difficulty of bringing forth a new *system of knowledge* [her italics] by just publishing bits and pieces of it among traditional work and then waiting until after it has all evolved and someone writes a book about it. What we decided to do," she concluded, "was to write about it while it is still happening." It would not be a typical text, encapsulating what was already known. "Our method was to project ourselves ahead ten years, and then write the book as though reflecting back upon a decade. Then we would let the people in the community critique it, and let the book itself become the focal point for the creation of methods."

Lest the community offer the ultimate critique of indifference, Mead began assigning chapters of the book to his classes as they were written, and Carlo Sequin of the University of California began teaching the chapters as well at Berkeley. Then, in the fall of 1978, just before Conway began teaching a class in circuit design that Ivan Sutherland had arranged for her at MIT, Mead and Conway finished the first draft. "It was she who drove it through," says Mead. It represented a full exposition of the state of the art in circuit design, complete with several examples of chips designed by various of Mead's students at Caltech using the new methods. It concluded with several prophetic chapters on the most promising future directions for computer architecture and with an explanation of Mead's entropy theory for judging design efficiency.

In one chapter, the manuscript introduced the novel idea of the multi-project chip—later to become the multi-project wafer. An innovation of Mead's, it aimed at drastically reducing the costs of creating prototypes of new circuit ideas. With the consistent techniques and design rules described in the book and the vast capacities of VLSI, several student design projects could be executed on one die. Afterward, each student could receive a packaged chip with only his own circuit wired in.

A key Conway contribution was the concept of "lambda"—a relative unit of measurement in chip geometries—that facilitated translation of designs between fabrication processes with different degrees of resolution (different minimal line sizes). The lambda approach freed designers of committing themselves to any particular fab until after they had finished the design. Thus it allowed "process independent" designs, emancipating their authors from bondage to any particular fab.

What Conway planned for her course at MIT was an audacious attack on all the practices and pretenses of the current design establishment. During one term at MIT, she would have her *students* draft an array of sophisticated chip designs. They would cover a gamut of projects, each of the complexity of a full board of TTL devices, each typically a year-long project that would involve several electrical engineers and layout personnel at one of the large semiconductor companies. The prototype chips would be delivered in six weeks. Using multi-project chips, they would cost less than $1,000 apiece. They would demonstrate conclusively the superiority in both cost and speed of the new design methodology.

Printing out three hundred copies of the finished manuscript at Xerox, Conway shipped some down to Mead in Pasadena, put the rest into her car, and drove off to Cambridge. *Electronics* magazine described the results in the cover story announcing its 1981 Annual Outstanding Achievement Award to MIT dropout Lynn Conway and the "compulsive prophet" Carver Mead:

> The MIT class was a smash hit. The students learned about the methodology in September, created their own designs in October and November, and handed them in by early December. Six weeks later, the masks for the multiproject chip design had been made by electron beam lithography at Micromask Inc. in Santa Clara, California, and the wafers had been processed at Hewlett Packard Co.'s IC processing lab in Palo Alto. The dice had been cut and packaged with custom wiring running from the 40 pin dual-in-line package to the internal pads for each separate circuit project within the chip. Each of the students received a silicon implementation of his design.

Most of the nineteen chips functioned as planned.

Returning to California, Conway resolved to expand her revolution into a national campaign against the silicon establishment. The time was ripe. Reports of the MIT course and Mead's feats at Caltech had reverberated through the nation's engineering campuses. Many professors were eager to conduct similar classes, introducing their students to the field in such a dramatic and intoxicating way. Conway would have hundreds of students, working separately at colleges across the country, execute designs in weeks through the hierarchical methodology.

Since most of the nation's technical institutions were tied into the Pentagon's interactive ARPANET (Advanced Research Projects Agency Network), Conway decided to have designs from all the classes transmitted to the computer at Xerox PARC. In the time-sharing system, queries about design approaches, scheduling, filing format, fabrication rules, multi-project wafer specifications, or bugs in the textbook could be entered into the network for a rapid response from the authors. Engineers at PARC would quickly integrate the submissions for multi-project chips on the multi-project wafers and convert them into pattern generator tapes for photomask production. Once again the photomasks would be inscribed on Micromask's new electron-beam writing system and the chips would be manufactured at Hewlett Packard.

To finish the elaborate software needed for such a massive project in interactive communications, with hundreds of packaging requirements, twelve different die types, and many complex layouts for the multi-project chips, required a crash program at Xerox PARC. But the job was completed in time and the classes across the country— some in universities that had never before run a course in VLSI design —ran off without a major hitch. The resulting chip set contained a total of 82 designs from 124 participating students and numerous faculty from MIT, Caltech, Stanford, Rochester, Carnegie Mellon, Berkeley, Yale, and the Universities of Illinois, Washington, Utah, Colorado, and Bristol (UK). Most were sent over ARPANET, but other networks were also used.

Included among the projects were a digital signal processor and several LISP microprocessors (for artificial intelligence applications). From the cutoff date for design submissions, the turnaround time for fabrication was twenty-nine days and the total cost per project was between $400 and $500. Less than the cost of TTL on a circuit board —even for the prototype—the system portended a drastic drop in the cost of designs produced in volume. Once again, most of the chips worked. The yield was especially high at MIT, which used a number of more advanced design and testing aids and benefited from Conway's class of the previous year. Conway and her PARC colleagues Alan Bell and Martin E. Newell reported that "we find many new applications are being explored, new architectural techniques discovered, and novel mappings of logic functions . . . are being invented by university participants. New and powerful design aids have been created and tested during efforts to create some of the projects."

Most important of all, thousands of students were experiencing a thrill unprecedented in the annals of the semiconductor industry. They were executing complex chip designs at low cost in a matter of weeks and seeing the fruits of their efforts a month later. Mead began traveling out across the country and giving similar courses at companies from Bell Labs and Digital Equipment to Boeing Aerospace, startling groups of executives ready for a series of lectures with the announcement: "What you are going to do in these three days is design a microprocessor."

Whether in school or business, most of these students would never again be satisfied with the old rhythms of chip design and fabrication. Scores would begin new companies to design VLSI chips or to create new computer-aided engineering packages. Three groups of students of Carver Mead would begin new companies to consummate his earlier concept of the chip compiler. A revolution was inexorably under way. It would permanently change the balance of power in world electronics and revitalize the industry for years to come.

A revolution does not finally triumph, however, until the establishment itself gives in, which rarely happens easily. The Mead-Conway text made its way into virtually every VLSI classroom and onto the desk of virtually every semiconductor executive in America. But most gratifying of all was the response of one of the leading established figures in chip design, one Masatoshi Shima. During 1979, Mead went to Zilog in Cupertino to visit with microprocessor pioneer Federico Faggin. Walking down one of the halls, the two men encountered Shima. "Hey, Carver," he said, "I want a copy of your book."

"Oh," said Mead bashfully, "it's really just for teaching students. It doesn't have anything to tell someone like you. It's not up to your standards."

But Shima insisted. "I've just finished the Z8000," he said. "It's the last of the big mothers. No way we can ever design a chip like that again. Send me your book!"

Shima knew that you couldn't keep an entire chip in your head any more. Mead was thrilled. "It was the first time," he said, "I thought there was any chance that the major people in the industry would accept my approach."

CHAPTER
15

The Silicon Compiler

The difference between history and life is perspective. From the historic point of view of the late 1980s, Carver Mead and Lynn Conway launched a juggernaut in 1978 that would sweep inexorably through the world of electronics. Their text would make its way into design courses around the globe and into the tiny office bookcase of nearly every electronics executive. Their students would begin new companies that would make "computer-aided engineering" and "hierarchical design" the buzz words of Silicon Valley and an open sesame for funds on Sand Hill Road, the Palo Alto promenade of the venture capitalists. Magazines and newsletters would proliferate to advance the revolution. Carver Mead and Lynn Conway would become two of the most charismatic names in the industry.

Eventually a key responsibility for the future of America's security would devolve onto the youthful shoulders of Lynn Conway. Still without a Ph.D., in 1984 she would be made chief scientist for strategic computing at the Advanced Research Projects Agency of the Department of Defense. Mead would become a world-ranging consultant parceling out his free minutes to corporations from Sweden to China. He would also serve on the boards of several vanguard firms in the movement toward computer-aided engineering and chip design.

That is all history. But in 1979 Conway was embattled at PARC, frustrated in most of her efforts to follow up on the successes of 1978. Teaching a summer course at the University of Washington, Mead felt his life was falling apart. The gap between history and life loomed like the difference between Seattle circled with snowy mountains on a clear day and Seattle bound in storm and fog.

Bert and Ivan Sutherland had been leaders and catalysts of the new movement in computer design. Ivan had shaped many of the central concepts and introduced Mead to Bert. Bert had brought Mead to Xerox PARC to teach his course and assigned Conway, Douglas Fairbairne, and Jim Rowson, among others, to take it. Fairbairne and Rowson would become key technical figures at VLSI Technology, or VTI, the first of several companies begun with an explicit mandate to exploit the Mead vision.

Beyond his efforts in VLSI design, Bert Sutherland had supported the work at Xerox PARC that led to the "windows" and the "mouse" on nearly every workstation and many personal computers, from Apple and Atari to Apollo and Sun. He formed the research department that made Ethernet the dominant small computer network and that conceived the "notebook" lap computer. Xerox's lead in IC design gave the company the tools—if the firm had only understood them—to lend new special features to every copier and printer and even to create the kind of electronic "personal copiers" later pioneered by Canon.

Bert Sutherland was the hero of Xerox PARC: that is history. But that was not life. In real life, Xerox fired him in 1979. While he worked day and night on the novel projects in Palo Alto that were to give Xerox an indelible role in the history of computer technology, jealous rivals conspired against him at headquarters. They said that his research, which would fuel the industry for a decade, was irrelevant to the needs of the company. In coming years, the research leadership that replaced him would make the company nearly irrelevant to the needs of the world.

Ivan Sutherland had been similarly important from the Caltech side. He transformed the Computer Science Department, once chiefly a skunkworks for Carver Mead and his students, into a vehicle for major industrial and defense appropriations. He co-authored a key paper with Mead that appeared in the famous microelectronics issue of *Scientific American* in September 1977. This major statement lent

Mead new authority beyond his established field of solid state circuits. Ivan introduced his brother to the Caltech projects and built up Caltech as the nation's leading center of what might be called solid state computer science, the source of most key innovations in the industry. That is history. But in real life, Sutherland left after two years and many of these accomplishments were allowed to slip away.

Then in 1981—in the midst of a thousand shards of sharp glass on a porch in Palo Alto—the real life of Carver Mead was bleeding away from a huge gash in the side of his skull. In real life, Carver Mead had been left by his wife, shorn of his laboratory by a new Caltech administration, and bereaved of his beloved eighteen-year-old daughter by a burglar with a gun who casually shot her in the head as she returned from a date. Carver Mead, once the Pied Piper of computer science, at one point was down to two graduate students in the corner of a basement at Caltech. Carver Mead, the master of several disciplines, all too often was losing control of himself, drunk and disorderly, maudlin and reckless, driving madly late at night into the mountains behind Caltech, or into the redwood forests, and ready to retire to his hazelnut farm in Oregon. From the euphoria of 1978 he plunged into a depression in 1979.

But history goes on. In 1978, at the behest of Ivan Sutherland, Mead had given a speech to the Defense Sciences Board, the Pentagon's advisory panel for research and development. In a scene that will affirm the worst fears of every feminist, the key contact was made in a Pentagon men's room. There Mead was accosted by Charles Boileaux, president of Boeing Aerospace, who asked him to "come up to Seattle and talk to some of his boys." Mead agreed. Arriving in Seattle, he found himself in the midst of a bitter debate that summed up the problems and promise of Mead's vision.

Today, in an aircraft company such as Boeing, most of the work is done on computers. Beginning with the stress charts and aerodynamic flow analysis that determine the form of the airplane, the vehicle emerges from concept to construction and into simulated flight on the screens of many workstations and terminals. On its powerful CAD systems, the company flies thousands of hypothetical planes for every one launched down the serpentine assembly lines in Boeing's colossal factories. Each part is digitally conceived and shaped and finally rendered as a tape that programs a numerically controlled machine tool or other automated metalworking device, or a plastic extruder or mold.

More and more of the features and functions of a Boeing jet, however, were no longer embodied in metal or plastic; they were inscribed in silicon. But the exhaustive exploration of hypotheses possible for airframes was impossible for integrated circuits. Designers tended to stick with old standby concepts such as memories, microprocessors, and logic families, and hope that process advances would improve their functions and performance. Unlikely though it might seem, a creator of Boeing's leviathan aircraft might have more freedom to explore new architectures than an author of Intel's tiny microprocessors.

In the early 1980s at Boeing, chip design was becoming a bottleneck. Yet the company had been burned in an expensive earlier venture in chip manufacturing and several Boeing executives were adamant against any further investments.

Mead gave his presentation and felt the group was listening attentively. But afterwards several of the men returned to the theme of their tribulations in wafer fab. Mead responded that the purpose of his system was to allow systems manufacturers to design chips, not make them. By using outside "foundries," they would never have to gaze upon or pay for another malfunctioning ion implanter as long as they lived. The Boeing officials then began listening more closely. Finally they contracted with Mead to give a course in chip design for the area's engineers the next summer at the University of Washington.

In 1979, Mead and his wife and her younger brother moved to Seattle for the summer. Mead invited the brother-in-law as company for his wife because he knew that the course would dominate his own time and concern. "But what could be lovelier than Seattle in the summer?" he asked. Mead was right about the course. Attended not by malleable students but by an assemblage of largely established figures from both businesses and universities, the classes were difficult and demanding, absorbing Mead day and night.

One afternoon during the second week, he said, "You must be getting onto this stuff by now." He assigned the group a short problem in circuit analysis, which he thought they should be able to solve in an hour or so. When he returned to the lab at his usual hour, five the next morning, he was startled to find the group still there. All barriers between the businessmen and academics had broken down and they were immersed together like any group of impassioned students in the adventure of VLSI design. A few more days and nights of similar adventures, however, and Mead went back to his apartment to find that his wife had left, apparently for good.

At the time, Mead believed that the course would be a crucial test of his system. With the usual students, the payoff would come in the future, as the students slowly gained influence in their companies. But this course was different. If it succeeded, not only would Boeing, Tektronics, Fluke, and the other Northwest sponsors adopt his methods, but the word would spread. Further companies would also join the movement. In his usual hyperbole, he felt the course and his own career were "right on the edge" poised between success and failure, and "could go either way."

Outside of his lab, Mead is an emotional man, eager for immediate gratification and prone to wild swings of mood. He had to choose between finishing the course and returning to California to put his family back together. He chose to finish the course, and though he succeeded, his family fell apart. He felt it would not matter what happened to him if there were enough students who comprehended his vision. It turned out that on this one point he was right, though it would not be easy.

Returning back to Pasadena from Seattle without a wife, Mead found his department in a shambles. A new administration at Caltech was attempting to downplay computer science and other multi-disciplinary fields in favor of rigorous pursuit of the established sciences. One of the first steps was to take away Mead's laboratory. With Ivan Sutherland leaving, Mead also saw he would have to assume increasing burdens of administration and he knew he was no administrator.

In the face of these problems, he received an invitation to take a new chair in computer science at the University of Washington, with a raise in salary and his own laboratory. He was sorely tempted. It would be a fresh start, an escape from all his personal and professional sloughs in Pasadena. But his mind kept returning to his students. In particular, he would think of David Johannsen, the best of them all.

Johannsen was the ultimate tall thin man, six feet eight inches high, with a childlike face, horn-rimmed glasses, and a unique ability to visualize all the levels of the design of an integrated circuit from device physics to computer architecture. At Caltech, Johannsen was a "townie," a Pasadena native educated in the local public schools. He also was a Christian, ready to reproach other students for disparagement of women and to debate them resourcefully on the origins of the universe. The son of an electronics teacher-inventor at Pasadena City College, Johannsen grew up in the midst of a household full of transistors, silicon-controlled rectifiers, diodes, and other circuitry all securely nested in a hierarchical structure, headed by the gentle authority of his father and the love of God. If you ask, David will give you detailed evidence that the dinosaurs were drowned in the Flood.

During that antediluvian period in late 1970s computer history, chip design in college was a heroic undertaking. At the time, the entire university—from the geologists to the astronomers—had to compete for precious moments of batch-processing time on the Caltech IBM 370. Laboriously keying in their advanced and complex designs on IBM punchcards, Mead and his students acquired a keen disdain for IBM's state of the art and contributions to it. It seemed to them that IBM made machines not for the people but for the priesthood of information systems. In its unending quest for the "gotcha," for example, IBM had created a massive and complex operating system for the 360 and 370 that was designed chiefly to prevent other companies from improving the machine with new software or peripherals.

Often access to Caltech's 370 was impossible until late at night. Mead vividly remembered one whole evening with his protégé toward the end of the school year of 1978, when they were processing the final layout of their main chip design project of the year: OM2 (our machine 2), the datapath processor that would comprise the central example for the Mead-Conway text.

Entering the chip design required more than an hour on the 370 CPU, a large cost in computer time. Then Mead and Johannsen proceeded across town in the darkness to have the layout printed on mylar sheets by the Calcomp plotter at Caltech's top-secret Jet Propulsion Laboratory. The Mutt and Jeff of Caltech computer science —one short and in his forties, one teen-aged geek; one lapsed Protestant, divorced, dissipated, and Darwinian, one fresh and bright-eyed Eagle Scout creationist—they passed through the guard gate clutching their computer tape from the 370.

Then Mead and Johannsen made their way into the inner sanctum of the government-sponsored lab. At any hour of the night, the place was hopping like a Watts police station in downtown LA. On one previous nocturnal visit, they remember seeing schematics for the space shuttle then being planned. On the current occasion, they submitted their precious tape of a year's labor and waited for the plots to be printed.

It didn't seem to be working out. The maximum time for a single job on the plotter was 99 CPU minutes, and when they finally got their tape into the machine at 11:00 P.M., it seemed unlikely to finish their project. The operation went down to the wire. One time the monitor revealed that less than 1 minute and 30 seconds remained before cancelation. The next report read five seconds before cancelation. At 4:00 A.M., the final signal revealed that the project was finished, but Mead and Johannsen had no idea at first whether it had merely run out of time or had completed the plot. It had completed the plot. They both whooped with joy.

Mead also remembered the limericks. Because the student projects were the very lowest priority at Intel and the other fabs they used, Mead began writing one line of a limerick for each of the five photomasks: the diffusion of dopants, the polysilicon gates, the ion implantation, the contact cuts, and the metalization. He thought the women working in the fabs would enjoy the entertainment and push through the Caltech designs to get to the punchline on the final wiring step. Although other more unsavory proposals had been offered, for OM Johannsen came through with a limerick from *Boys Life:* "There once was a man from Spokane/ Whose verses would never scan/ When told this was so/ He said, Yes I know/ But I always try to get as many words into the last line as I possibly can."

Then in 1978 came the low point. Johannsen graduated with both bachelor's and master's degrees ("I had been taking a few extra courses"). In a rare decision, Caltech decided to keep him for a doctorate as well, partly to finish the OM project. But the project came to a sudden halt. As Johannsen recalls: "Our tools got pulled out from under us." When the Computer Science Department acquired a DEC-20 minicomputer, the university banned the bit-gorging Mead classes from the IBM 370. The problem was that all the chip design tools used by the group—relatively primitive but indispensable software packages for graphics and wiring—ran only on the IBM mainframe. Not only would the class be unable to complete the peripherals for the OM project but they would not even be able to debug it.

At the time, Johannsen's wife Janet was doing the graveyard shift at work from 11:00 P.M. to 7:00 A.M., and Johannsen began to use his time alone to finish the OM project. But he quickly decided that he would first need to build some new tools that would work on the DEC-20. Other students were already making impressive contributions. Ronald Ayres, a software expert pulled into IC design by Mead, developed a new programming language with embedded graphics.

Johannsen used the new language to extend Mead's most advanced idea—his software program of 1971, created to design programmable logic arrays (PLAs) and rejected by the ISSCC. Years ahead of all the other pioneers in computer-aided engineering, Johannsen created a full general-purpose chip compiler program, called "Bristle Blocks," that could design a wide range of datapath CPUs from architecture to bonding pads.

Like most inventors, Johannsen did not set out with his invention as an explicit goal. "I didn't say, 'What the world needs now is a silicon compiler.' I said, 'What *I* need now is a silicon compiler.' " To get an idea of Johannsen's problem in designing the chip, imagine that you need to design a town from scratch. A datapath chip like Johannsen's presents a problem in complexity roughly comparable to designing a town for five thousand people.

Like an old-fashioned chip designer, with the usual design tools, you decide to make it simple and bottom up. You base the town as much as possible on a single fungible component: bricks, tiles, or some other building block that can be made from sand. You can start by designing the bricks, as chip designers meted out transistors one by one, each one slightly different. But chances are you will give up the freedom of having a variety of different-shaped bricks in favor of a design tool that makes all the bricks the same. (Chip designers would resist long and hard before giving up the flexibility of their uniquely shaped and twisted transistors.)

You also may have a room generator that takes the brick designs and puts them together with rooms. You then find you can design a house—possibly a bunch of houses. Even a child could conceive of such a town with blocks.

This in fact was more or less how designers created chips. It worked for very simple designs such as TTL components. The computer designers could wire them together later on printed circuit boards. But for the town design, there is no further step; the town is the system. Once it is designed, the houses cannot be moved around on the map as if they were so many ICs on a printed circuit board. Similarly, a complex chip design is the system.

In the case of the town designed as if by a child with blocks, you would then have to go back and place all the structures and route all the links between them: the roads, power lines, telephone wires, water and sewer pipes. The connections turn out to be the essence of the design and to work them out will take 90 percent of the time. Back to the drawing board. To route these interconnections in any town of significant size, you will essentially have to start over.

You discover that you can design simple systems bottom up, but with large numbers of houses the complexity explodes far beyond the power of your mind to encompass it. You find you cannot start with sand or bricks or even houses. You must begin hierarchically with an overall town design optimized for the ganglia of pipes, wires, and roads that will make it work. This is what Joahnnsen discovered.

At this point, you see you need a town planner, and an array of architects, electricians, plumbers, and other experts working within a top-down model. These are the short fat men of chip design. Such an aggregation of experts will work in town design. In any particular year very few large towns are designed. But Mead envisaged hundreds of thousands of chips to be designed every year. Johannsen invented a silicon compiler to automate these chip designs.

Starting top down, you begin with the town plan and embed in it plans for roads, pipes, wires, and houses. In the house model you embed rules for creating rooms and you would integrate piping and wiring information with every room generator. You would develop design rules so that you do not connect the septic system with the water pipes or water pipes with wires, assuring that the sewage does not run into the dishwasher or the electricity into the bathtub. You will want to include cost and time constraints.

If you are lucky, an arduous discipline of design rules maintained throughout the process will make the town design correct by construction. The rules embodied in the town compiler will relieve you of the need to simulate and test every part of the design under every conceivable pattern of usage. Such simulation would be impossible; no computer model could encompass the complexity of life in a town. Similarly, the complexity of electrical activity on a complex chip could be simulated only if the entire chip was designed in a thoroughly regular and symmetrical way, with rigorous design rules that make the system correct by construction. This is what Johannsen did in Bristle Blocks.

The crux of compiler design is to partition the overall job into a group of problems of manageable size, in which each part consists of a countable and intelligible number of entities. Making one brick, or one brickmaking machine, or one room design, or one house design, or one town's street design, are all manageable problems. So is designing a transistor, a logic gate, a functional block, a computer architecture. Designing thousands of houses and millions of bricks and billions of electrical nodes and wires are beyond reach.

Throwing more people into the effort will not solve a complexity problem but make it worse. A complexity problem must be solved hierarchically, with the solution to the more detailed, lower-level expression nested in the higher-level abstraction. The town compiler problem cannot be solved as a problem in shaping and distributing sand and copper any more than the silicon compiler problem can be defined as a question of the distribution of silicon, its dopant chemicals, and its aluminum connections.

Designing a compiler is the opposite of using one. The purpose of the compiler is to shield the user from the details. But to shield the user, the designer must immerse himself in these very details. At the same time, he must constantly relate the details to larger structures; he must optimize at the right level. If he decides to pave the streets with gold or locate all the toilets in one place, he will optimize in the wrong way. Similarly, the silicon compiler must forgo many attractive local optimizations that lead to global problems and complexities. To economize on silicon usage, a designer may wish to distribute memory cells in every unused nook and cranny of the chip; but he had better have a good idea of how to interconnect them later.

As a programmer since his early teens, Johannsen had a deep-seated hierarchical impulse. Every time he set out to do a functional cell or block of the chip, it always seemed more attractive to him to do a program first. As he proceeded, from the registers, buses, and ALU of the datapath core, out to the buffers that drove the timing and control lines, on to the instruction decoder that managed and sequenced the datapath operations, and finally on to the pads that connected the device to the outside world, he found he was building what would now be called a series of block compilers. This is a fine achievement. But as Johannsen commented at the time, such aids were only one step beyond "fancy computer filing cabinet design aids." With these, 90 percent of the design is completed in 10 percent of the time, but the final 10 percent of the design (the final wiring step, placing and interconnecting the blocks) takes 90 percent of the time.

He addressed the routing problem with several algorithms that stretch the wires around the blocks without touching other wires. But a successful device must also connect the wires to the blocks in ways that accurately collect the correct signals with the right timing. Johannsen began incorporating a field of logical and timing information in each of the connections—bristles for each block—so that the routing program could locate and address each of the cells of the device, and all the outputs of the datapath and its parts could be interconnected without error.

With this step, the program began to attain global powers beyond compiling particular blocks. It also could interconnect them. Johannsen discovered that he was building something far more important than a particular chip for a classroom project. He was building a system for constructing any datapath chip in NMOS. Since it seemed to him comparable to the compilers he had used in writing software, he named the device a "silicon compiler."

Both silicon compilers and software compilers perform the crucial function of allowing the designer, whether of chips or programs, to avoid the multifarious intricacies of the lower-level implementation. When writing software, the programmer does not want to become involved with the location of each physical address in the code or to see which registers the machine will use or even the exact instruction set of the target machine. Freed from the details, the programmer can address the specific needs of the user.

Similarly in creating a chip, a systems designer wants to operate on the architectural or functional level rather than on the transistor level. The key difference is that the two-dimensional locations, three-dimensional physics, and multi-dimensional timing and electrical parameters of an integrated circuit create a problem of complexity far beyond the constraints of software code. For example, in a software compiler, all locations are essentially equal; they are merely addresses in various forms of memory and can be interconnected mostly at will. In a silicon compiler, by contrast, different addresses represent real locations on a silicon chip, and interconnections between distant transistors pose huge complications not only of the expense of real silicon real estate but also of power usage, signal deterioration, timing, and heat dissipation—essentially all the parameters of Mead's Boltzmann model.

The writer of a silicon compiler cannot be simply a software programmer; he must be to some important degree a master of device physics and electronic circuitry. As a student of Carver Mead, David Johannsen had long ago learned to "listen to the silicon." As a programmer since boyhood, he had mastered the art of generalized structures of software.

So history went on. While Carver Mead was stewing about the politics of his department and the problems and passions of his private life, Johannsen took the technology a massive step forward. In 1979, Mead began to make the exhilarating discovery that the movement had grown beyond his reach and control. In January, his colleague Charles Seitz organized Caltech's first conference on VLSI design and assembled speakers from around the country. The most important address, given on the second day, was David Johannsen's paper on "Bristle Blocks," which announced the age of the silicon compiler: "The goal . . . is to produce an entire LSI mask set from a single-page, high-level description of the integrated circuit."

Johannsen's achievement provided a new stimulus to Carver Mead's sense of possibilities. Asked in 1980 to give a course for designers at Bell Laboratories, he decided to make it from the outset a class on silicon compilation as much as a class on structured design. "It was the first course I could give free of previous lore," Mead says. "All the people attending were systems designers rather than chip designers. I could start fresh, with the right conceptual underpinnings, and give a clear picture of the prospects for compilation." Among the some thirty participants were two leading figures at Bell Labs: Hal Alles and Misha Burich. A typical electronics team, one was a Midwestern farmboy from a strict religious home and the other a refugee from socialism. They were both enthralled by the Mead presentations, two more converts of Carver Mead.

As a small child on a farm in Greeley, Colorado, Alles says, he was providentially dropped on his head. As a consequence, he cannot spell to save his life ("It is a small mind that can think of only one way to spell a word"), but he can see an information structure, a logical system, or an electronic circuit entire in his brain.

This ability eventually attracted the attention of Bell Laboratories, which hired him out of school. At thirty-five, he was one of the youngest division heads in the history of the company. In this role, Alles found himself in direct contemplation of Mead's gap: needing scores of custom designs but without the resources to create them. Carver Mead's concept seemed the solution to the problem. In Mead's design course, Alles did not rest on his laurels. He won one of the two prizes for best design. The other was won by Misha Burich.

Misha Burich was born in Belgrade, Yugoslavia, in 1947, and grew up with a yen for the microcosm. His idol as a child was Nikola Tesla, the inventor of alternating current, also born in Belgrade, who ended up as an embattled American entrepreneur.

Burich threw himself with increasing energy into his studies of English and began poring through magazines in technical libraries, focusing on the fascinating reports on semiconductor technology from the West. At Belgrade University, he wrote his thesis in 1971 on semiconductor memories (just then being created by Intel and a few other companies), and he affronted his professor by arguing fervently that they would usurp cores and become the prevailing form of working storage in all computers, a view not yet widely accepted in the West. When American astronauts landed on the moon, Burich resolved, one way or another, to come to America. Eventually arriving at the University of Minnesota, he turned out to be one of the more valuable prizes of the space effort.

In his studies, he pursued computer science, mathematics, and electrical engineering, but he earned his doctorate in the mathematics of multi-dimensional spaces. The ability to organize complexity would serve him well later. In December 1977, Burich came to Bell Labs. Working in the field of speech recognition, he found he would need a specialized collection of custom chips. But the chip design priesthood of Bell had time to work on only the most urgent projects. Like Alles, he was facing the Mead gap and was eager to join Mead's course on VLSI design.

These were just a few in the growing ranks of the students of Carver Mead. Among them they already assured him a prominent place in the industry. But Mead felt he had been working on these matters as long as he could remember and the idea that merely giving more classes would suffice seemed a delusion to him.

Then, in 1980, Mead had a fateful lunch at an Embarcadero restaurant in San Francisco with two leading venture capitalists. One was John Doerr, formerly one of Intel's best-looking and fastest-talking marketeers, and the other was Tom Perkins of Kleiner-Perkins, the inventor of the $100 laser who sold out to Spectra-Physics and became a leading venturer. Both of them had attended the Caltech conference on VLSI the year before. Perkins brought up the subject of computer-aided design: "There's something wrong with computer-aided design," he said. "I see all these CAD companies and we haven't invested in any of them. Who is going to do it right?"

"Well," Mead responded, "*we* are." In fact, Mead went on, Dave Johannsen and some of his other students were already doing it right, but "it will take another couple years of research to get it to the point where any venture guy would want to fund it."

"I'll fund it now," barked Perkins. "So it's clear who owns the rights. It's worth the extra half a million or whatever to get it right." Perhaps the Valley's leading venture capitalist, Perkins had spoken and blessed the project. Mead was suddenly on the verge of becoming an entrepreneur.

Partly to consider these heady possibilities, Mead decided to visit a beautiful and brilliant Chinese American physicist named Cecilia Shen. He had first seen her but not met her years before at a sparsely attended lecture he gave at Carnegie Mellon in Pittsburgh. Among her many attainments had been a stint as a consultant for MITI in

Japan, when her former husband was assigned to Tokyo. There she made the Japanese secretaries—protesting bitterly at such a violation of custom—serve her tea in her office in the afternoon when they served the male engineers. When her husband found other interests in Japan, she returned to the United States and consulted on several major projects for Bell Laboratories. Yet for all her successes in a male world of engineering, she retained a vivid sense of the differences between the sexes and an intense feminine warmth and luminosity. All of it was radiated at Carver Mead.

At Cecilia's place, he would sit on a chair by the swimming pool in the warm night, under a low-branching fir, and feel content. His trials at Caltech, the departure of his wife, the murder of his daughter, the resistance of the industry all still pained him grievously. But Cecilia gave him perspective and peace. He began to see that most of what he had achieved at Caltech would be lost under the new administration if he left. He saw that the university was his life and he couldn't leave it: "I was a lifer at Caltech." He got up to go inside to discuss his feelings with Cecilia.

Mead leaned forward to step up into the lighted room and crashed into the invisible glass door. In a true outrage of American technology —these accidents happen repeatedly across the country—the door shattered. It cut into the side of his skull and sliced off a six-inch swathe of skin, his entire left cheek, just missing his jugular and the node of his optic and other facial nerves.

Cecilia came down and looked through the glass in horror at the gushing blood, but Mead, still unaware of the extent of his injury, began reaching in to stop her from stepping on the glass. She rushed for towels to stanch the bleeding and at midnight managed to get him to one of the state's leading cosmetic surgeons. He remarked on Mead's arrival: "Well, I'm glad we got all the pieces." When the three hours of surgery were finished under tranquilizers and local anesthesia, he felt strangely exhilarated and began flirting extravagantly with the nurse.

"What would your wife say if she knew you were behaving this way?" she asked. Mead at first was taken aback. His wife? Then he realized the nurse meant Cecilia. "Oh, she would just be glad I'm still normal," he said. Later, the nurse went out and told Cecilia that her husband was behaving like a mischievous teenager. "I'm just glad he's still okay," she said. Mead remained under her care for five days before he could return to Pasadena.

Over the next few years, in times of trouble, Mead would often take Cecilia over the mountains from Palo Alto, veering dangerously through the slalom turns on Route 15, and show her the redwoods in the state forest near Santa Cruz. They would walk under the tall trees through the filigreed light and talk of the pangs of ambition and love. The redwoods spoke deeply to him. Even his stationery has a redwood design running the length of it.

Once when he had been drinking, he told her the trees were sheathed in hard, fire-resistant bark and could survive flames and grow back. "I have been burned many times," he said, "but I have hard bark and can grow back. And the new trees are my students. They keep coming and they will grow very tall and strong." That would be history and it would be life.

CHAPTER

16

Competing Visions

The invention of the silicon compiler was the climax of a series of smaller advances in computer-aided engineering which were rippling through the industry in the wake of the Mead-Conway démarche. The first development was the emergence of the workstation firms: Daisy, Mentor, and Valid. These companies provided a "turnkey" package: a minicomputer and software system which gave chip designers an array of new tools for accelerating every step of IC design, from logic to layout and simulation.

By greatly facilitating the job of conventional chip design, these companies tended to favor firms such as Intel and AMD which produced a variety of standard chips. In themselves, the workstations did not greatly change the balance of power between the established firms and the new firms. The workstations merely allowed both to continue designing chips in the new age of VLSI when otherwise the design bottleneck would have strangled progress. The workstations, in themselves, were a conservative force in the industry.

Even the gate array, a semi-custom device, was conservative in nature. An array of unconnected logic "gates"—linked or "programmed" in a final metallization step in accordance with the customer's design—the gate array provided a way for established firms to preserve their logic business—"glue" chips—against the threat of full-custom solutions. Gate arrays conformed to the competitive strengths of the established firms. Produced on fabrication lines honed by the production of commodity memories and customized in those same fabs according to designs developed on computer workstations, these commodity devices represent the evolutionary response by the silicon establishment to the new challenges of a revolutionary era.

The problem is that the order which these technologies conserve favors the Japanese conglomerates. The evolutionary trend in the industry brings ever greater densities in both logic and memory, achieved through an ever-advancing technology of production, ever more centered in Japan. To meet the challenge from Japan required a radical move toward VLSI: fast turnaround of specialized chips which fulfill the customers' needs far better than any combination of commodity devices.

The major innovation arrived just in time. It was a true American breakthrough that potentially struck at the heart of the Japanese advantage. The great Japanese conglomerates had been investing as much as 30 percent of their sales in a massive campaign to create the leading fabrication facilities in the industry. The silicon compiler promised to move the center of value added from the fab area to the functional design and then move design capabilities away from the established semiconductor firms and make them available to the world. The established firms, long accustomed to regarding their design expertise as a strategic asset, saw the compiler as a threat. Ironically, so did some of the leaders at Silicon Compilers, Inc.

The transition from the old to the new in any industry is far more difficult than it appears. The hard fact is that innovation always hurts. If a technology is truly new, its masters also will tend to be new. Like the gaggle of callow programmers—many of them teenagers—who took over the fastest markets for personal computer software, the masters of the silicon compiler were young geeks scarcely out of school. The experts on the old system resented these newcomers who dismissed their hard-won expertise as "polygon pushing," spoke of current design methods as "archaic," and espoused new forms of knowledge, such as complexity theory, for which the elders had no particular aptitude. As in most industries, they tried to twist the new technology to match the old expertise. Such efforts usually produced the semiconductor equivalent of the prop jet.

Silicon Compilers was fraught with such tensions of transition be-

tween old and new. It began in Pasadena, run by two young men
scarcely out of graduate school, Dave Johannsen and Ronald Ayres,
under the supervision of a visionary and sometimes abrasive professor.
Both the venture capitalists and the principals felt a need for some
Silicon Valley expertise. Carver Mead hired Edmund Cheng, one of
his first students and one of Intel's best chip designers. Cheng brought
on board much of the Intel design team that had created the 286
microprocessor in every IBM AT or clone computer. The first president would be John Doerr, an Intel star who would return to his
venture capital responsibilities after one year, and the second president would be Phillip Kaufman, also an eminent Intel manager.

Commanding some of the industry's highest-powered chip design
expertise, its most prestigious venture capital, and the silicon compiler's very inventor, Silicon Compilers was perhaps the most promising company to emerge in Silicon Valley since Intel itself. It had a
three-year lead on the rest of the industry in the most important new
product of the epoch. Its every move created a splash of color on the
covers of electronics journals. But everything did not go well.

Johannsen's "Bristle Blocks" was a brilliant beginning, and students
following him at Caltech used the program to design many new datapath devices and build several working chips. But it was out of date,
devoted to NMOS in the CMOS era, devoted to datapaths in a time
of entire CPUs on a chip, and most important it lacked a method of
automatically generating test patterns to exercise the circuitry and
discover any production defects. The design might be "correct by
construction," but in any fab many actual chips would be incorrect by
production. No other design tool had even addressed this problem;
but then no other design tool purported to achieve complete design
automation.

An untested chip is like an untested airplane. No one will use it,
regardless of how perfect its schematics and mathematical simulation
or how beautiful it looks on the ground. Working in Pasadena, Johannsen was busily trying to rectify these problems. In two years, he
thought he could do it.

Summoned to Santa Clara, however, Johannsen found himself embroiled instead in a two-year effort to provide tools for the creation
of the custom chips in which Cheng and his Intel associates excelled.
The company began with a chip for SEEQ to connect computers to
the Ethernet networking system invented by Robert Metcalfe at Xerox
PARC. Silicon Compilers treated the project exactly as an establish-

ment firm would have—giving all credit to the designers rather than the tools—and then moved on to a second great success: its Microvax chip done for Digital Equipment Corporation. Meanwhile, the compiler was almost forgotten. For a while Silicon Compiler's designers were even drawing layouts on mylar. The former Intel experts simply didn't believe that compilation could work. They were right. It had to be perfected first by a company entirely committed to the goal. As long as Silicon Compilers, Inc., devoted most of its limited resources to chip design rather than to compiler design, the invention would languish.

When the Microvax was finished—after intense efforts by both the chip designers in Santa Clara and the tool creators in Pasadena—Phil Kaufman awarded special stock bonuses to the chip designers and not to the tool creators. The flagship company of the revolution was telling its followers and indirectly proclaiming to the world what the silicon establishment wanted to hear: that silicon compilation was mostly hype.

Then, in September 1983, at an international conference on circuit design held in a large auditorium in Santa Clara, Ed Cheng received a serious shock. Addressing an audience of several hundred, Tom Matheson, a young colleague of Hal Alles and Misha Burich, described the new silicon compiler they had built at Bell following Mead's course at the laboratories. Called PLEX, it could rapidly execute a complete custom microcomputer design from a functional description of the device. Matheson described a RISC (reduced instruction set) architecture created by the compiler that performed 8 million instructions per second, faster than the Microvax for many applications.

Cheng rushed forward from the crowd to intercept Matheson and take him to Silicon Compilers. They had lunch together and then visited the SCI offices. Before the end of the afternoon, Cheng had offered him a job and a piece of the company. But as far as Matheson could tell, Silicon Compilers at the time had no coherent vision of silicon compilation. Johannsen, whom Matheson met briefly, had not been able to advance the technology far beyond Bristle Blocks. . . .

Dick Gossen, meanwhile, manager of TI's DRAM designs into the megabit generation, resolved to fly the coop. In early March of 1983, he announced his resignation from the firm he had served for seventeen years. Then he ordered some four hundred engraved invitations

from a printer, to be sent to the potential witnesses. His flight—
literally an airplane flight—was scheduled for June 4, three months
after his resignation, and was to be recorded by two television crews
and celebrated with hundreds of friends and some $2,000 worth of
barbecue and beer. Like most of what he did in his life, Gossen's blast-
off from TI would show a sense of drama and metaphorical flair. In
mid-March, he mailed off the invitations. Then he set out to finish his
airplane.

He had been working on this craft—a Christen Eagle designed for
aerobatics—since the grim days of 1978, when his 64K DRAM design was yielding wafers one after another each as dead as a discus.
He had crafted nearly every one of the five thousand separate parts of
the plane himself, from airfoil to fuselage. Then in exact accord with
the instructions in the thousands of pages of some twenty-eight separate manuals, he had assembled the parts. Physically taxing but men-
tally routine, working on his plane was the antithesis of his work at
TI. The handiwork helped him relax and allowed his mind to wander.
Putting parts together in accord with some twenty-eight manuals
might be diverting for a hobbyist. But it occurred to Gossen that TI
had put together its chip designs in much the same way. He had
scarcely more latitude to experiment on his chip designs than he had
in putting together his aircraft.

As a director of new product development at TI, he had an avid
interest in the development of better design tools. But the company
was still behind in the field. One TI engineering manager used to
boast that "our automatic test pattern generator takes only ten times
longer to generate vectors than our worst test engineer takes to do it
manually." That about summed it up. When Gossen left the company,
he hoped to join a firm that could do the design side right. His first
choice was VLSI Technology, best known as VTI, the company
begun by Mead students Fairbairne, Rowson, Steven Trimberger,
and others to exploit new technologies of structured design.

With Mead on the board and Evans & Sutherland as an early investor, VTI had raised a record-breaking semiconductor bonanza of $90
million in an initial offering. VTI appealed to Gossen as a way to
catch the new wave of silicon art. Gossen was especially pleased that
the new president of the firm was Alfred Stein, formerly a manufacturing star at TI and then the savior of Motorola in Phoenix. Combining many Caltech design aids with a top silicon foundry run by the
industry's leading production men, VTI was then the most prominent
force in structured design. The company turned out to be as eager to

hire Gossen as he was to enter the firm. He joined VTI in March 1983 and gave up the idea of a party in June.

Perhaps Gossen had exaggerated the degree to which seventeen years of experience at TI and four thousand hours in his garage would prepare him for rapid ascent in this new semiconductor specialty. In any case, no sooner had he moved into his new office in Santa Clara than he discovered an irreconcilable conflict between his job as he understood it and his job as his bosses defined it. He spent two weeks at the Red Lion Inn in San Jose trying to work out the problem, and then, in the beginning of April, he returned to Houston to work full time on his airplane. It was a setback for his career, but at least the party would go on.

At 2:00 A.M. on June 4, he was still at work doing final "design rule checks" on the plane. Sure enough, it was "correct by construction." He had followed the rules and put every part in its place. He didn't make it to bed until after four. Nonetheless, the next day he was out at Wolfe Field ready to go.

Gossen pressed the ignition, and it ground the engine like a Model T on a cold day. Finally it caught, sputtered, revved up, and began to roar. A burst of air from the propeller splashed a Coke in the faces of his family. He made a last-minute check of the panel, waved smiling to the crowd, disentangled a scarf caught on a hinge, and then released the plane down the runway. It took crisply off and soared up over the fields and away, aiming toward the patches of blue in the gray Texas sky.

Free at last, Gossen could go anywhere he wanted, sprinting off over the Texas prairies or down over the Houston marshes. Within weeks, he would be using his plane for aerobatics, reveling in the freedom of the sky, "drilling holes in the air," zooming up and plunging back toward the ground, right side up and upside down, breaking the molds of his mind. . . .

By early in 1984, Gossen was ensconced as president of a new firm called Silicon Design Labs, begun by a group of scientists from Bell Laboratories, and led by Alles and Burich. Mead's star students of three years before, they had built the microprocessor compiler that Matheson had described in Santa Clara, and had hired Matheson to help create a full silicon compiler.

In the effort, Burich invented a new language for silicon compilation that combined geometrical and electrical information in every line of code. Rather than integrating these data in separate subroutines of "C"—the Bell Labs Unix language—Burich achieved a further step of synthesis. It conferred several efficiency benefits and allowed graphics to be directly edited without need to refer to a "C" program.

Silicon Design Labs (SDL) was a firm nearly devoid of the kind of detailed circuit design and layout experience in which Silicon Compilers, Inc., excelled. But with Gossen's help the company developed a way both to overcome the resistance of the design culture from which he had come and to disprove the establishment view that a compiler could be no better than the chip designers who built it.

Gossen summed up the strategy in early 1985. "All semiconductor companies believe they have the world's best IC design talent, and they're right! Each company is an expert at some segment of the IC market. But as the demand for semicustom and custom ICs expands, manufacturers of semiconductors will need a way to reduce the time and expense in the design process. But they will not want to lose control of the process by using the design expertise of a third party's silicon compiler. They will demand the ability to develop their own proprietary compilers or generators to automatically capture their design expertise and reuse it in a way that provides hand-crafted quality. We view GDT (SDL's 'cell generator' system for automatically creating cell compilers) as the first and only such tool. These generators will allow each manufacturer to maintain its competitive edge by continuing to link the best design talent to the best process."

Although Johannsen had failed to induce Silicon Compilers' own chip designers to contribute to their own compiler, Gossen had arrived at a strategy to induce chip designers around the world to contribute to Silicon Design Labs's compiler. If a designer developed a new cell for the system, SDL would compensate his company. It was an impressive scheme. Within months, Silicon Compilers adopted the same strategy, launching what it called Genesis, a compiler development tool. But Mead's flagship firm, once three years ahead, had become a follower. . . .

In Seattle, meanwhile, shortly after Charles Boileaux retired from Boeing, the company canceled its semiconductor design program. The new president was adamantly against "competing with its IC suppliers." He would not listen to protests from Boeing's semiconductor labs that for small runs of specialized chips, each to go in a few hundred planes, there were then no suppliers at all. Nonetheless, under the leadership of Mead students Dick Oettel and Vince Corbin, the CAD group had already reduced the time needed to turn around a new design at Boeing from nine months to five weeks.

When Boeing canceled the program, Corbin, Oettel, and several other engineers found themselves with a partly developed silicon compiler but without a company willing to use it. Teaming with Gordon Kuenster, a leading Seattle venture capitalist who previously had launched or rescued several small systems companies, Oettel and Corbin founded Seattle Silicon in April 1983. Like Silicon Design Labs, they lacked rigorous IC design expertise. But also like SDL later, they created a conduit for the expertise of others. Within two years Seattle Silicon had their Concorde program running not only on Mentor workstations but also on Valid, Metheus, VTI, and Tektronics CAE systems. Following the Seattle Silicon example, Silicon Compilers in 1985 began negotiating with Daisy, the one CAE company Seattle Silicon had missed. Silicon Compilers was no longer clearly preeminent.

By 1988, Silicon Compilers and Silicon Design Labs—both shaped and inspired by the classes of Carver Mead—were in the forefront of the most important development in present-day semiconductor technology. Also guided by Mead students, VLSI Technology (VTI) had survived a period of turbulence that ended with Mead's resignation from the board. Under Al Stein, the company had swept to near $300 million in revenues in 1988. With a strategy of coupling an array of compiling tools with a leading edge CMOS fab, VTI became the first fully conceived and executed silicon foundry.

Nonetheless, no company had yet effectively addressed the testing issue. Johannsen decided to bring it to the fore at an engineering meeting at Silicon Compilers' Los Gatos headquarters. Twenty of the firm's engineers crowded into the room to consider the problem of generating test patterns or vectors for a chip. The problem that confounded the design aid division at TI and most other companies, test pattern generation is fundamental to the development of silicon compilers.

The testing problem is so central to silicon compilation that the lack of test vectors might ultimately relegate compilers to a marginal role in the industry. The one thing gained from a hand design is the intimate knowledge needed to test the chip. The purpose of a compiler is to shield the designer from the very details he needs to know in order to create a testing program. Without a generator of test vectors, the designer may lose in the testing phase much of the time he gained in the design phase. In the issue of testing, the very future of Silicon Compilers both as a concept and a company was at stake. Yet the experts at the company—the inner sanctum of the flagship firm in the field—were busily declaring the impossibility of test vector generation.

No sooner had Johannsen raised the question than heads began shaking and the chorus came forth. "Can't do test. . . . People have tried and died in the attempt to computerize testing. . . . It's impossible. . . . Let the customer do it; he's the only one who really knows the functions."

While the agitated talk continued, Johannsen went quietly to the blackboard and made a list: RAM, ROM, Datapath, FIFO, PLA, pads, Random Logic. "I agree," he said disarmingly, "it is very hard to do test. But what about a RAM? Can you make a program to do test vectors for a RAM?"

"Sure," one of the engineers conceded, "a RAM is different. It's a regular structure. You could write a program that could do test vectors for a RAM." As the discussion again surged forth in opposition to the idea of testing, Johannsen quietly checked RAM off the blackboard list.

A minute later, he asked about a ROM. Could you do test vectors for a ROM? Again a similar reply. By the end of the meeting, Johannsen had checked off all the various cells he had listed on the board, which collectively represented nearly all the key functions of any compiled microprocessor. But even at the end of the meeting most of the chip designers were deeply dubious of the possibility of including a test vector generator as part of the compiler. "Couldn't ever trust it," one said. "It won't work globally," said another. The meeting broke up, with most of the designers going off to develop newer and better chips. Johannsen returned to Pasadena to develop new proposals for a test vector generator. This advance, together with other verification tools, promised to give Silicon Compilers its major asset in the industry: ways to reassure its customers that compiled chips would actually work as specified.

Then, early in 1987, a deputation from Silicon Design Labs in New Jersey arrived at Silicon Compilers in California and proposed a merger of the two firms. A dream come true for Mead as a director of Silicon Compilers and as a teacher of Burich and Alles, the new firm would bring together most of his best students in compilation in one company. The new firm, to be called Silicon Compilers Systems Corporation (SCSC), would end the confusion in the industry over the two key silicon compiler products by integrating them into one system as some customers had already been trying to do.

The new merged company would use the New Jersey firm's layout language and technology-independent circuit design tools and would use Silicon Compilers' functional and behavioral systems for product definition and development. Crucial to the firm's success would be testing and verification tools. In 1987, compiler sales in the industry had risen nearly 100 percent, to close to $200 million, and Andrew Rappaport of the Technology Research Group expected still faster growth in 1988. A $500 million market was foreseen in the early 1990s. Accelerating the acceptance of the technology by systems designers, the new integrated tool would spur on the revolution in the industry.

Impelled by the logic of the microcosm, the silicon compiler represents a further shift of the sphere. Like the integrated circuit that displaced all the producers of discrete devices, semiconductor memories that laid waste to cores, and microprocessors that destroyed the economies of scale in computing, it is an innovation with the ability to hurt. Carver Mead and David Johannsen were determined to computerize the design of chips. Chips would indeed be shaped "automagically," as Ed Cheng had disdainfully said. The compiler is not merely a fancy tool to install in design centers at Texas Instruments and Intel. It is a weapon of creative destruction that will ultimately force these established firms to change the way they run their businesses.

From the establishment point of view the compiler would lead to "bad designs," with layouts far less than optimal. But Mead and Johannsen were saying that in the usual sense, for most chips, *quality didn't matter*. They were claiming that for many purposes the elegance of the actual circuit topography—the citadel of expertise of every semiconductor firm—was a needless luxury. In Drucker's terms, what matters is not designing right but choosing the right designs and getting them to market at the right time. Mead didn't even believe that it was important to produce a superb new general-purpose microprocessor. They were in the business of ultimately eliminating most general-purpose microprocessors. They thought they could tell that to a group of Intel graduates and microprocessor designers. Who was crazy anyway?

Nonetheless, the compiler paradigm would ultimately prevail regardless of the views at the established firms. It would move the power to design chips—and to determine which chips get designed—away from the semiconductor firms and toward the systems companies. Most portentously, the silicon compiler would move industrial power away from commodity chip designers and toward systems architects. By 1988, for example, four of the five most important new RISC chip architectures had been introduced by systems firms: MIPS Computer Systems, Sun Microsystems, Hewlett Packard, and IBM.

As Andrew Rappaport summed up the arithmetic, around 9,500 chips were designed in 1986, approximately half by the some 5,000 chip designers and half by systems designers. But there are some 295,000 more systems designers than chip designers. Most of the systems architects were still working with commodity logic devices on printed circuit boards. As late as 1988, over 90 percent of logic designs were still done with primitive standard components on printed circuit boards. But as systems designers master the use of computer-aided engineering and move toward silicon compilation, they will increasingly reduce those boardsful of logic to single VLSI chips. Based on the current difffusion of design tools, a tenfold increase in the number of designs is predictable by the early 1990s.

This development is by far the most important event in the industry. It means that unless the semiconductor firms take action, their share of designs may drop some 90 percent. Since the majority of systems designers are in the United States, however, compilers will tend to move the technology away from Japan and toward America, from the Japanese commodity manufacturing plants to U.S. flexible semiconductor fabs.

This dramatic potential shift in favor of the U.S. industry can only happen, however, if U.S. semiconductor firms act decisively to exploit the opportunity. They must build fast-turnaround, flexible manufacturing systems—that is, fabs made for custom chips—to accommodate the upsurge of design creativity at computer and other systems houses. They must learn to handle hundreds rather than tens of mask sets or must dispense with masks altogether through direct write on wafer technologies. And if they disdain the foundry role, they can even retain a large share of design revenues by themselves aggressively adopting silicon compilation and other new design technologies.

Using silicon compilers more aggressively than anyone else, chip firms might even keep much of their present market share of designs. Compilers have been shown to improve design productivity by a factor of near 10 to 1, which is Rappaport's estimate of the likely increase in designs.

The irony, however, was that most of the large U.S. semiconductor firms seemed to care more about solving the problem of EPROM pricing than about pursuing this opportunity to usurp EPROM software carriers with special-purpose processors. They cared more about their own values than their customers' values. Thus the semiconductor companies actually became a threat to the systems designers in the American computer industry. They even supported a trade agreement with Japan that forced most U.S. systems firms to pay twice as much for memory chips as their Japanese rivals which produced them in house.

The large American firms saw their problem as the Japanese, but were nearly blind to the opportunity of nullifying the Japanese strategy. But with them or without them, the balance of power was inexorably shifting, and they would ultimately have to shift as well.

CHAPTER

17

The New Balance of Power

The descent into the microcosm had begun with a new scientific balance of power between man and matter. What was once opaque and unintelligible became a transparent tracery of physical laws. Gaining mastery of these laws, human beings could achieve new mastery over nature. From that point on, a great cleavage had opened between peoples and institutions that could exploit this new human advantage and groups that remained in the darkness of the previous materialism.

The first crude breakthrough—the atomic bomb—radically redistributed power from territorially and mechanically oriented nations to mind-centered free peoples. But it also distracted the postwar world from the richer and subtler promise of the microcosm. Nations that targeted the nucleus alone tended to adopt yet another new materialism: worshipping raw energy and control rather than fathoming the ever-widening ken of free minds.

Even within the domains of information technology, power constantly shifted according to the Mead paradigm. From those who sought a haven in some comfortable waystation of the technology— whether metal gate circuits or specific commodity designs—the microcosm moved power to those who grasped the immensity of the promise still untapped, and pressed on deeper into the quantum economy.

With the revelation of the huge benefits of ever denser circuitry in ever smaller spaces, the end was in sight for all companies or engineers who insisted on unaided human design of chips. There would still be a place for a few relatively simple, hand-hewn designs of specific new memory cell structures or logic gates. But even memories succumbed to the possibility of more complex and specialized features. There emerged a stream of new memory types—from video RAMs to DRAMs with on-board error-correct to charge coupled devices for cameras—nearly as diverse as the old catalogue of commodity logic TTL. These designer memories again tended to enhance the power of the inventors of new circuitry and processes against the owners of large fabs and other physical capital. Pioneering most of the new memory technologies were new entrepreneurial companies.

The single most important new storage device was the non-volatile RAM called the E-square (for electrically erasable programmable read-only memory). Of all memory types, it came closest to repeating all the functions of the ancient magnetic core. The E-square is important because it offers the eventual promise of solid state disks.

Why, after all, in the quantum age, does anyone consign precious data to a box containing an electric motor revolving a rigid disk at tremendous speed to be read by a drive head flying microns above the surface, and vulnerable to catastrophic crashes, killing all on board? Why tolerate these mechanical media, all whirling and wobbling, swishing and clicking, like so much breakfast cereal, interconnected to the world of silicon by an array of expensive controllers, comparators, and converters, and backed up with a technology called "streaming tape"? Isn't this the age of solid state?

The only reason anyone tolerates these electromechanical memories is the inability of the semiconductor industry to create a cheap and reliable form of solid state storage that keeps its contents when the power goes off. In the drive to collapse the computer into the microcosm, the creation of cheap and dense non-volatile memories is near the top of the industry's agenda.

Richard Kahn at Bell, Dov Frohman at Intel, and Shumpei Yamazaki in Japan had all outlined the crucial concept: the loading of a floating gate memory cell by quantum tunneling across a thin oxide. Virtually every major firm in the United States and Japan attempted to create a workable E-square by this means. A cluster of new firms, propelled by a total of some $200 million in venture funds, emerged in pursuit of the goal. But the clear winner, with some 55 percent of the market, was Xicor, which boasted no venture sponsorship and used an approach repeatedly spurned by Frohman and other Intel experts.

Xicor won by taking the technology deeper into the microcosm. Using a bizarre quantum effect discovered by Richard Simko at Intel, Xicor loads its floating gates by tunneling through oxides six times as thick as Frohman's devices. The result is a chip far easier to miniaturize and manufacture than the thin oxide E-squares made by all other companies. In 1989, Xicor introduced a one-megabit E-square that for the first time offers a realistic prospect of collapsing at least some disk drives into the microcosm. As the price drops over coming years, these solid state devices will steadily encroach on mechanical media for all but long-term storage of substantial databases (probably the domain of optical disks).

Led by Israeli immigrant Raphael Klein, Xicor demonstrates that even in the field of commodity memories, ideas and designs are still more important than large fabs and funding. Moving from the realm of memories to the application-specific technologies promoted by Carver Mead, however, the shift in the balance of power is still more dramatic.

For the last five years, companies that aggressively pursued the Mead mandate for silicon compilers and similar tools have been gaining ground against firms that resisted the new regime. The result was as Andrew Rappaport had predicted: the number of designs produced in the industry shot up explosively—from 9,500 in 1986 to 25,000 in 1988, and toward 100,000 in the early 1990s.

Although some American chip firms protested mightily, the microcosmic insights of Carver Mead were transforming the industry in a way that deeply favored the United States. The new tools shifted power toward systems design; the United States commanded nearly half of all the world systems designers, four times the share of Japan. The devaluation of entrenched physical capital reduced the barriers to entrepreneurship; the open U.S. environment and culture uniquely favors entrepreneurship. The new balance of power between mind and matter was yielding a new balance of industrial might and new opportunities for nimble entrepreneurs. . . .

Among the entrepreneurs of the microcosm, none were nimbler than Gordon Campbell, the former founder and president of SEEQ. Taking Phillip Salsbury and other non-volatile memory stars out of

Intel in 1981, Campbell had begun meteorically. But after a few years, SEEQ's E-square technology had slipped against Xicor and the industry went into its mid-eighties slump. While many experts bogged down in the problems of transition, however, Campbell seized the opportunities. In a new firm, he would demonstrate beyond cavil the new balance of power in electronics.

He left SEEQ in 1984 and at once steered his Ferrari back into the semiconductor fray. But few observers favored his prospects. If the truth be known, many semiconductor people thought they had already seen plenty of Gordon Campbell, company president.

Campbell is a complex man, with a rich fund of ego and a boyish look that belies his shrewd sense of strategy and technology. To a strong-minded venture capitalist such as Frank Caulfield of Kleiner, Perkins, Caulfield, & Byers—or even to a smooth operator such as John Doerr—Campbell appeared to be a pushover. A man with no money, no social ivy, no advanced professional degrees, no obvious scientific mastery, he was a disposable tool: some kid who had snuck into the E-square huddle at Intel and popped out into the end zone just in time to make a miracle catch of several million dollars in venture capital.

When Campbell used the money to give SEEQ the Valley's most elegant new headquarters and one of the best-designed fabs—and began solemnly comparing himself to Gordon Moore—many observers thought it was all going to his head. When SEEQ's fab manager B. K. Marya defected to form Exel, the venture community exulted, Hambrecht & Quist in the lead, and scrambled with unseemly haste to give Marya funds. Campbell would be easy to beat in E-squares; so would Raffi Klein. But with E-squares going slowly everywhere except at Xicor, Marya stewed at Exel for a year, then sold out the technology to the Koreans and Japanese, and began a new firm called Catalyst to enter the smart card market.

The real smart card, however, turned out to be Campbell. He surprised everyone by bursting into the booming market for EPROMs and revving up SEEQ to a year of $43 million in sales, $60 million in backlog, a profitable quarter, and a market valuation of $200 million. This was bad enough. What was worse, he fell in love with his financial officer, Maria Ligeti. Then, shocking everyone from Fortune Drive to Sand Hill Road, he turned rumor into reality by marrying her. With his new family responsibilities, he began complaining about the heavy bias toward the venture capitalists in the apportionment of equity in the company. A few weeks later he was out on his tail on

Fortune Drive, the venturers were running SEEQ, and the industry was plunging deep into a slump. The market revalued SEEQ's now diluted stock down to one seventh of its previous high.

Campbell, however, is not a man to sit around contemplating last year's news. While doing an Ethernet chip project with SEEQ and Silicon Compilers, he had gained an acute sense of the Mead paradigm. Visitors to Campbell's open office at SEEQ would learn of the liabilities of random logic, the promise of hierarchical design techniques, and the opportunities created by the new design tools. At the time he was fired, he was attempting to move SEEQ toward the use of silicon compilation to reduce industry standard boards full of commodity chips to a few VLSI devices.

He saw that compiler technology had radically changed the shape of the industry. Beyond a few major products such as memories and microprocessor CPUs, commodity chips were no longer state of the art. Yet IBM, among other firms, did not seem to know it yet. Its AT personal computer—a new industry standard replacing the PC—contained two boards full of old commodity chips interconnected with "glue logic." Campbell's idea for a company was to use a silicon compiler to put those boards into custom silicon and to provide a means by which scores of companies could produce AT clones faster, cheaper, better, and more reliable than IBM's.

Campbell drew up his business plan and brought it to some fifty venture capitalists. A moneyed yawn issued from Sand Hill Road, echoed down the canyons of San Francisco's financial district, and reechoed through downtown Manhattan. A jaded group that had funded some forty very hard disk projects and some fifty rather floppy computer firms within the previous two years, venture capitalists eyed Campbell's boyish manner and lightweight look and they contemplated his business plan (a personal computer chip project during a PC and semiconductor depression), and they identified the heart of his overall strategy (compete with IBM). They rolled the firm's proposed name over their tongues: Chips & Technologies. Wouldn't Microtech be better? Then they laughed nervously. Not this time, Gordy.

Finally, Campbell found a friend: Bill Marocco, who had built the SEEQ headquarters, and had once offered to support a future project. Marocco put up $1 million, and Chips & Technologies was off the ground. But Campbell did not have a silicon compiler, and Silicon Compilers, Inc., was not ready to provide the powerful equipment he needed (though later it would unwillingly provide two key designers for Campbell's team building their own silicon compiler tools.)

Campbell then focused on the other critical change in the industry, the dissolution of the old boundaries between mainframes and desktop computers. The expertise he needed was available not chiefly in personal computer firms such as Apple and Compaq famously competing with IBM in small computers. The crucial design expertise was in Amdahl, the maker of plug-compatible mainframes. Campbell saw that the greatest gains would come not merely from redesigning IBM chips but from reconfiguring the entire AT system. It was Amdahl that had by far the most experience in this pursuit.

Just before Campbell left and the company crashed, he had lured Amdahl's CAD-CAM chief, Morris Jones, to SEEQ. Now Campbell returned to Jones and said: "Boy, have I got *another* deal for you, Morris." Jones was intrigued by Campbell's business plan and uncomfortable with the turmoil at SEEQ. He joined the new venture.

With Dado Banatao also leaving SEEQ to come on board as engineering director, Campbell leased a top of the line Amdahl mainframe and set Jones to work creating a compiler that could restructure and repartition the entire AT system outside the central 286 microprocessor. Then Jones and the company's other designers would collect the boards of AT logic and glue, controllers, input-output chips, and other devices into a few custom silicon packages.

For Jones, experienced with condensing IBM mainframes, the AT seemed to be a relatively simple system. He proposed to reduce the total number of support chips in the machine from sixty-three to five. The new designs would allow the creation of a complete AT system using little more than one third the board space of IBM's, consuming about 40 percent of the power, and using just 47 chips compared to IBM's 130. Moving further into the microcosm, the new AT would be smaller, cheaper, more reliable, and faster.

While Jones worked on the new design tools, Campbell decided to finance his company by reducing IBM's graphics enhancement board (EGA), first to several gate arrays and eventually to one chip. Finally, the firm would render the next-generation PC, based on Intel's 386 microprocessor, in an amazing seven chips before IBM even came to market.

If Chips & Technologies could succeed in this series of projects, the company would overwhelmingly prove the powers of silicon compilation. In 1986, Morris Jones might need a mainframe. But mainframe powers were rapidly devolving into desktop machines. Soon any designer with a workstation would be able to create chips that for specific purposes excel all the standard devices of IBM and Intel. The customers would regain control. A wafer fab could never again be confused with a throne.

No one would ever again be able to design a major project in the old way. Companies could not prevail by providing inferior technology and trapping their customers with "gotchas." Intel's practice of forcing purchasers of its microprocessor to buy an array of inferior support chips at monopoly prices would no longer work. The gotcha approach—a negative value added scheme long favored by some Intel strategists—had always been a loser. It entirely failed Intel in Japan. It would no longer work even with small U.S. companies. It would no longer be possible to sell microprocessors and other designs as loss leaders and make money on glue logic. It would no longer be possible to milk old technologies while better designs were feasible. From then on, every chip producer and systems firm would increasingly have to put the design needs of its customers first.

To any semiconductor firm that understood the change, the news was good. A proliferation of proprietary designs would free semiconductor producers from the commodity manufacturing treadmill. A robust and flexible process would prevail over an automated juggernaut. No reduction in device geometries for a commodity product could compete with a special design exactly adapted to the customers' needs. With proprietary designs that responded much better to the needs of the customers, firms could make far more money on each wafer because the customers could make more on each final product.

By 1986, this message was clear to all close observers. Using similar design tools and compilers to improve on IBM architectures, Compaq Computer, Ben Rosen's startup, had become the first company in history to have a first year of $100 million in sales and then to more than triple that record in its second year. But Compaq was only the most luminous of the new firms to exploit the new design technologies with unexpected results.

In 1981, for example, two Chinese American defectors from Hewlett Packard formed a firm called Weitek (Microtech in Chinese), using the new design tools to break into major markets without purchasing a fab. Instead, they farmed out designs to foundries, as Mead had urged. The Weitek founders—Chi Tzun Wong, Ed Sun, and Dwan Fan Fong, all former Mead students or associates—noticed that the floating point, fast mathematical units produced by Intel and

Motorola were "gotchas"; they could be used only with their own microprocessor families. The three young Orientals announced what must have been Silicon Valley's simplest and most comprehensive business plan: "Add, Subtract, Multiply, Divide." Do math for anyone, fast as a tunneled electron, regardless of age, sex, microprocessor family, or company affiliation.

The market for unaffiliated math processors—designed in a matter of months—turned out to be about twice as large as the market for "gotcha" devices. In 1987, AMD relied on Weitek for the floating point unit for its 29,000 RISC microprocessor, at 25 MIPS the fastest of all the new central processing units, and according to Sanders the most important product in the history of AMD.

Similar breakthroughs by Mead students and associates include the Brooktree offshoot from North American Rockwell. Using design ingenuity to compete with manufacturing prowess, Brooktree produces 16-bit digital-analog converters that can be made without laser trimming or other expensive processing. Fujitsu licensed one of their first products. Once again, the chip designers have broken away from the domination of the commodity chip manufacturers.

A further example is Silicon Solutions, Inc. (now part of Zycad), started by hierarchical design pioneers John Newkirk and Robert Matthews (readily nicknamed NewMath), who proceeded to use advanced design tools to launch still more advanced design tools: chiefly an accelerator for simulation. A prime obstacle to the custom chip revolution, simulation is at the heart of the design process. It is the way the designer can tell if he is devising a workable chip, and it is crucial to the creation of test vectors. Simulation and test preparation take more than two thirds of the design cycle. By using custom-designed parallel architectures to create simulators ten times faster for many designs, NewMath greatly advanced the age of the silicon compiler.

Exploiting the advances in design automation, scores of new firms were successfully competing with much larger firms that used designs to limit customers rather than liberate them. Gordon Campbell thought there would be many similar ways of competing with Intel and IBM. "Those companies' strategies cry out: come and get me." The only unanswerable gotcha is providing more of the functions and features desired by the customer at ever lower prices. "SEEQ was profitable in three years," Campbell said, without a smile; "Chips & Technologies will be profitable in a year and one half."

Oh-oh. It's June 1985 and the EGA graphics chip set is not ready and there is no money in the bank and twenty-three people are on the payroll. Campbell remembered the words of Nolan Bushnell: "You are not a real entrepreneur until you've got to meet a payroll from your own bank account." There was truth in those words. There was a sense in which Gordon Campbell was still not a real entrepreneur.

If you are a real entrepreneurial hero, you do not get your start by rolling out of bed one morning in rumpled pajamas to answer the telephone at Oakmead Plaza and find that it's the man from Kleiner-Perkins announcing you've won the lottery (for spinning out of Intel with Dr. Salsbury and the rest). Real entrepreneurs do not usually become paper millionaires and Ferrari corsairs in a public offering without ever experiencing the warm sensation of a profitable year. Raphael Klein had put up his house to save Xicor; he was an entrepreneur. In the desperate silicon panic of the summer of 1985, Gordy Campbell too was going to join the club.

The venture capitalists were all waiting for Campbell to fail. He had no chance of money from them. But other sources would also be difficult. Campbell had been careful to buy no real assets and channel all his money into intellectual capital. Morris Jones's Amdahl 470—a powerful mainframe that ran the company's CAE programs—was a second-hand machine, leased by the month. The rest of their CAD and CAE equipment was either designed by Jones and his team, including two defectors from Silicon Compilers, or it consisted of various IBM workstations. The company's most valuable asset, beyond its ideas, was a compaction algorithm that Jones had developed from a Bell Labs model. It allowed the scaling down of CMOS technology into difficult non-linear volt warps near 1-micron geometries. Couldn't mortgage that at a bank.

Campbell could scarcely believe what was happening to him. There was nothing to do but use his own personal money to keep the company afloat. But if the truth be known, his personal funds were running a bit low. It was out of the question, of course, to sell the Ferrari. He could hardly putter forth onto Route 280 and down toward Sand Hill Road like a beggar with some tin cup from Toyota. Campbell's other wealth, though, was mostly in SEEQ stock that was then selling at $2 per share and going down.

Campbell would have to sell at the very bottom of the market and use his own last personal wealth to finance a company with no revenues and a burn rate of some $4,000 a day. He gasped and did it. He went through a couple of cliff-hanging months, with shortened fingernails. But the act of personal sacrifice was catalytic. Within a few weeks, several of the employees and other friends also put up some money, including $200,000 from his financial officer, Gary Martin. Before the year was out he had raised another indispensable $1.5 million from a number of companies in Japan, including Kyocera, Mitsui, Yamaha, and Ascii, Kay Nishi's PC software firm that represented Chips in Asia. By July, the IBM graphics enhancement chip set was finished and Chips & Technologies was a company almost fully owned and controlled by its employees.

By July 1986, when the chip set for the IBM AT computer was finished, most of the world had decided that the AT would be the next major personal computer standard. In the United States, Tandy, PC's Limited (now Dell), and several other then unknown manufacturers bought the Chips & Technologies set. Tandy became the leading AT compatible producer, assembling the computers in a factory in Fort Worth manned by immigrants from twenty countries led by an immigrant from Japan. Among the purchasers of the Chips set in Europe were Olivetti, Apricot, Siemens, and Bull. Nishi signed up NEC, Sony, Epson, and Mitsubishi in Japan; Goldstar, Samsung, Daewoo, and Hyundai in Korea; a number of companies in Taiwan; and the Great Wall Computer Company of China. Most of these firms —plus Compaq and a slew of producers of IBM add-in graphics gear —also were buying the graphics enhancement chip set.

At the outset, Campbell had boldly predicted profitability in a year and a half. In fact, the firm was profitable by the last quarter of the first year. In the midst of a worldwide recession in the industry, Chips brought in $12.7 million of revenues in its first fiscal year and $70 million in its second (which ended an annualized run rate of over $100 million) and $12.4 million in profits. It was the fastest-growing chip firm in history. In 1989, Chips reached a run rate of nearly $200 million, mostly from scores of companies launching computers using the Intel 386 microprocessor to run an array of advanced software for multi-tasking and networks.

Chips & Technologies is not only a design center. Its business is to sell chips. But Campbell early decided he would not build a fab. He reasoned that with original designs produced in a timely way, he would not need access to the most advanced processing lines. But just as he was pleasantly surprised by early profits, he was delighted to discover the new balance of power in semiconductors. He found that he could produce his chips on the world's best 1.2 micron processing lines. As Mead had predicted, many semiconductor firms were happy to serve as foundries. Not only would he produce better designs than IBM but he would be able to manufacture his chips on equally efficient fabrication lines. His AT chips would be fabricated in 1.2 micron CMOS fabs owned by Fujitsu, Oki, Toshiba, Yamaha, NEC, National Semiconductor, and AT&T. With 115 employees, 70 devoted to R&D, Chips could send forth more than $300,000 worth of chips every day from its tiny shipping area with virtually no manual laborers.

Weitek—the Add, Subtract, Multiply, Divide people—had a similar experience. At first they went to Intel with their design, but the company refused to open its foundry to a competitor. In the short-sighted policy of many established American firms, they refuse to "help the competition." So many of America's best new designs are being manufactured in foundries in Japan instead. But Intel is learning. Soon Moore's protégé, David House, watching the Weitek "letchas" outsell the Intel "gotchas" two to one, offered to serve the company as a foundry. By creating all-purpose multipliers and signal processors, at home in both the IBM and Motorola domains, Weitek became a sudden success.

Even Tateyama—Takami Takahashi's automated line for flattened ball bearings—dramatized the new balance of power. Fulfilling the overall prophetic scheme of Carver Mead, NMB semiconductor made Tateyama a silicon foundry for the world's best chip designers. NMB bought rights to a 64K fast static RAM from Lattice, a small design house in Portland, Oregon. It bought a 256K DRAM design from Inmos. It bought a slow but cheap megabit DRAM design from Vitalic and a fast DRAM design from Alliance, a startup by the brilliant Reddy brothers, Indian designers formerly of Texas Instruments, who in 1989 took over AT&T's DRAM fab in Kansas City. Finally, NMB contracted to produce the new ferroelectric four-megabit design from Ramtron, a firm also in pursuit of the solid state disk.

The Japanese had solved the problem of automating the semiconductor production process at the very time that the balance of power in the industry was shifting away from process automation and toward product innovation. At a time when the industry was bidding feverishly for new designs that use fewer wafers and less silicon, the Japanese were feverishly building new capacity.

In part, the bidding for designs was an effect of the recession in the industry, with all innovators enjoying a seller's market while commodity producers could not fill their fabs. But the late 1980s were also very different from previous recessionary crises. The industry was undergoing a technical transition unprecedented since the invention of the integrated circuit.

Prestigious analysts, from Gordon Moore to the business press, predicted in the late 1970s that the number of designs would have to decline because of the greater cost and complexity of product development. Instead, the number of designs exploded. As Rappaport had predicted, an IC fab with just 1 percent of the market would have to produce some 1,500 mask sets a year.

These mask sets would arrive in largely unpredictable patterns from hundreds of designers. The chief problem of the industry would not be fixed automation for the most advanced commodity chips. The problem would be creating flexible manufacturing systems that could run robustly reliable processes for a variety of chip types. As the 1980s proceeded, hundreds of companies were experimenting with scores of techniques for solving this problem.

What had happened was a general shift of the balance of power in semiconductors toward the United States. As value added shifted toward design and software, the United States found itself in the lead in the CAE revolution spurred by the students of Carver Mead. The Japanese had some superb facilities for CAE. But they all ran on mainframe computers in large design centers. It was in the United States that every engineer's desk was becoming a design center and new designs were being produced in unprecedented numbers.

The Japanese would command the bulk of the commodity semiconductor market. They would produce the chips with ever greater efficiency and lower their prices with ever greater decisiveness whenever politics permitted. As a result, the price of computing would continue to decline and the market would continue to expand. But the greatest beneficiaries would be the American chip and computer firms who take advantage of the new capabilities of scores of thousands of chip designers. They are using the new technologies to create ever more attractive features and ever more efficient architectures for the computer industry in all its forms. Manufacturing efficiencies may still move toward Asia for many products; but value added will gravitate toward the most effective users of the new technologies of design, led by the silicon compiler.

The climax of the paradigm shift came in 1988, when the law of the microcosm yielded yet another breakthrough from the students of Carver Mead. An Egyptian immigrant, Amr Mohsen, a Mead Ph.D. in 1971, took a radical further step in silicon compilation. He launched what might be called a system of desktop printing of chips.

With Mead on the board and Andrew Rappaport as a key consultant, Mohsen's firm, Actel, developed a system for using 386-based PCs to configure or program gate arrays with tens of thousands of transistors. Suddenly, systems designers would be able to design, produce, and test prototypes on their own desks. Rather than waiting weeks and paying $10,000 to $20,000 in non-recurring costs to program an array at a semiconductor fab, the designer can program all but the most advanced array in minutes for a cost of pennies.

Following Jerry Sanders's rule of not introducing a technology until all its parts are ready, Actel had eighteen patents either approved or pending when the company made its stunning announcement at the ISSCC in San Francisco in February 1988. In July, it introduced both the programming box and a set of design tools. While Silicon Compilers would provide systems designers the gear to design complex microprocessors, Actel would give them the power to define and produce all the smaller glue and accessory chips in house. Using Actel equipment, designers could customize cheaply produceable arrays of logic gates with a small box costing a few thousand dollars rather than a multi-million-dollar gate array fab.

This new shift in the sphere sharply accelerates the advance of the Mead vision. It emancipates creative engineers from the rhythms and requirements of the silicon wafer printing plant and further shifts the power of design from the 4,000 masters of printer technology to the world's some 400,000 experts in electronic systems. This change will unleash new energies of creation throughout the world economy.

The larger issue, however, remains: not what new design tools have been invented but what new designs will be created. Chips & Technology is a virtuoso in shrinking computer CPUs and peripherals, putting ever more processing power on fewer and fewer chips taking up less and less space on printed circuit boards. But the computer's collapse into the microcosm had now reached a more difficult challenge: the keyboard and screen.

The next step for the quantum computer was microcosmic input and output. The industry would need compact chip sets for cheap and reliable speech recognition and synthesis, for vision and projection. Many resourceful firms and researchers have pursued these elusive goals for more than a decade. Requiring numerous massively parallel application-specific chips, their achievement will entail not only mastery of the digital electronics of computation but also a revival of analog sensory systems—yet another route into the microcosm.

PART

**The Imperial
Computer**

IV

CHAPTER
18

Patterns and Analogies

Profound study of nature is the most fertile source of mathematical discoveries. . . . It brings together phenomena the most diverse and discovers the hidden analogies which unite them.
—JEAN-BAPTISTE JOSEPH FOURIER

After triumphs in logic and memory, the continuing drive to collapse the computer hits a wall when it comes to sensory input. Except for limited gains in speech recognition and primitive systems of robotic vision, the computer industry mostly relies on humans at keyboards or at screens to connect its machines to the world. This limitation means that people still have to sit—or kneel—before computers to put them to use.

It was to overcome this bondage of humans to machines that Mead first embarked on his drive to provide new tools for designing new computer architectures. But in order to apply those architectures to the cause of emancipating people from keyboards and kneelers, the industry would have to give its machines sensory powers.

The problem is that the computer is still primarily a digital symbol shuffler, and the world does not present itself in discrete or digital symbols. Outside the quantum domain, the world is mostly continuous, multifarious, murky, merged, and unmeaning. A human being can descry the shape of a horse on a hill, whether in a line drawing, a color photograph, an Impressionist painting, or a Cubist sculpture, or in a field, and he can link it to a neigh or a word or a clatter of hoofs or a pile of dung or a carpenter's tool or a broomstick. A human being can identify the sound of a cry in a crowded room, or the shadow of a smile on the face of a cat, or the meaning of a word in the speech of a child, and link it to the same word gasped in the voice of a crone. But a computer does not even know where to begin. Even if—with appropriate transducers—it can capture large amounts of information about the waves and particles of the world, the machine is confounded by the webs and layers of meaning, abstraction, and association as they are filtered by the human eye and ear and mind.

The gulf between the sensory powers of humans and the input apparatus of computers is the chief challenge in the further collapse of the computer into the microcosm. To overcome it will demand a reopening of some of the most fundamental and most firmly resolved issues of computer logic and architecture, beginning with the question of how information will be represented within the machine.

When Carver Mead predicted a computer system on a single chip, he did not define what kind of computer system he had in mind. The entire world knew that computers were digital: they consisted of sets of switches with only two states: on or off. Out of these binary elements, usually denoted one or zero, can be formed any digit or group of digits, number or symbolic notation. As long as the phenomena represented by the binary symbols are clearly defined, separable from one another, and expressible in mathematical terms, the digital system can process them with tremendous speed and accuracy according to a specified program of logical steps. The digital computer has become so completely dominant in the field that few people are even aware of any alternative model of computation.

Yet all digital systems are entirely dependent on what are called analog systems. Digital devices represent numerical quantities by sets of switches. Analog devices represent real-world flows and forces by electrical flows and forces. Currents and voltages serve as analogs for such continuously varying phenomena as pressure, sound, and heat. Linking the palpable world to the digital world—mediating between mass and microcosm—these circuits comprise the computer's sensory apparatus. Analog devices convert inputs, such as pressure on a keyboard or temperature in an oven, into information for the computer. Then they transform computed results into output: whether a screen display, a printout, or a mechanical movement.

As the body incarnates the mind, analog converters surround the microcosmic machine. It is thrilling in the microcosm. But a man has got to eat; a digital chip has got to connect. Like a tourist in Tasmania, unless a digital chip has access to interpreters, currency exchangers, maps, taxis, telephones, and buses, it is lost. Every process by which a computer couples its digital action to an input or output depends on analog circuitry. If computers are going to drive cars, cook meals, input handwritten material, inform the blind, run flexible assembly lines, find submarines and missiles, or identify terrorists, they can no longer function as blind, deaf, and dumb devices almost entirely dependent on human input.

Most of all, if computers are going to overcome the software slough that currently slows the industry's growth, they must be given the power to learn from experience. Learning is in large degree a sensory process. A computer that can learn from its inputs is in effect creating new software itself. Mead believed that new analog powers would be crucial to a new learning process for machines.

Learning has always been a target of what is known as artificial intelligence: AI, the simulation of human mental functions by a computer. Launched by Marvin Minsky, John McCarthy, and Herbert Simon in 1956 at a conference at Dartmouth, the AI movement began with a series of bold predictions of powerful thinking machines. But AI champions always assumed that thinking and learning were functions of pure logic. Captivated by the materialist superstition, they envisaged the laws of logic in the terms of mathematical and Newtonian causality. As software engineers, they saw both the computer and the brain as feats of software engineering. A mutual glorification society arose between computer scientists exalting the human brain as a computer and computers performing ever more dazzling logical feats for their masters. For some time, no one noticed that in essence the AI movement was playing a game with mirrors.

Logic is usually the outcome of a learning process, not the process itself. People can think logically about a problem only after they understand it. Understanding comes from an analogical or intuitive process, heavily based on the engagement of the senses, rather than a mechanical or digital engagement of pure intellect. It is an elusive process of pattern matching and association. Acquiring sensory powers, Mead believed, is a crucial step toward functioning artificial intelligence.

Mead saw that to meet this challenge would entail new architectural designs, new silicon circuitry, that could better accommodate an analog world. The new hardware could not be more of the same. Increasing the speed of CPUs by another ten MIPS or megahertz or reducing the size of lithographic geometries by another fifth of a micron or expanding the capacity of memory chips by another factor of four could yield only small gains comparèd to the achievement of genuine perceptual powers. Without transcending logic, the computer could never escape its house of mirrors.

Many computer experts resist these ideas. They routinely say that by digital methods any phenomenon can be captured to any arbitrary degree of precision—or imprecision. This is true enough once the phenomenon has been defined and structured logically. But in perceptual roles, it is the structure and definition itself that is at issue: you wonder who or what rather than how many.

Today, analog devices use voltages and currents, band gaps and temperature effects, among other physical analogs of reality. But the ultimate analog computer—the human brain—uses the entire flux and physiology, wavescape and chemistry of human neural wetware as analogs for both the large structure and analytical detail of sensory phenomena. Rather than a binary vessel to catch the world, analog electronics potentially offers a broad palette of physical properties as correlates of perceived experience. As the computer continues to collapse into the microcosm, some of the next key steps are likely to be analog. Just as Mead provided crucial tools for the creation of better digital designs, he would provide key technologies for the journey toward a neural computer.

In achieving the benefits of the analog vision, Mead and other pioneers would follow two different strategies. They would simulate analog functions with digital electronics, based on binary two-state functions, and they would create real analog devices, based on voltages and currents. In both these pursuits, they would gain a deeper sense of the nature of computation in both its abstract and physical aspects and a truer grasp of the future of computer architectures.

The Von Neumann machine was totally swamped by the challenge of performing such analog functions as hearing, vision, music, and imaging in real time. To deal with the onrush of analog reality, whether as sounds or sights, pioneers such as Raymond Kurzweil and Fred Jellinek in speech recognition and Lawrence Ryan and Arthur Kaiman in image processing found themselves forced to use silicon compilation to create new parallel architectures using largely digital circuitry. In the process, they enriched computer science and technology with an ever deeper and more fruitful interplay between the analog and digital realms.

At the same time that computer industry pioneers were using digital processes to simulate analog functions, analog pioneers were taking real analog devices ever deeper into the microcosm. Heading this enterprise was one of the leading inventors and most intriguing personalities in the history of electronics, Robert Widlar, the inventor of the first analog integrated circuit. Following in his footsteps was Carver Mead, who would create for the first time a fully microcosmic mode of analog design.

Out of this prolific union of digital and analog visions and devices came a spate of new technologies that would transform our lives, creating new visions and tools for office, outer space, highway, and living room. Before the turn of the century, such familiar and apparently impregnable technologies as computer keyboards, symphony orchestras, and high-definition television sets would fall into deep shadows of obsolescence.

The basic analog function today is the amplifier, which allows a small force on a control circuit to regulate a large force on a working circuit. In an amplifier, the ratio between the small force and the larger force is called the "gain." In the digital world, transistors use the small force to switch larger currents off and on. In the analog world, a continuous flux of small force can continuously vary a large working force.

Such valve effects, as they are called, are crucial in virtually all technologies. Typical is a microphone, in which sound waves are converted to small electrical voltages and currents and then amplified into larger sound waves. Other "amplifiers" of small forces include the accelerator on your car and the faucet in your kitchen sink.

The faucet, in fact, provides a fair model of an electrical amplifier, with the water pressure corresponding to the voltage, water current to electrical current, and the force on the handle to the force on the grid or gate.

Simply by reversing the analogy, we get an analog computer. Rather than using the faucet to understand the electrical amplifier, the analog computer uses the electrical amplifier to understand the faucet. In fact, the electrical amplifier can be used in a computer to model any plumbing function involving water pressure, current, and valve effects. By manipulating the amplifier, we could gain a precise notion of the performance of a hydraulic system. In more general terms, an amplifier is an analog multiplier and can perform that vital mathematical function in any analog system. The gain is the multiplier of the input.

We have seen that gain is just as important in digital electronics. A typical chip has just forty input-output pins but a hundred thousand or more interior devices and interconnections, all with resistance and capacitance leaching away power. A typical computation—taking millions of steps—would be impossible without constant renewal of the signals by transistors functioning as amplifiers.

In both digital and analog systems, therefore, the fundamental electronic function is the amplifier. But amplifiers are worthless unless they are stable and predictable. Amplification must be controlled to be useful. Unless you know exactly what the gain is, you have no way of knowing what is being multiplied by what. You don't know whether the car will lurch or ooze forward when you press the accelerator.

The key to stable amplification is the negative feedback loop. It solves a major problem of most systems: how to get them back on track when, as is inevitable in a world of flawed men and materials, they stray. The problem is even worse when during their deviations, they get positive feedback—good vibrations—and stray more, until they crash and all the good vibrations turn into the bad kind. For example, a thermostat is a negative feedback device. If instead it gave the furnace positive feedback, the more heat it provided the more the thermostat would demand, until the furnace either exhausted its fuel or exploded. This phenomenon of positive feedback loops or oscillations is familiar among drunks, addicts, computer users, and analog designers (some of whom are also notable authorities on feedback cycles in alcohol and drugs).

Like valves, therefore, negative feedback is fundamental to machinery, from James Watt's steam engine, which had a feedback governor to keep it from the extremes of expiring or exploding, to Harold A. Wheeler's automatic volume control, which keeps AM radio output from constantly spiking up and down when the signals shift as you drive your car through a city. Essentially, any system converts some cause (steam, radio waves, voltage) into a desired effect (power, or sound, or signals).

Consisting of an effect sensor and a cause adjustor, the negative feedback loop monitors the output and feeds back undesired variations—in inverse proportion—to the input. Again, a thermostat provides a simple example. If the output is too large (the room too hot), the feedback is proportionately small and turns down the furnace; if the output is too small (the room too cold), the feedback is large and turns up the furnace. Thus the loop keeps the overall system steady.

The idea of a negative feedback loop was first applied to amplifiers in 1928 by Harold S. Black of AT&T, as he rode on the ferry from

New Jersey toward the West Side Manhattan docks where Bell Labs were then located. He drew the circuit diagram on his *New York Times,* took it to the lab, and essentially solved the problem of drift and distortion in long-distance phone calls. Long-distance signals must be amplified repeatedly as they are transmitted through telephone circuits across the country. With the flaws in each amplifier corrected through negative feedback, the noise and distortion could not accumulate and thus would little affect the overall performance of the system.

Many engineers, including key figures at Bell and at an indifferent U.S. Patent Office, long failed to recognize the significance of Black's insight. But Frederick Terman, the great professor and inspirational pioneer of electronics at Stanford, understood its epochal importance and communicated his enthusiasm in the late 1940s to two of his students, William Hewlett and David Packard. Thus were laid the foundations for the flagship company of Silicon Valley. As Hewlett remembers, "Fred was greatly intrigued with [negative feedback] and really fired up the class on everything that could be done with this thing." Hewlett used the concept to assure that his first major product, the audio oscillator, would not deviate from its specified electrical frequencies. It didn't, and became the key product of Hewlett Packard's industrial empire of electronic instruments controlled through negative feedback.

The leading analog device using this principle was the operational amplifier, or op amp. Produced by George A. Philbrick Researches, Inc., off Route 128 in Dedham, Massachusetts, the op amp might be termed the basic analog microprocessor. First manufactured in 1952 and made with vacuum tubes, Philbrick's device was designed to be the basic cell of an analog computer. In that era, the world of computing was split between the advocates of analog systems and the pioneers of the digital switch and the Boolean code. Since Vannevar Bush had created a series of increasingly sophisticated analog computers at MIT, many experts failed to see the need to go digital. Why give up the beautifully flexible analog ballet of currents and voltages and translate them into crude binary switches, only to translate them back into the real world again when the computation was over?

Even then, linear people were different. Philbrick, for example, doggedly spelled it "Analogue Computor." He also periodically published elegant essays under the title *The Lightening Empiricist* and distributed more than a million copies of a *Palimpsest on the Electronic*

Analog Art and subsequent catalogues. He liked to quote Kepler: ". . . and I cherish more than anything else the Analogies, my most trustworthy masters. They know all the secrets of Nature. . . ."

The elegance of the analog computer stems from these natural "Analogies" among mathematics, electronics, mechanics, chemistry, and physics. For example, if a series of conducting wires comes together, the current in the single wire will equal the sum of the currents in the branches; hence, with a few wires, an adder can be created that in digital form would require hundreds or even thousands of Boolean switches. If a current is split evenly between a pair of resistors, their voltages will be in proportion to their resistances. Hence, by applying resistors, the most easily available and precisely defined component, a voltage signal may be enhanced or diminished by a set proportion (or using variable resistors, called potentiometers, by a variable proportion as in radio tuners).

By manipulating voltages and currents in resistors or amplifiers, a designer can create a voltage divider, an accurate multiplier, or a feedback loop that precisely inverts the deviations in the output. Moreover, using capacitors, which collect and then discharge current over time, he can perform all the integrations and differentiations of the calculus. Far more elegantly than the Rube Goldberg contraptions of digital logic—with their binary numbers and Boolean mazes—the analog computer could theoretically perform any mathematical function simply and directly.

Better still for many purposes, analog devices could simulate the dynamics of mechanical processes without translating them first into mathematical terms. For example, voltage, current, and inductance are perfect analogs for the interplay of mass and velocity. The operation of a capacitor nicely recapitulates the depression and release of a spring. And resistance precisely correlates with damping or friction.

The advocates of analog computing contrived endless tables of such correspondences in every field of science, from geology to hydraulics, and they were beautiful to behold. To simulate any dynamic process, one merely translated these mechanical, chemical, or hydraulic functions and effects into electronic components and turned on the device. Analog computers, in particular, could use op amps to simulate precisely any industrial control system with a feedback loop. To simulate the dynamics of any change, one merely changed the settings of a potentiometer or variable resistor on the computer.

By simulating nuclear behavior in the Manhattan Project and by aiming artillery, analog computers helped win World War II. They were useful too in many peacetime applications, from developing aircraft autopilots to controlling continuous manufacturing processes or laying out power grids. Nonetheless, as the computations became more complex, the "Analogue Computor" faced a fundamental difficulty.

In any complex computation, the same distortions that afflicted the phone company's long-distance calls before the era of Harold Black also eroded the accuracy of the computer. Unlike the on-off signals of digital systems that could be transmitted flawlessly through billions of switches, very slight defects, drifts, noises, or distortions in the tubes or wires of an analog system would accumulate into major errors during the course of mere thousands of operations. The operational amplifier was designed to overcome this problem.

It was a heroic effort, and though it did not stem the historic tide of binary computation, the op amp became one of the most useful of all electronic devices. An amplifier that eats its tail, it suggests both in symbol and in sales, and in apparent uselessness, the alligator of Lacoste. In more specific terms, the device normally has a very high open loop gain, or amplification effect, and a feedback loop that cancels out most of it. It essentially achieves incredible accuracy by blowing itself up to a tremendous size, only to shrink itself down again by a factor of thousands. Just as chip designers originally achieved their submicron accuracies by first drawing their layouts at five hundred times their final size and then photographically shrinking them, op amps attain precision by blowing up and then shrinking their inputs.

Shrinking in electronic feedback loops is done by passive components such as resistors. Because these devices are far more accurate than active processes like amplification, the result is an amplifier that maintains a very stable gain. The residual output of the entire system is almost entirely determined by the size of the resistors in the feedback loop: that is, by reliable passive devices rather than by the complex and skittish vacuum tubes (or in later years, bipolar transistors) in the open loop of the op amp itself.

Let's say the amplifier controls an automobile motor. Without the negative feedback, a distortion in the device that doubled gain would double the speed of the car. But if the output gain is reduced 90 percent by feedback resistors, this otherwise catastrophic error will lead to only a 10 percent rise in the speed of the car. It is by that principle that the op amp achieves its stability and accuracy.

Op amps, however, can be far more accurate than the automotive example suggests. For instance, assume that an op amp initially magnifies its input by a factor of 100,000, and then reduces it by the feedback loop to a closed loop gain of 10. Any errors, distortions, noises, or temperature changes afflicting the initial amplification of each op amp are virtually eliminated in its feedback reduction process. The ratio of the initial op amp gain to the overall system gain is 100,000 to 10, or 10,000 to 1. Thus distortions of the op amp gain will have 10,000 times less impact on the overall gain at the output.

Even if, by some spike of charge or noise or some major change in temperature, the op amp's gain doubles, the effect of this cataclysm on the overall system is virtually undetectable for most purposes: just 100 percent divided by 10,000, or 0.01 percent. Since op amps can be easily cascaded one after another, just three of these devices with a gain of ten could create a gain of 10 to the third power, or a thousand: ample magnification for most purposes. Thus the op amp offers amazing power and stability of function to any system, whether a microphone or a phonograph, a precision instrument or an active radio frequency (RF) filter.

The problem for analog computing, however, is that even the infinitesimal remaining inaccuracy accumulates to monstrous errors in typical computing tasks involving billions of serial operations. So Philbrick's "Analogue Computor" became an archaism. Digital systems are everywhere putting analog to rout. Today even sound recording, manipulation, and transmission—the classic analog functions—are increasingly done in digital forms, albeit in systems containing scores of analog chips.

A compact digital disk is theoretically far less faithful to the original performance than an analog vinyl LP. The vinyl LP actually simulates the same sound waves that entered the microphones, while the digital system first converts these waves to off-on switches merely conveying mathematical data about the waves. But when digital systems can convey millions of bits a second, without distortion, the overall effect is higher in fidelity than a perfect reproduction of the actual sound waves run down fallible wires in analog systems.

Philbrick's "Analogue Computor" was a beautiful idea too good for the real world at the time. The digital computer might have been less elegant and might require an endless rigamarole of A-D and D-A translations. It may entail endless scurrying back and forth to memory to pick up data and instructions for the dogged one-by-one, clock-pulsed, serial access to the central processing unit. But all its flaws were dwarfed by two huge virtues: it was incredibly fast and awesomely accurate.

Nonetheless, do not forget the analog computer. Because it is conceptually so compelling, it has a way of always coming back. It offers a standard by which to judge all computers and to find them sadly wanting. It offers a way to look at all electronic equipment and clarify its nature. In a sense, moreover, the analog computer died only by becoming ubiquitous.

Consider, for example, the common radio. Even if it has a digital tuner, a radio must detect, filter, and amplify analog signals, control their volume with variable resistors, transduce them into sound waves with other analog devices. From the feedback loop AM volume stabilizer to the phase lock loop FM signal interpreter, a radio is crammed with analog circuitry that manipulates signals in ways inherently far more complicated than a digital calculator. The calculator just seems complex because it is such a digital kludge, using thousands of gates to do what an analog machine could theoretically accomplish with a few wires, resistors, capacitors, and op amps. The radio, in fact, represents an inherently more impressive and versatile feat of electronic manipulation. And what is the radio? In essence it can be identified as nothing less than an "Analogue Computor" dedicated to the formidable task of the analysis and synthesis of sound.

Most important of all, however, if we are to understand the challenge of artificial intelligence, we must address that supreme integrated system: the human mind. Applying our mostly analog brain, we might shake our assurance that the analog computer is obsolete. The secret of the triumph of digital was the materialist superstition: the belief that thoughts followed the same chains of cause and effect that were seen to characterize things. Like many superstitions, it led to magnificent towers of achievement. The computer could use the output of one transistor as the input of another in a chain of virtually infinite length and flawless accuracy. Yet because the machine was ultimately based on a superstition, an infinitude of flawless computations ended with a monstrous imbalance. Derived from false physical models of reality, chains of serial logic bound the computer off from the real world.

Now the industry is facing new challenges of sensory input. Sensation is not a serial and digital but a parallel and analog process. What matters in vision, for example, is not the absolutely precise accumulation of inputs over time but the immediate results of innumerable photons arriving at one instant. As Raymond Kurzweil estimates, "We would need about a billion personal computers to match the edge detection capability of human vision. And that's just for one eye.

. . .[However] even the hundred trillion multiplications per second required for human vision, while out of the question using digital circuits, is not altogether impractical using analog techniques. After all, the human brain accomplishes image filtering tasks using precisely this combination of methods—massively parallel processing and analog computation."

The preoccupation with serial logic led to the phenomenal capabilities of the digital computer, but sealed it off from raw experience and from a more authentic model of the brain. To link the computer to the flux and flow of reality would require the kind of investment in analog capabilities that has driven the industry to its present digital peak, centered on the digital microprocessor. In analog, we begin with the op amp.

CHAPTER

19

Analog People

Herbert Philbrick's vacuum tube operational amplifier—priced at under $100 and used in scores of thousands of measuring instruments, servo-mechanisms, control systems, and simulators—failed to make him a computer magnate but did make him a millionaire. When Ray Stata at Analog Devices made op amp modules out of discrete germanium transistors and resistors, the function became still smaller and more useful, and fueled a handsome business. When op amps were put by Robert Widlar onto a tiny piece of silicon in an eight-pin package and sold for $5 apiece, it created a multi-billion-dollar industry in tiny "Analogue Computor" components.

Comparable to the microprocessor in the digital realm, the op amp is one of the most important chips in the entire industry. In 1986, National Semiconductor and Fairchild between them sold more than half a billion op amps based on Widlar designs. Now the two companies are one and the Widlar legacy continues to expand its influence. The op amp—a tiny "Analogue Computor"—is the ubiquitous manifestation of the analog side of the microcosm. It is used in microphones, radios, disk drives, keyboards, sensors, and anywhere else in electronics requiring the reliable conversion of relatively tiny real-world disturbances (radio waves, sound waves, magnetic fluxes) into useful electrical signals.

Today, when Mead's prophecy has come true and the logic in a computer's central processing unit (CPU) is routinely inscribed on single chips, the next step is to create entire computer systems on single chips. This step will mean integrating the sensory system as well as the cognitive functions; for many purposes collapsing the keyboard and screen into the microcosmic maw. This will mean combining digital and analog (or linear) elements on single chips.

Just as the industry long believed that it was impossible to put resistors and capacitors on integrated circuits with transistors, today many observers believe that it is cumbersome and unnecessary to combine digital logic with analog-to-digital and digital-to-analog converters and op amps on chips. But the pressure of the microcosm is relentless. Already some 40 percent of all custom devices contain both digital and analog elements. No account of the history and prospects of information systems—or of chip technology—can any longer ignore analog concepts and devices. In the information age, we must deal with both forms of information, just as the computer must and most chips eventually will.

Widlar was the pioneer who took the initial key steps in bringing analog into the microcosm. Like Robert Noyce and Jack Kilby in the digital realm, Widlar created the first integrated circuits in analog. Thus, using the same simple ingredients of silicon, aluminum, and air, he made it possible to link the digital miracles of the microcosm to the real world.

But Widlar cannot be treated simply as an ingenious chip designer. He also incarnated the spirit of an industry.

"An American inventor," it says here, "creator of the IC op amp, born in Ohio, famous around San Francisco, disappeared into Mexico. After his disappearance, his legend grew."

Why does he have an ax in his office? A wild sheep in his Mercedes? A bale of hay on his desk? A bottle of champagne at the conference center? If he is going to give a lecture, why is he drunk? "I have to be drunk to get down to your level," he says. Why does he have a loaded pistol in his attaché? "In case anyone falls asleep." Sorry you asked?

But why is he appearing in competitor's ads, urging customers to spurn his chips?

You don't believe it? Well, here's the picture of National's star— see him, the bushy hair, the rounded beard, the drink in his hand, the smirk on his face: "What national semiconductor firm would you turn to to fill the op amp price/performance gap? The challenger, Teledyne." In the picture, Widlar is raising a glass, saying, "I'll drink to that. I'll drink to anything."

Most important of all, why do you say that Widlar is a portent of things to come in the American semiconductor industry? Are we all going to end up over the hill and on the bottle?

"You want to find him," said his old colleague David Talbert, "call the cops. . . . Or else," Talbert adds, "you could try the bar at the Hilton in Puerto Vallarta." Widlar says: "If you want to find Talbert . . . and you gotta find Talbert and his old lady (she actually laid out most of the circuits). . . ." Well, finding David Talbert is another story. He runs a ranch in Paradise, they say.

Is Widlar crazy? Sure he is. *Affected with a high degree of intellectual independence; not conforming to standards of thought and speech and action derived by the conformants from the study of themselves.—Ambrose Bierce.* And who is this Ambrose Bierce? "An American writer," it says here, "famous for *The Devil's Dictionary* and the chip on his shoulder, who was born in Ohio, worked around San Francisco, and disappeared into the wilds of Mexico. After his disappearance, his legend grew."

Why does Widlar so often quote Bierce? Could it be because, in elegant ingenuity and precision, analog circuits are silicon epigrams?

For some people, epigrams seemed too limited a form. Too much of it was in the material world, merely on the thresholds of the microcosm. Digital heroes could break away into the vastness of microcosmic space, where you can escape matter altogether, achieve "process independence," relegating the details of execution to some "technology file" in the computer—and fab engineer on the line. But there is no way the analog designer can reduce his system to an abstraction of binary switches and achieve independence from process or technology.

Rather than routing and switching millions of bits, a linear device must accurately amplify one signal, define a voltage, compare two currents, or set a timing pulse. The technology—its material substance, its physical properties, its band gap, its conductance and resistance, its thermal coefficient—is the essence of the analog device. The detailed characteristics of the circuit are its data sheet, its truth table.

Compared to digital layouts, linear features and geometries tend to be "large," ranging from 6- to 10-micron dimensions. The devices have to be "large" to handle the outside world with all its voltage spikes and temperature changes: to get the electrical charges and currents ready for the microcosm. Analog chips actually perform worse, much of the time, as they get smaller. To digital heroes, analog people seemed better than mechanics only because they have to confront the microcosm as well as the world.

Nevertheless, as the industry fulfills the prophecies of Carver Mead —collapsing the computer and reshaping it—the analog interfaces are now moving to the forefront as a focus of innovation. The chief obstacle to further extending the reach of information technology is not the speed of the internal logic of microchips but the rigidity of the interfaces to the world. A new set of analog inventions, extending the insights of Widlar, will be needed to consummate the new era in electronics. And yet, there would be problems. . . .

Here is Widlar at one of Peter Sprague's musical salons, citing Bierce again, to the elegant young lady in velvet about to assume her seat at the piano. "The piano, ah yes . . . a parlor utensil," he quoted in his George C. Scott rasp. "It is operated by depressing the keys of the machine and the spirits of the audience." That is the way that Widlar talks, when he is quoting Ambrose Bierce, at the house of the man who hired him first from Fairchild and still was chairman of National and, so to speak, his ultimate boss.

It is not clear that Widlar has ever been managed by anyone. For example, Widlar appreciates female companionship *(Woman—The most widely distributed of all beasts of prey . . . lithe and graceful in its movements, especially the American variety, is omnivorous and can be taught not to talk)*. However, he has never married *(Marriage—The state or condition of a community consisting of a master, a mistress, and two slaves, making in all, two)*.

Of Widlar's youth, little is known and he is not talking. *(Infancy— The period in our lives when, according to Wordsworth, "Heaven lies about us." The world begins lying about us pretty soon afterward.)* But everybody agrees that Widlar was "absolutely uncontrollable" even as a teenager doing radio pranks on the police in Cleveland, even in college as a math whiz and classroom bully at the University of Colorado, even as a young engineer playing cat-and-souse games with chip salesmen in Boulder.

In Boulder, Widlar was working on board-level designs for Ball Brothers, Inc., which had been inspired by John Kennedy to extend its line from Mason jars into aerospace electronics. One day, Jerry Sanders, then selling chips for Fairchild, arrived in town looking sleek, tan, and terrific, and wondering where the girls were. A rising star in Fairchild sales, Sanders eyed Widlar, short, stocky, sartorially tacky, and thought he could take him. Widlar appraised Sanders carefully and escorted him down to a "local gin mill." Widlar "drank him under the table, no sweat." Then he took all Sanders's chip samples—"double diffused planar pnp transistors, the original 1613s, great stuff"—and walked him out none the wiser. No one else ever did that to Jerry Sanders, before or since.

That kind of story contributed to the legend but it worried Robert Swanson, the young Bostonian then in charge of the linear division at National. Widlar was going to be the key man in his department and they said he drank . . . to excess sometimes. He gets in bad fights. They said he breaks down doors to get supplies. Uses explosives in the office to attract the attention of his bosses (always works, he says). At a certain point, Swanson thought, someone is going to have to draw the line. Lots of luck, Swanson (you sell the stuff, fine), but it's Widlar who draws the lines. Swanson would need to be trained to handle the famous elite National team, called by *Fortune* "the animals of Silicon Valley."

Swanson, though, was not sure he could ever be trained to run a zoo. "I mean I don't know if I can handle these guys. These linear guys are different," Swanson said. He noticed it right away.

The first difference is their intense individuality. Analog designers must learn their trade mostly on their own. In reconciling macrocosm with microcosm, they cannot rely on some prefabricated system of knowledge. Neither digital logic nor electrical engineering will teach them how to bridge the gap—to create an amplifier that operates partly in solid state and partly in the state of California, that is ruled on one level by the microcosm and on the other by Governor Deukmejian. It is lore and intuition that the creators of new circuits must learn, mostly by themselves. There is no mold by which to multiply their numbers; they threw away the mold when they made Bob Widlar.

For example, if you want to catch Barrie Gilbert—he's another one, best speech, 1967 International Solid State Circuits Conference, a list of major credits only a little shorter than Widlar's—you can try the woods of Oregon, where he went to join Tektronics and now works virtually alone for Analog Devices, a company in Norwood, Massachusetts. An Englishman, he lives on a farm in Forest Grove between the Cascade Mountains and the coast. By discovering the laws governing currents in linked bipolar transistors, Gilbert made much of analog electronics possible. Or take Paul Brokaw, back home in Analog's design center. In a peripatetic early life, in the Air Force and out, "wiring filaments" for vacuum tubes, he carried along, creased and torn, like a letter from a long-lost love, a schematic of a transistor radio that he had clipped from a magazine at age fifteen. Now age fifty, a master's triathlon competitor, lean and gray, he is one of the world's leading circuit designers with some thirty patents and many more original designs.

In an era when many analysts believe that all progress comes from intimately coordinated teams of engineers—great gray armies from IBM and Bell—linear circuitry offers as an alternative caricature a zoo controlled by the animals. In part, analog people are different from digital people because the business is different. The analog market comprises thousands of individual designs sold through distributors rather than a few team products marketed by the company sales force. Linear success depends chiefly on personal inventiveness rather than on orchestrating groups of designers, reducing a computer schematic or memory to silicon, ekeing out angstroms with electron beams, and ramping up a fab line to awesome yields.

As Widlar says, "You can't force linear designs, each one is a separate invention of its own." Japan is learning fast in many areas of electronics. But so far the individualistic American culture, favoring design inventiveness over manufacturing teamwork, has given the United States a lead, and even a trade surplus, in advanced analog chips and the instruments they make possible.

Linear devices come in many forms. Voltage and current levels in most electronic devices need to be precise, stable, and predictable. For stabilizing functions, engineers use linear chips: voltage references, voltage regulators, comparators, and most of all, operational amplifiers.

Every system also needs some device, some *transducer,* to translate pressures, as of fingers on a keyboard, or levels—of temperature, moisture, light, fluid, and sound—into electrical signals that a digital device can use. Every digital system needs A-D converters to transform analog electrical signals, consisting of currents and voltages, into the digital language of binary bits and bytes. Then it needs D-A converters to translate the bits and bytes back into the language of the real world: into voltages and currents. It may need transducers also to translate the voltages and currents into lights, pressures, temperatures again for the actual operation of machines. In all its many dimensions, the converter business is making Analog Devices into one of the world's major semiconductor companies.

A sophisticated personal computer will have voltage regulators and fast digital-to-analog converters in its CRT; voltage regulators and high-speed comparators in its disk drive; voltage regulators and analog-to-digital converters in its mouse or joysticks; voltage regulators and telecom devices in its modem; voltage regulators, comparators, and drivers in its printer; scores of similar devices in its CPU; and op amps everywhere.

Finally, even digital chips are full of analog devices. From the perspective of the user, for example, a memory chip is chiefly a system of switches containing and conveying data in binary form. But at the lowest level of abstraction, every memory cell is a small linear system dealing in tiny voltages and currents, and storing them in a capacitor. From there a charge must be impelled down polysilicon wires by line drivers, to be registered by a sense amplifier, measured against a voltage reference by a comparator (by which it is identified as a digital "bit"), and finally sent down wires and through more complex linear circuitry to the system bus. These are linear devices all, even if, needing only to approximate two states, they are largely standardized and relatively imprecise in their operations.

Widlar created an analog IC and brought the op amp into the microcosm. Yet in the early 1960s at Fairchild, Widlar's proposed analog IC seemed even more improbable than the digital IC at the time of Kilby and Noyce.

In the first place, the passive components that are most crucial in an op amp are the very weakest link in a silicon chip. Resistors hogged space and capacitors leaked like sieves. It was the difficulty of putting these components on silicon that almost halted Kilby and Noyce. Yet it is the passive components, particularly the resistors, that give an op amp its accuracy.

In addition, unlike digital chips which merely have to switch—and gain accuracy only by switching millions of times—each linear device is useless if it is not exactly accurate and stable. Individual digital ICs have a "noise margin" of as much as half a volt. Linear ICs operate internally in microvolts (millionths of volts). While digital designers can use computers to cram millions of transistors on a chip, linear designers place hundreds one at a time.

Each linear transistor is laid out not to maximize density but to optimize the overall balance of heat and voltage across the chip—to control both the movement of molecules (heat) and of electrons (electricity). All of hundreds of specifications must be worked out with the highest precision. All this was perfectly possible using separate components ("discretes"). But it was simply impossible to translate all the features and functions of these discrete devices onto a single silicon chip.

Widlar, however, was one of the men who from the beginning understood the rules of the microcosm. Many engineers have always regarded integrated circuits as a vessel for replicating printed circuit boards in miniature through clever advances in process technology. (Widlar might quote Bierce: *Craft—A fool's substitute for brains.*) But Widlar always knew that ICs would require completely original modes of thought. As Barrie Gilbert puts it: "He understood the medium."

Widlar outlined his strategy: "There are far better approaches available than directly adapting discrete component designs to microcircuits. Many of the restrictions imposed by monolithic construction can be overcome on a circuit design level. This is of particular practical significance because circuit design is a nonrecurring cost in a particular microcircuit while restrictive component tolerances or extra processing steps represent a continuing expense in manufacture."

Widlar set out to create entirely new ways of implementing linear circuits on silicon and off. He attacked the resistor problem first by allowing the user to attach any desired resistors outside the op amp itself. Then he began using elongated transistors as resistors; and finally he resorted to pinch resistors, achieved by channeling current through thin passages of wire or polysilicon.

Theoretically, only the outside resistors could duplicate the performance of the discrete devices. But in the microcosm, as Widlar showed, wonderful things can happen. It turned out that the on-chip resistors actually outperformed discretes in many applications because being on the same chip with the amplifier, they responded the same way to changes in temperatures and conditions.

Still, Widlar had to work with a drastically limited palette of transistors. One of the key obstacles to a flexible and efficient linear IC was the absence of pnp transistors: devices that used holes rather than electrons as majority carriers of charge. But Widlar knew that CMOS IC designers spent much of their time trying to prevent the emergence of "latchup": unwanted parasitic pnp transistor effects between the positive and negative transistors. Turning this problem into a solution, Widlar managed to create usable pnp transistors on his linear ICs. This was a major coup that greatly enhanced his design resources and made it possible to solve a number of other problems as well.

Finally, the key to an op amp is a powerful amplifier to magnify the voltage before it is reduced by the feedback circuit. To get high-gain amplifiers for his op amps, he resorted to an idea from the reject bins. He had noticed that when transistors are left in the furnace too long, their base widths shrink to the point of vulnerability to short circuits, "punching through" even at relatively low voltages. These chips are thrown out. Nevertheless, the narrow base did result in far higher amplification gain for the device.

By deliberate overcooking of the transistors on his IC and by use of careful circuit design to shield them from untoward voltages, he managed to create bipolar devices with gains of 3,000 to 10,000, compared to the usual 50 to 300 on IC transistors. This super-beta transistor—an extraordinarily high-gain device—provided the foundation for the 108A, still one of the best-selling op amps. In the end, Widlar's op amps outperformed the best modules made with discretes.

In all these inventive solutions, Widlar relied on David Talbert and his wife Dolores for the actual fabrication. In recruiting Robert Dobkin, his successor, Widlar stressed that no one could compete with National in linear chips because only Talbert could process the strange new circuitry that was needed. Talbert used relatively old technology, but he applied it in unique ways. Filtering photoresists through a stocking, tuning his epitaxial reactor with a pair of pliers, risking the usual catastrophes with arsine gas and no safety interrupt gear, Talbert complemented Widlar's conceptual brilliance with an abstruse lore of process tweaks and intuitions that made it all work. "But if the room begins to smell like garlic [i.e., arsenic]," he said, "scram."

Nonetheless, the conceptual advances were Widlar's. In the end, Widlar created a whole new vocabulary and syntax of linear integrated circuitry that permanently emancipated the designer from most of the rules and limits of the old Philbrick op amps. Toward the end of his first stay at National, Widlar explained the result of his efforts: "It is now virtually impossible to draw a schematic diagram of a modern linear circuit. Usually, several pages of notes must be attached to describe components that the schematic cannot."

Widlar's analog chips were also crucial in financing the digital advance into the microcosm. By putting linear ICs ahead of digital ICs in cost-performance ratios, Widlar led the way for digital electronics. While digital IC designers were still merely reiterating the simplest discrete designs in homely logic families and asking the users to apply them creatively, Widlar incorporated the creativity in the IC itself. While digital designers were creating integrated circuits that performed less well than board-level designs at Digital Equipment or IBM, Widlar was offering completely novel functions at a lower price. He both gave the customer devices cheaper and more effective than the modules of discretes and invented new devices such as the voltage comparator and voltage reference that were not executed in modules at all.

For this reason, linear chips took off faster than digital chips and financed the early gains of several companies. More than military contracts, Widlar's creativity financed the first years of the key entrepreneurial firms of the Valley. One of his most important contributions, however, may have been to prompt the dismantling of Fairchild and the spawning of these firms that made Silicon Valley an entrepreneurial hotbed.

After several years at Fairchild, there dawned on Widlar—he is a genius, but he drinks a lot—a compelling new thought. He concluded that everyone was getting rich off his designs but himself. A head-thumping recognition, it drove him and Talbert to issue an unconditional demand to Sporck: Fairchild must make them millionaires within four years. "That's *after* taxes," Talbert said. "I want to get rich," Widlar explained.

Charles Sporck delivered the demand to corporate headquarters in Long Island. In the first of several similarly obtuse decisions that finally dismantled the firm, Sherman Fairchild said no (*Fools—The once and forever rulers of all creation*). Waving the wand now known near Lawrence Station as Vulture Capital, Peter Sprague said he thought Widlar could achieve his goals through the magic of *equity participation*. And so he did.

It was by hiring Widlar and Talbert that Sprague broke open the ranks of Fairchild and a year later lured away the manufacturing master Sporck. Sporck couldn't handle Widlar any better than anyone else could. But after deliberation (*The act of examining one's bread to determine which side it is buttered on*), Sporck decided to join Widlar at National rather than compete with him at Fairchild. Largely by recognizing the worth of human capital—making Bob Widlar a millionaire in 1969—Sprague and Sporck launched a company that would harvest billions in revenues from Widlar's designs. Fifteen years later, in 1984, long after Widlar had left the company, the volume of their sales was still increasing.

Widlar had also been indispensable to luring Robert Dobkin from the East. Among Dobkin's many inventions under Widlar's direction was an op amp that overcame the speed limit imposed by his ingenious but sluggish lateral pnp transistor. With a "feedforward" loop circumventing the slow-moving positive "holes," Dobkin accelerated the new op amp's speed to a level fifteen times faster than Widlar's 101.

Even Fairchild continued to reap huge benefits from Widlar's inspiration long after his departure. David Fullagar created an improved version of the 101 called the 741. Proliferating through the designs of electronics hobbyists and high fliers alike, it now may be National's biggest selling single product.

At the end of 1970, exactly four years after their arrival, Widlar and Talbert could exercise their options and collect their million dollars apiece. In relation to their contributions and in relation to the stock's value soon after they sold it, a million dollars, even an after-tax million, was a pathetically small amount. Still, Widlar felt rich at the time and summarily stopped working for five years. He left for Mexico, where he was unreachable by phone and was said to enjoy "the beach, the bimbos, and an occasional exotic smoke" (*Absentee—A person with an income who has had the forethought to remove himself from the sphere of exaction*). But there is reason to believe that his life in Puerto Vallarta also fulfilled Bierce's doubtful view of *emancipation (i.e., a bondman's change from the tyranny of another to the despotism of himself)*.

Talbert meanwhile put his after-tax million in tax-free bonds and calculated on a map the point in middle California where the smog of Los Angeles and San Francisco would be last to reach. The township would be called Paradise, California, but, needless to say, there is no town in Paradise. There he bought a ranch for $70,000 and settled in with his wife Dolores. He became a real cowboy at last. His number is unlisted. To get to Paradise, you take a flight in a small plane from San Francisco over the mountains to San Luis Obispo. Then you rent a car and drive eighty miles to Paso Robles. A friendly agent in a real estate office will tell you how to reach his ranch high in the hills, in the shadow of Fairoaks Mountain, where he lives with Dolores, big dogs, and a gun. He does not want to see you.

A young designer hired by National in late 1968, Dobkin was Widlar's accomplice in many practical jokes. He was also his successor as chief linear designer at National, chief affliction of National production chief Pierre Lamond, and leading figure in world linear design. In 1976, he would be named man of the year by *Electronics* magazine, "for leading a small group of designers at National Semiconductor to an impressive series of successful linear product designs encompassing both circuit and process innovations."

Of the early linear elite, Dobkin is one of the few still on the case, still producing three or four circuits a year and running the technology side of Linear Technology, Inc., Bob Swanson's splitoff from National. Dobkin has brought to Linear Technology, Inc., five out of the ten or so leading linear circuit designers in the business, including himself, Carl Nelson, James Williams, and George Erdi, the founder of a leading competitor, Precision Monolithics. And Widlar. Ah yes, Widlar.

Arriving to work for Swanson at Linear Technology, Widlar confided, over a drink: "I've been burnt out for ten years."

Swanson gulped. He had just granted Widlar 5 percent of the gross and 15 percent of the equity in his firm. "Hey, that's great, Bob. Now you tell me," Swanson muttered under his breath. Widlar then returned to Puerto Vallarta to work on his designs. "You don't invent with a computer," he said.

A couple months later, though, he cruised in, demanded a lab and special assistants for himself. Then he left again. He would produce a design once every two or three years. In his prime he used to put out a design every month. (*Achievement—The death of endeavor and the birth of disgust.*)

Finally Dobkin had to fire Widlar, which was very like firing his father, but he managed. Widlar would be bitter. But Linear Technology, Inc., was not going to house the Smithsonian Patent Model collection or compete with Lawrence Station or IBM. Widlar should have known; he partly owned Linear. But he had read too much Bierce and apparently regarded corporate property as "an ingenius device for obtaining individual profit without individual responsibility." He returned to the capacious arms of National, which could still afford his working style.

Still, he was not through. Watch Widlar, in 1985, there on the stage at the International Solid State Circuits Conference at the New York Hilton. He doesn't fit in—even after he gets through complaining about the lack of vodka on the podium. He is too gray and American and establishmentarian. He follows a long polyglot Babel of dapper geeks and teen-aged boffins with Adam's apples popping, and Chinese designers wrestling with consonants to the death, and Indians speaking of artificial intelligence chips in the tones of Oxford dons, and throngs of Japanese going through the motions of speech as they brandish 4-megabit RAM cells on their Samurai slides, and an occasional Intel American from Israel or Portland spieling out the specs of a fast microprocessor floating point unit.

Widlar, however, is serious, going for another best paper prize, rasping out the details of his unique new heat-compensated super-power 150-watt op amp. Still speaking in precisely the tones of George C. Scott, he outlines his new LM-12, which will be useful as an image enhancer, a motor driver, a high-power audio amplifier, or, as he hisses, a missile controller. He is looking spruce, too, with a trim gray beard, a crisp blue suit, and a big glossy tie with a bulging red Windsor knot. He gave the concept of the new chip to a young Japanese American named Minoru Yamatake and the young man produced the silicon prototype. Perhaps that is the way an elder statesman should work. That is one way to see Widlar as a portent of the American future in semiconductors.

In 1985, he succeeded. At the industry's leading conference, he remained king of the hill. Twenty-five years after his first breakthrough and after a flood of intoxicants had washed through his system, he once again won the award for the year's best paper and most ingenious invention. It was a dire setback for Alcoholics Anonymous.

Widlar's career, however, and indeed the entire history of analog computing, offers deeper lessons. In entering the microcosm, the industry made a deal. In exchange for giving up the testimony of human senses, electronic engineers could revel in the magic of an invisible world, achieving speeds and spaces of a nearly metaphysical cast. But the deal was always based on a deception. For the world needs computers that work not only in schematics but also in silicon, not only in oscilloscopes but also in offices. In real applications, analog circuits are often more useful than digital logic.

Analog, by definition, works by analogy with real events. Digital works through abstraction, by reducing all the murky physicality of solid state phenomena into binary Boolean codes. The analog computer is a true physical machine that must work materially.

In creating digital computers, computer scientists often forget the physical model in favor of the mathematical figment of a Turing machine or a truth table. But all computers are ultimately physical devices and thus in a sense analog machines. Hovering on the edge of an industry of software hackers and mathematical gurus therefore were always the hands-on harbingers of analog—the ultimately irrepressible intimations of the flesh beyond the word. Widlar was an avatar of what the industry often tried to forget or ignore: the analog dream that could be shunted aside but would always recur.

Not only is the op amp a rudimentary analog computer on a chip.

Large computers as well are increasingly adopting a concept of computer architecture derived, whether consciously or not, from the analog ideal. While Philbrick's "Analogue Computor" is as dead as any obsolete technology, its Platonic idea continues to exert a magnetism in the world of computing.

In the late 1980s, scores of companies are arising to apply the most advanced ideas in computer architecture. They are all doggedly digital firms, on the frontiers of the field. They include IBM and Teradata, B.B.&N, N-Cube and Thinking Machines, Sequent and Alliant, and AT&T. At long last, the computer industry is facing up to its basic problem of excessive reliance on serial logic. The solution seems obvious. The words buzz from Silicon Valley to Route 128, and with the usual lags, through Tokyo's Fifth Generation gap, and even arrive in due course as part of some European Eureka or Esprit. The words are "parallelism," "concurrency," "asynchronicity," "applications specificity." They sum up a growing consensus on the new signs and architectures needed to overcome the Von Neumann bottleneck. And when you investigate these brave new architectures, these buzzing new concepts, they turn out merely to represent a return to the grizzly old signs and architectures of the "Analogue Computor" that the Great Von Neumann banished to oblivion.

In Philbrick's machine, all is asynchronous (throw away the clock), all is parallel (no central processing unit here), all is concurrent (the entire machine is the CPU). All logic is combinatorial rather than memory-based. The algorithm is in the architecture itself. The problem comes in changing the architecture of the analog flow and in organizing data to take advantage of the asynchronous parallel processors. These problems have only been partly solved. But the solution most widely entertained comes in those cherished "Analogies," the symbolic regresses of Bush and Philbrick, and in the acceptance of the specificity of particular functions.

No longer is it taken for granted that the dominant computers of the future will be general-purpose processors. They may well resemble more closely the application-specific devices of the past. Forty years after the heyday of analog, analog concepts return on every side: in signal processors, in multipliers, in artifical intelligence chips, in multiprocessor machines, in parallel database servers, none of which will know and do everything, like some Fifth Generation son of God, but all of which will perform intelligently in their particular roles. This is what the original analog devices could do: each solved one particular set of problems. From the flow of force around an airplane wing to the flow of power in a grid, from the dynamics of chemical reactions to the operations of a secret code, the "Analogue Computor," slow and inaccurate though it often was, performed in elegance and truth.

The final irony is that executing these new analog architectures are mostly digital devices, being mobilized to simulate the integrators and operational amplifiers of the Philbrick era. Even at Barrie Gilbert's lab in the woods of rural Oregon, they are using digital microprocessors and teaching them to function in a Philbrick machine to simulate the dynamics of signal processors. With some success, the array of processors in essence simulates the actions of the Philbrick array of operational amplifiers to perform summing, multiplication, and integration. The direction seems clear. Ultimately, the "Analogue Computor" will return. It will be called the Fifth Generation super-computer. And with the needed arrays of analog supports, it will be made of digital chips, scrupulously imitating the old analog schematics and symbolic codes that Von Neumann threw away.

The nth-generation machine, however, the Platonic machine to which all computer architects aspire, may well represent a more exact embodiment of the analog ideal. At least it may become possible to eschew the baroque loops of the operational amplifier, eating its tail to keep its footing and find itself. Photonic or optic computers today are conceived as digital devices, with the designers laboriously trying to create switches out of light. But is it not likely that in due course some new Widlar or Widlaser mind will recognize the meaning of the new medium? He will see that the genius of light is its immunity to noise, drift, and physical defects, and he will create photonic integrators and differentiators that fulfill at last the Philbrick dream.

Listen to the technology, says Carver Mead, and you will find yourself returning to the analog inspiration. Using application-specific massively parallel machines, it may soon be possible to achieve the remaining goals of the collapsing computer, readily recognizing patterns in the world. The leading pioneer and practitioner in this field has long been Raymond Kurzweil.

CHAPTER
20

The Age of Intelligent Machines

Whether finding words in a stream of sound or reading them in a rush of text—or descrying objects in a passing tray, or missiles in a cluttered sky, or faces in a bustling crowd, or waves in a brain—the microcosmic computer demands new powers of pattern recognition. Raymond Kurzweil has devoted his entire career, in advanced research and practical technology, to the epochal enigmas of this field.

As a student at MIT in the mid-1960s, he discovered some initial clues in *Cybernetics,* Norbert Wiener's classic work on control or piloting theory (the word comes from the Greek *cyber* for pilot or governor). Pattern recognition is essentially the finding of significance in murky phenomena, or pure notes in noise. You may protest that the world does not seem so murky to you. That is because you see and hear its welter of signals only after they have been processed by the greatest of all pattern recognizers: the human brain. Ending his book with a contemplation of brain waves, Wiener suggested an agenda of work in pattern recognition that would imitate these powers of human beings.

A pioneer of the microcosm, Wiener saw that pattern seeking had to transcend conscious logical sifting of material shapes and sizes. The eyes and ears do not initially take in the world as a system of logically connected geometries located by material coordinates of time and space. For most pattern recognition, it is necessary to enter the microcosm and work with waves as well as masses, thoughts as well as things.

To Wiener, the supreme breakthrough in the field came from Jean-Baptiste Joseph Fourier. In exploring patterns in the movements of heated molecules, Fourier discovered that all periodic or repetitive mathematical functions, however complex, can be expressed as the sum of certain simple functions. In an inspiring poem of microcosmic harmony—integrating mathematics, physics, and aesthetics—these simple functions turned out to be the sine waves of pure tones in music or pure colors in light. By Fourier's math, anything that recurs regularly, that oscillates—whether earthquakes, electrical currents, sounds, or light waves—can be distilled to simple curves or synthesized to full complexity.

When a prism—or a raindrop—breaks up light into a rainbow, it is performing a Fourier analysis; when the cochlea in the ear transforms complex sound waves into simple sounds, it is performing a Fourier analysis. Prisms and cochlea and quantum phenomena already know the Fourier process; by nature, they perform Fourier processes all the time in what we call analog form. Computers and other pattern recognizers must usually learn Fourier analysis in the form of complex digital equations using numbers. But when they master it, they gain the key to such feats as identifying words and simulating violins.

Wiener's second contribution to the Kurzweil portfolio of tools was the concept of feedback as the basis of learning systems. As World War II approached, Wiener had worked with Vannevar Bush on a project designed to counteract German air superiority by improving air-defense. It was a problem comparable to the current enigmas of missile defense. Since planes could fly at a speed comparable to the speed of antiaircraft missiles, the guns would have to aim far ahead of the target. In order to shoot down a plane, it would be necessary to anticipate its future path and its possible evasive maneuvers and to find an optimal point at which to aim. It would be necessary to recognize the pattern of its flight.

From this wartime research, Wiener developed his theory of cybernetics. To perform any purposeful work—from forging a machine part to maintaining room temperature—a device must have a way of adjusting performance to purpose: appraising the current output and altering inputs when the output strays from the desired path. In this familiar feedback process—crucial to most manaufacturing control systems—Wiener saw the key to machines that could *learn* from their mistakes and constantly improve their performance.

Learning by feedback, a machine could achieve a further form of pattern recognition. As through a feedback loop a device homed in on a target—whether a room temperature, a missile, or a set of specified sine waves—in effect, it was learning to recognize the object. Such learning behavior he also identified as the crux of intelligence in seeking patterns.

A third Wiener lesson was breaking the mold. Most manufacturing systems reproduce a form by the use of molds or templates—static renditions of the original pattern. Similarly, most pattern recognition equipment has always worked by comparing a phenomenon to a template. The problem is that providing a template for every pattern requires immense amounts of expensive memory that eventually gluts the machine and prevents real-time operation. In a world full of surprises—spikes, spillage, smears, specks, dirt—template systems also limit recognition to anticipated phenomena and thus tend to work only in fully controlled environments. By seeking dynamic patterns that could change in response to changing conditions, Wiener broke the mold and released a new spirit of creativity in the conception of machines. Kurzweil continued in the new dynamic mode of pattern-recognizing devices.

After a series of intellectual oscillations—from poetry to philosophy—Kurzweil homed in on Wiener's idea of using new machines to overcome blindness or deafness "by replacing the information normally conveyed through a lost sense by information through another sense still available." Kurzweil envisaged machines that could generally convert type to talk and talk to type, machines that could make the computer generally accessible to the world. By pattern recognition, he would launch and name the age of intelligent machines.

The history began long ago in 1976 in Cambridge, Massachusetts, in a shabby little one-story structure at 68 Rogers Street. Part of a cluster of industrial buildings off Kendall Square, it had been upgraded in the late 1960s for the National Aeronautics and Space Administration. When Massachusetts voted for McGovern in 1972—so the story goes—President Nixon canceled the NASA project and doomed the area to dingy decay. Through most of the 1970s, the square declined from an aspiring mecca of space research, on the threshold of the cosmos, to a center of manual manufacturing, on the threshold of the Third World.

On the snowy morning of January 13, 1976, however, there was unusual traffic on Rogers Street. Outside the gray one-story buildings with their clouded tilt-out windows, vans from various television channels maneuvered to park. A man from the National Federation of the Blind struggled over a snow bank onto the sidewalk and began tapping earnestly to get his bearings. A dark-haired young man set out on a three-block trek to the nearest vendor of coffee and donuts for the gathering media. In the room at number 68, two engineers poked at a gray box that looked like a mimeograph machine sprouting wires to a Digital Equipment Corporation computer. Several intense young men in their early twenties debated when to begin a demonstration of the device. The short, curly-haired leader of the group, twenty-seven-year-old Raymond Kurzweil, refused to start until the arrival of a reporter from *The New York Times*.

The event was a press conference announcing the first breakthrough product in the field of artificial intelligence: a reader for the blind. Described as an "omnifont character recognition device" linked to a synthetic voice, the machine could read nearly any kind of book or document laid face down on its glass lens. With a learning faculty that improved the device's performance as it proceeded through blurred, faded, or otherwise illegible print, the machine solved problems of pattern recognition and synthesis that had long confounded IBM, Xerox, and the Japanese conglomerates, as well as thousands of university researchers.

Directly performing a mental task never before done without human aid, the Kurzweil reader would achieve the distinction, amazing in this fast-moving field, of being unequaled by any other firm for a decade and more. It would survive as the only product from the 1970s to appear in a show of the most advanced AI technology, entitled "The Age of Intelligent Machines," at the Boston Museum of Science in 1987. In its crisp replication of an act of human cognition, it excelled most of the other devices on display. By then improvements in the machine had raised the speed of its voice synthesizer to some 250 words a minute—nearly twice as fast as normal human speech—allowing the blind to read extended works or scan documents at a pace about equal to an average sight reader. But in all essentials it was the same device that Kurzweil had created in just eighteen months and was launching at 68 Rogers Street on January 13, 1976.

So where was the reporter from *The New York Times*? Brought up in a Jewish family in Queens, Kurzweil wanted the paper of record to certify his achievement. ABC, NBC, and CBS, plus the two wire services, had arrived and all were pressing Kurzweil to begin. But he had been told the *Times* would appear and he would wait for the *Times*. Finally the man clumped in, a new science specialist named Victor McElheny. Kurzweil sighed with relief and the show went on.

Kurzweil's prototype worked flawlessly on all three networks and created a sensation for this tiny, nearly bankrupt firm. When the synthetic voice read Walter Cronkite's signoff from the air, the device captured some of the magic of the anchorman himself. An appearance on the "Today Show" followed. Everywhere, the machine was described as the most important breakthrough in vision since the invention of Braille. As James Gashel of the National Federation of the Blind put it, after testing a machine at the New York Public Library, "Our major barrier in life has always been the print barrier. . . . Until [the Kurzweil] arrived, most of the books in the world's largest public library were off-limits to blind people."

Victor McElheny, however, was not impressed. He had stopped first at the AI Lab at nearby MIT headed by Marvin Minsky, the leader in the field, and had been told that Kurzweil's machine was of no special significance. Cobbled together with generally available components, it was inferior to many other projects across the country.

Nonetheless, the next day's *New York Times* did publish on page 10 a photograph of Gashel using the Kurzweil machine. The caption falsely explained that the device "translates ordinary reading material into Braille." McElheny's story quoted MIT scientists who predicted better machines within "a few months." Other artificial intelligence researchers dismissed Kurzweil's achievement as "mostly hype" and said it lacked all the crucial features of an AI product. Kurzweil was furious, and wrote a long letter to the *Times* listing numerous errors in the article.

In the following days, however, he was appeased by a more encouraging response. Stevie Wonder, the great blind musician, called. He had heard about the device after its appearance on the "Today Show" and it seemed a lifelong dream come true. He headed up to Cambridge to meet with Kurzweil.

As the son of a prominent pianist and conductor who had fled Austria and the Nazis before World War II, Kurzweil had a lifelong interest in music. He took Wonder out to lunch at Legal Seafood near Inman Square and the two young men found they shared a commanding musical intelligence and a fervent interest in electronics.

As Kurzweil remembers, "He was very excited about it and wanted one right away, so we actually turned the factory upside down and produced a unit that day. We showed him how to hook it up himself. He left with it practically under his arm. I understand he took it straight to his hotel room, set it up, and read all night." As Wonder said, the technology has been "a brother and a friend . . . without question, another sunshine of my life." Wonder stayed in touch with Kurzweil over the years and would play a key role in conceiving and launching a second major Kurzweil product.

A few days after the visit of the great blind musician came a call from a still more sophisticated source, the Boston Computer Society (BCS). The Society wanted a presentation at the group's March meeting at the Commonwealth School in Boston. Kurzweil readily agreed. Arriving early at the venerable school building, Kurzweil's vice president, Aaron Kleiner, was introduced to the Society's president and founder, Jonathan Rotenberg, who had contacted Kurzweil. Kleiner looked around in confusion. He saw no one in the room but a young boy. "Where is he?" Kleiner asked.

Rotenberg was twelve years old at the time. He had started the Society when he discovered that local computer stores could tell him nothing about how to purchase the personal computer kits that were then being created by a few small firms. Purchasing impressive new stationery and practicing a deeper voice for the phone, Rotenberg became a Compleat Society President. He immediately recognized the importance of the Kurzweil machine. During the subsequent ten years the Society grew into the industry's largest and most prestigious open membership group, and Kurzweil's companies grew apace. At the tenth anniversary meeting early in 1987, the twenty-two-year-old Jonathan Rotenberg, president of an organization of some 23,000 members, introduced Kurzweil, then thirty-eight, to present the premiere of a film on artificial intelligence and deliver the evening's main address.

Kurzweil had reason to listen to blind musicians and twelve-year-old seers. By the age of five, so he remembers, Kurzweil knew that he would grow up to be a brilliant scientist. According to his family, the computer soon became his "best friend." At the age of twelve, he wrote his first major software program and saw it adopted by IBM. At thirteen, he contrived an ingenious new database system that won him a prize. He built his first computer in 1962 at the age of fourteen. At sixteen, with a program that could create original musical works in the style of designated great composers, he came in first among some 65,000 high school entrants to win the International Science Fair.

Entering MIT, he became known as the "phantom" for his long entrepreneurial absences. Among his early feats, he wrote a computer program that given a student's record and preferences could name the most suitable college or university to attend. The program was purchased for $100,000 plus royalties by Harcourt Brace Publishing Company and Kurzweil, still a teenager, made his first appearance in the pages of *Time*.

If Kurzweil's openness to the wisdom of children was understandable, his reasons for learning from the blind were more complex. He had no blind relatives or close acquaintances. But perhaps, as the "friend" and creator of computers, he sensed that their tremendous capability—performing millions of operations a second, recalling data from huge memories, tirelessly repeating specific tasks—obscured some grave handicaps. After all, the computer functions under a triple affliction; it is deaf, dumb, and blind.

Any device that allowed blind humans to read would also bring sight to the blind machines that Kurzweil loved. Following the launch of the reader came application of the same technology to the problem of data entry into computers. In 1978, the firm succeeded in introducing an optical scanner that could read data into a computer at more than double the pace of a fast human typist, and Kurzweil set out to raise the money to produce and market it. But his main goal was still more ambitious. He wanted not only to convert type to talk but also to convert talk to type: a voice-actuated word processor that when reduced to a few custom chips would finally bring computer input into the microcosm. This remains one of the central challenges in all computer science.

He would begin with simpler sounds than speech. He would provide what many humans regard as the most precious analogies of all: the tones and timbres of musical instruments.

Martin Schneider, the media guru who pulled off the three-network coup for Kurzweil's reading machine, remembers visiting the young inventor in 1982 in his home. Kurzweil told him he had a new product and not to question him about it until he had said his piece. He launched into a long history of the piano, the early analog synthesizers of Robert Moog, even the Theramin, an ill-fated early effort at electrical synthesis of voice sounds. Finally he described the Kurzweil 250.

It captured perfectly the sounds of scores of acoustic instruments; it allowed endless layering, chorusing, sequencing, and other manipulation of music; it was based on pattern recognition and learning technologies and on the generation of wave forms by electronic oscillators. It achieved unprecedented computational speed and power, using a high degree of parallelism and scores of new specialized chips. It would come to dominate the world of musical instruments.

Schneider was entranced. "That's great," he said. "Where is it? Can I hear it?"

At the time, the machine did not exist and the company to create it had not yet been formed. But in the mind of Raymond Kurzweil, the synthesizer existed with all the power and authority of a primal truth. He had conceived it and reified it in his mind. He would will it into the world.

He took Schneider up to his studio over the garage and showed him two computers rigged up. He described some new insights into the contouring and compressing of musical sounds and timbres. He mentioned new computational strategies that could make the machine possible. These concepts would suffice. His first chief engineer, Al Becker, who had worked on the reading machine, put it succinctly: "Ray believed that once the concept of a machine had crystallized in his mind, it existed. Everything else was just technical details." Kurzweil was right; provided the mental reification is sufficiently thorough and powerful, mind can generate matter. The word can launch the thing. "Let there be music," he said, and there was.

It had all begun with Stevie Wonder. After the lunch at Inman Square, their friendship deepened and Kurzweil visited Wonder at his studios. During the visit, the musician described a vexing problem. As a composer and songwriter, Wonder reveled in the mastery—the power to mold and manipulate sounds—that he could achieve through use of synthesizers. But he felt the synthesizers' sounds were thin and shallow, and clearly artificial.

On the other hand, although savoring the rich timbres and colors of acoustic instruments, Wonder found that in using them he lost all control to conductors, ensembles, and even arrangers. Wonder challenged Kurzweil to design a synthesizer that could capture the rich tapestry of acoustical instruments and give the performance back to the composer. Kurzweil said he thought he could do it.

Himself a sometime composer and a performer of music, Kurzweil exactly understood the problem. Watching his father perform on the piano and with the orchestra, the boy Kurzweil had long sensed that conventional instruments were at once the consummation and the stalemate of a technology. In the creation of musical instruments, there was too much of the bondage and expense of rare materials and laborious craft and too little of the magic of mind. The world of music needed to enter the age of the microcosm.

Consider the concert grand piano, perfected more than three centuries ago and distilling the accumulated craft of hundreds of thousands of artisans. Each of some 12,000 parts—10,000 moving, 90 for each note—was originally wrought by hand. Walruses and elephants keeled over in bloody humps to yield ivory for the keys; rosewood veneer glows from the surface; the sounding board—an expanse of spruce some 25 square feet in area and millimeters thick—reverberates under perfect control; 250 pins on a ponderous cast-iron block hold the 88 strings, vibrating in endless interwoven waves of sound. It is a fabulous creation, the summit of the mechanical and cabinetworking technology of its time. And yet, in another sense, it is a ponderous weight of centuries on the back of the musician.

Each of the great composers of the past had risen to dominate and exploit the new technology of a particular era. The organ, the clavichord, the viol, the chamber orchestra, the pianoforte, the symphony orchestra, each occasioned an efflorescence of powerful compositions, nearly all created by master performers. Rather than operating from ivory towers, composers used the most advanced technical resources of their time to launch the most enthralling works of their day.

Perhaps because of the biases against business and science in the domains of high art, modern composers have largely lacked access to the leading technologies of the age. If they wanted to create new live music, they found only the plangency of electric guitars and a tinny tintinabulation of synthesizers as new resources for their palette. While the great composers of previous eras commanded the supreme mechanical achievements of their epoch, the modern composer has used relatively second-rate offerings from the hugely more powerful technologies of today. What is needed for a triumphant new music may well be less a cultural breakthrough than a technological innovation, less a new access to symphony orchestras in the inner city than access to microcosmic instruments that will soon allow inner-city musicians to outperform the symphony.

Kurzweil set out to achieve it, and in the process outdo the thousands of companies then in pursuit of the $8 billion (now some $11 billion) market for musical instruments. With neither hardware nor software in hand, he committed his company to an appearance at a music equipment trade show scheduled for early in the summer of 1983. Then he put so much pressure on the software team that several quit. Having sold his other company to Xerox but still running it,

Kurzweil himself pitched in with sixty-hour weeks writing code himself. By June, just twelve months after the conversation with Stevie Wonder, Kurzweil unveiled the prototype on schedule at the exposition.

Potentially allowing one person to simultaneously perform scores of musical instruments in thousands of combinations, the Kurzweil machine was acknowledged to represent "an order of magnitude improvement" over previous keyboards. Whatever its flaws, the acoustic qualities of the Kurzweil 250 tolled the end of the line for six centuries of handcrafted instruments.

The facts are simple. Every year the essential electronic components in the synthesizer drop in price or rise in efficiency by about 35 percent. Every year the essential cost of the skilled labor and rare materials in handcrafted acoustic instruments tends to rise and their quality to decline. Between 1980 and 1986, electronic keyboard instruments have risen from 9.4 percent to 55.2 percent of the American market. It is just a matter of time before the mechanical piano will go the way of the harpsichord, if not of the mechanical calculator.

There are many lamentations and regrets. Many connoisseurs of acoustic devices detect flaws and distortions and audible inaccuracies in the synthetic sounds. Those defects will be readily corrected as the technology changes and improves. Then the connoisseurs will speak of the dehumanization of music and the usurpation of art and craft by disembodied mechanism. As Kurzweil understands—in his own living room his father's Baldwin Grand stands next to his own 250—the critics will be partly right. Progress imposes inevitable costs.

Nonetheless, the central fact remains. Based on the capabilities and potential of modern microelectronics, the Kurzweil machine is not only a better, incomparably more powerful and resourceful instrument than the acoustic piano or any other acoustic device, but also a fuller and richer expression of the true genius of human beings.

Every era has its pinnacle. In medieval Europe, it was the Gothic cathedral. In the late twentieth century, it is the very large scale integrated circuit, the Gothic cathedral of modern America. Like a cathedral, it is wrought of common elements—glass, sand, and metal—transfigured by human genius and art. Viewed through a microscope, the chip unfolds as a panorama of infinitesimal tapestries, as intricate and variegated, efflorescent and symmetrical, radiant and colorful, as all the glass and stone of Notre Dame or Chartres. Passing over for a time the issue of whether it will be comparably given to God, the feat of designing and producing a modern microchip entails craft, devotion, diligence, genius, and vision comparable to any artistic pinnacle of ages past.

Not only are the quantum technologies that made the 250 and other synthesizers possible the highest attainment of human art and creativity—containing not mere thousands of components but millions of transistors and trillions of managed moving parts (or particles)—but the musical resources represented by the 250 can enrich and invigorate the world of music. Allowing one creative artist to far excel the musical capabilities of entire orchestras of the past at a fraction of the cost of a single piano, the synthesizer can revitalize and democratize serious music. Not only is the composer united again with the performer but the performer faces a new and stimulating challenge of technique. Far from withering in some electronic moonscape, his human capacities as an artist and composer can find fuller and richer modes of expression, as well as new markets for live music of all kinds.

We can gain confidence in the fate of the 250 and other synthesizers, moreover, by contemplating a similar rite of passage that afflicted the music of an earlier era. The piano itself wreaked creative destruction comparable to anything we can expect from the Kurzweil. With the simplest motions of the hand and fingers, the pianist could command seven octaves of sound and entire new dimensions of touch and dynamics, playing harmonies and melodies at once, trilling notes at inhuman speeds, achieving effects of counterpoint and convergence that would require entire orchestras of previous instruments. In its time a marvel of innovation, the piano laid waste to a world of lutes, harps, guitars, harpsichords, and other plucked devices. If anything, the electronic synthesizer represents a less drastic change. But its future promise is far greater because its cost will eventually make it available not to an elite but to the world. In 1988, Kurzweil's 250 technology was purchasable in a keyboard costing less than $1,000.

Raymond Kurzweil says his father anticipated some such achievement from his son and would have been pleased. Perhaps one reason for the note of sorrow in Ray's brown eyes is his regret that Frederic Kurzweil did not live to escape the cages of mechanical craft into the microcosmic spaces of the Kurzweil.

The synthesizer, however, was in part a digression in Kurzweil's career. Before the product was fully on the market, he began a new company, Kurzweil Applied Intelligence, to achieve the second and most formidable half of his agenda for an accessible computer: a computer that can take direct human dictation at the pace of natural human speech and recognize a full human vocabulary. This product has the potential to oust the computer keyboard from the offices and production lines of the world; banish the humbling hassle of hunt and peck and the need for stenography; create the possibility of a phone for the deaf; and enable fully voice-controlled robots and voice-accessible databases. By drastically improving the responsiveness and accessibility of every computer, the voicewriter would do for the computer industry what the synthesizer could do for music.

This new step in collapsing the computer and allowing the spread of intelligent machines was quite simply the most important current goal in all commercial high technology. From Japan, with its continuing problem of entering thousands of Kangi characters, to IBM, facing ever more telling competition across its product line, and to scores of other companies and countries looking for a technical edge, speech recognition is the most inviting target. It is the imperative next step in computer technology needed to launch a new surge of growth both in the electronics industry and in the world economy. Literally hundreds of companies and university laboratories have been pursuing the technology for twenty years. The murky multiplicity of human speech makes it a fiendishly difficult challenge.

Consider how human beings recognize words. In general, they depend heavily on meaning and context. The error rate for word identification rises sharply, to levels that would be unacceptable in office equipment, when words are recited in nonsense order. It seems that the human brain is incapable of recognizing words consistently on the basis of their phonemes or sounds alone. Humans need to know the meaning and context. Yet even the most ambitious proponents of artificial intelligence cannot claim they have the slightest idea how to teach computers the meanings of 20,000 words and their endless applications in context.

The voicewriter thus cannot succeed simply by equaling human capabilities in sound analysis; in some ways it may have to excel them. Yet the seventy phonemes, or voice sounds, will be rendered in thousands of different voices, accents, dialects, timbres, and intensities. Even particular individuals frequently change their pronunciation. Programming a conventional computer to identify these protean patterns in real time is impossible. Indeed, using conventional spectrum analysis, a Cray supercomputer would take minutes for each word, and in the many cases in which humans elide or slur syllables, the computer could not come up with the solution at all. The experts agree that a truly robust and effective voicewriter is still impossible.

Regardless of these perplexities, in August 1986, Ray Kurzweil described his new voicewriter. It could take dictation at an average pace of 150 words a minute; it could recognize an active vocabulary of 10,000 words out of an available vocabulary of some 200,000 words; it was even then being demonstrated at "beta" sites; and major firms would soon put it on the market for $5,000.

A visit to the offices of Kurzweil Applied Intelligence, however, yielded no evidence of this miraculous device that would transform the world of word processing. Neither the head of the semiconductor division, nor the linguistic consultant, nor the research director, nor the manager of human factors design could verify the truth of any of those statements by their boss. One hundred and fifty words a minute? Aaron Kleiner looked doubtful. Possibly in "burst mode."

A vocabulary of 20,000 words? Well, Kurzweil engineers thought it would be closer to 7,000, but research had shown this number to be ample for all office uses. After all, Shakespeare used a total of some 30,000 words, but only about half of them appeared more than once. Executives could function on half the working vocabulary of the Bard. In fact, James Joyce snuck by with 7,000 words in *Ulysses.* Demonstrations at beta sites? Perhaps in a year or so. As for large firms, many were interested but had yet to design any product that used the voicewriter. The price, moreover, at the outset would be closer to $20,000. "Ray must have been talking about the ultimate price."

The men at Kurzweil Applied Intelligence were trying to get off the hook. But Kurzweil wanted them on it. The facts of the world are less important to him than the force of his will. He knew that all the conceptual problems of the product had been solved, all its key components had been demonstrated in one form or another, and so it "existed"—and his staff had better believe it too. In a sense the product had existed ever since Kurzweil as a student at MIT read Norbert Wiener's prediction of it in *Cybernetics,* the prophetic volume of artificial intelligence.

But Kurzweil's staff were not yet sure. They would presumably be the last to know. Only Kurzweil knew for sure.

Kurzweil's optical character recognition technology, mostly invented a decade ago, is still a leading force in the industry. Using conventional techniques of digital sampling, the music synthesizer was simply impossible except at a prohibitive price. For example, to create a high-fidelity sample of each note on the piano would take some 30 billion bits, or more than 30,000 memory chips of a million bits apiece (costing a total of some $300,000 for memory alone). Speech recognition imposes far larger burdens. Even Kurzweil's 1,000-word machine—widely used by hospital x-ray and emergency room personnel—was yet to be equaled in the marketplace by 1988.

Some general principles of how to create machines for recognizing speech were widely known. The problem was that the machines did not work. Kurzweil's genius was to embody these concepts in working products that could be sold at an acceptable price. For this goal, Wiener's vision did not suffice. Also necessary was Mead's technology.

Kurzweil's secret is that he is the first man to exploit simultaneously both the cybernetics of pattern recognition and the new balance of power in the industry. In particular, he used the three microcosmic breakthroughs of the last decade promoted by Carver Mead—artificial intelligence, parallel processing, and silicon compilation. Fully committed to a digital strategy, Kurzweil took full advantage of the accelerating pace of microcosmic progress to digitize and manipulate huge arrays of analog information in music, voice, and print.

Many companies have pursued one or two of these concepts. Such companies as Texas Instruments and Symbolics have used the techniques of artificial intelligence to enhance the problem-solving capabilities of conventional computer architectures. Goodyear Aerospace, Bolt, Beranek, & Newman, Thinking Machines, N-Cube, and several other firms have used high levels of parallelism to accelerate the speed of general-purpose computers. Scores of companies have used silicon compilation to create new commodity chips or to integrate the functions of conventional computers. But as Mead insisted from the beginning, the full power of the three technologies emerges only when they are used together. Kurzweil has proved that Mead was right.

To operate in real time, processing the enormous flood and flux of speech signals, the system would have to mobilize massive digital computing power. Kurzweil estimated that it would have to operate at a pace of at least 2,000 MIPS or 2 billion instructions per second. In other words, it would require the equivalent of some $17 million worth of available computing equipment. Kurzweil resolved to create this equivalent for a few thousand dollars per machine by using silicon compilers to design massively parallel special-purpose chips.

His rationale was simple. Although the function might require 2 billion instructions per second on general purpose supercomputers using very high speed serial processing and costing millions of dollars, the specific voice recognition algorithms could run digitally at similar effective speeds on far less expensive custom circuits running in parallel. Although the voicewriter imposed special demands, Kurzweil was confident that this approach would work. Both the latest version of the omnifont optical character recognition system and the portable "1000" music synthesizer would concentrate their power in highly parallel custom chips designed on silicon compilers from VTI. He was ready to pursue a similar approach also with the voicewriter.

At the time Kurzweil assembled his nine-man team in offices next to the Charles River in Newton in mid-1982, both parallel computing and silicon compilation were far below the necessary level. Of the leading compiler firms, only VLSI Technology (VTI) was in the field with useful products. Parallel computing remained a highly abstruse and abstract domain pursued chiefly by firms with government contracts and with products scheduled to sell for close to a million dollars a unit. However ardently executives might lust to talk to their typewriters, few were thought to be willing to pay more than a few thousand dollars for the privilege.

Kurzweil's key insight was that by the time the large vocabulary voicewriter was ready for silicon, the silicon would be ready for it: the needed parallel architectures could be integrated on a manageable number of custom devices. His decision to go ahead before the technology was ready was critical for the success of the firm.

Since no one had ever written large vocabulary speech recognition software to run in parallel, the Kurzweil project would have to break new ground. Nearly the first project of the company therefore was to build a parallel computer on which to develop the necessary software. The Kurzweil team created a system with an array of 64 Motorola 68000 microprocessors running together, each with its own memory. This machine inevitably became known as the "Ray processor."

Although these portions of the actual voicewriter would eventually demand the equivalent of more than a thousand 68000s, sixty-four sufficed to develop and test the new parallel algorithms. By the time the company was ready to go into silicon—and VTI was ready to build the chips—all the necessary software would be written and tested in parallel form. Then the software algorithms, largely developed and tested on the parallel computer, could be embedded in dedicated silicon processors running in parallel.

For the final hardware designs, Kurzweil resorted to a little "new math" from John Newkirk and Robert Matthews, who had introduced the Mead system at Stanford and were authors of compiler tools to accelerate parallel processes. The new math helped Kurzweil chip designer Amnon Aliphas to reduce the thousand parallel 68000s to just eight 24-bit digital filter-processing chips. Of course, a thousand 68000s would have far greater general-purpose processing power. But the Kurzweil chip, customized through silicon compilation, was dedicated to just one task. Fabricated at VTI, the Kurzweil chip could filter and analyze voice signals at supercomputer speeds. Unique in the industry, the chip won an award as an *Electronics Products* magazine design of the year. Yet the chips cost not millions but under $10 apiece.

Such approaches, used widely in the machine, turned the impossible dream of the voicewriter into a small box easily employed with an IBM AT personal computer. Kurzweil's correct vision that the hardware would be available in time allowed him to begin developing algorithms in 1982 that took advantage of the state of the art in 1986. By 1988, the voicewriter contained more than a thousand separate parallel processes and custom chips controlled the critical datapaths.

In the new order, as Kurzweil comprehended, algorithms—the basic logical concepts in a program—did not merely consist of software written to run on the best general-purpose microprocessors available. With the customizing power of silicon compilation, algorithms had become architecture. Embedded in silicon, the software became part of the very definition and structure of the product.

In Kurzweil's view, however, parallelism and silicon compilation alone could never create a voicewriter without the third key theme of the new order: artificial intelligence. Artificial intelligence, the use of human paradigms of information processing, was central to the entire project. Taking the human voice and translating it into human language required a comprehension of human mental processes. For example, even the front end—processing the voice spectrograms themselves—was based on a detailed model of the electronics of human hearing. The award-winning filter chips executed an algorithm derived from an in-depth study by Rich Goldhor of the cochlea and auditory nerve in the human ear. Similar to the more detailed model being developed by Carver Mead, this design immediately adapted the signals for later interpretation by expert systems also based on concepts of artificial intelligence.

Filtered into sixteen channels of auditory information, the signals rush on to the central pattern recognition board of the system. This board comprises seven "experts," each of which attempts to identify the word. The phonetic expert considers its sound; the syntactical expert offers a guess on the basis of its function in the sentence; the semantic expert judges the context and makes a stab; an expert on psychoacoustics appraises the particular accent of the speaker; and the other experts analyze other aspects of the waveforms and arrive at a conclusion. Each expert combines its guess as to the word with a weight of probability. Finally a managing 68000 family processor brings to bear a model of the likely strengths of the different experts in judging different kinds and sizes of words. Its output is a list of the possibilities in order of their probability. If there is a clear winner, it registers alone on the screen; if not, all the possibilities are listed.

Crucial to the success of the voicewriter is its learning faculty. Each user begins with a two-hour initiation, which creates a personal voice profile on a disk. From then on, the machine continues to learn, adding new words, adapting itself to a particular job, and constantly refining its models. After a few words, for example, the machine will discover that the user has a cold or is speaking in a noisy room. This process of adaptation resembles the operation of Kurzweil's optical character recognition device, which will drastically improve its accuracy after just one page of experience with a difficult stretch of type (for example, a fourth-generation photocopy).

The voicewriter, finally released as "Voiceworks," is a true vessel of artificial intelligence, learning through feedback loops and recognizing patterns through dynamic models. Like most expert systems, the bulk of the work and the contents are specific to the task. Kurzweil could not simply use the same people who created the optical scanner or the music synthesizer. He had to turn to the universities to find specific experts in linguistics, snytax, acoustics, signal processing, auditory physiology, and other disciplines.

Ultimately Kurzweil's success stemmed from his mastery of Wiener's early insight in *Cybernetics:*

> The most fruitful areas for the growth of the sciences were those which had been neglected as a no-man's land between the various established fields. Since Leibniz there has perhaps been no man who has had a full command of all the intellectual activity of his day. Since that time, science has been increasingly the task of specialists, in fields which show a tendency to grow progressively narrower. . . . Today there are few scholars who can call themselves mathematicians or physicists or biologists without restriction. A man may be a topologist or an acoustician or a coleopterist. He will be full of the jargon of his field, and will know all its literature and all its ramifications, but, more frequently than not, he will regard the next subject as something belonging to his colleague three doors down the

corridor, and will consider any interest in it on his own part as an unwarrantable breach of privacy.

In other words, the average scientist may be what Ortega y Gasset called a "barbarian of specialization." Kurzweil's immediate domains of knowledge were pattern recognition, computer science, and artificial intelligence. But like Carver Mead, Kurzweil found a way to transcend such specializations with a powerful vision of general truths. By learning from others and venturing readily into new disciplines, each man gained a mastery of the scientific and technical culture of his day as effective as that of any of the great scientists of the past. The great unifying power of the microcosmic laws—from Fourier transforms and information theory to quantum physics and biochemistry—is creating a new age of Renaissance men.

By the measure of bringing the fruits of this microcosmic knowledge into the marketplace, in the form of new products that change the world, Kurzweil is unsurpassed in his time. Exploiting the tremendous opportunities in those three linked fields expounded by Carver Mead—artificial intelligence, parallel processing, and silicon compilation—he is fulfilling the promise of the quantum era.

Kurzweil's products were launched directly into the path of the giants, particularly IBM. Kurzweil recognized the new balance of power in the industry better than any other entrepreneur of systems. IBM and other companies had assigned the highest priority to similar projects and had pursued them for decades. Yet it took Kurzweil just two years to introduce his omnifont reader and scanner, a year and a half to develop his music synthesizer, two years to launch a 1,000-word voice-activated control system, and three years more for a voicewriter with a 15,000-word vocabulary and discontinuous speech. Even if he should falter in coming years in his effort at continuous speech recognition, his strengths and strategies, plans and products, have brilliantly illuminated the path ahead.

Nonetheless, IBM chose a different course.

CHAPTER
21

The IBM Machine

"Please write to Mrs. Wright right now." Oh, yes, Miss Right, whatever you say.

Zoom in on sultry red lips and come-hither smile. No more big blue pinstripes. Zoom back for feminine curves, big in blue cashmere, and eyes welling with warmth and promise and playfulness. No question, this company really knows how to sell sweaters. But no. Back to that ovoid mouth, hinting kisses to dream by. . . .

Listen, the lips are caressing a message: "Please write to Mrs. Wright right now." Then cut to a logo of IBM. For this is an ad not for cashmere or cosmetics but for voice-activated word processing. Right now. Three homonyms in one sentence: write, Wright, right. Claim that IBM has the most advanced system of its kind, even if it is actually not ready "right now" and even if the words are not displayed "right now" but only after a few seconds delay. After all, it is an experimental device. Zoom all around the issue of when the product will be available; the girl anyway is sure available right now. Perhaps the world will believe that IBM is still in the lead in speech recognition. Right.

What has happened to Charlie Chaplin? Can IBM be breaking its long taboo against using sex to sell a product? Can Big Blue be using a television blitz to announce an experimental prototype? At Kurzweil Applied Intelligence, where the ad is available for display at a moment's notice, there is much amusement and speculation. Some observers believe that IBM's campaign is an overwrought reaction to Kurzweil's premature announcement of a voicewriter with a vocabulary of 15,000 words that takes dictation at a pace of 80 words per minute and offers its results in milliseconds and is available if not "right now," sometime soon.

The first of a series of full vocabulary speech processors, the Kurzweil Voiceworks and its kin will drastically transform the office environment. Within fifteen years the only keyboards with a growing market will be on music synthesizers. In offices, hand entry will be as archaic as mechanical typewriting today. Unless IBM can respond successfully, such products could menace the company.

IBM clearly can respond. Big Blue spent as much on that one-minute advertisement, shown repeatedly in prime time over several weeks, as Kurzweil Applied Intelligence has invested altogether in the entire history of the company, some $15 million. But the consensus at IBM is that the ad reflects not the development of a salable machine but the display of IBM's prowess as an innovator. IBM was not flaunting an impending product; it was flaunting the potential to create a product. Yet things are rarely as they seem at IBM.

Zoom in again on those delectable lips. This time they hover motionless on a screen above a long metal table in a conference center at IBM Research Headquarters. A huge complex of blue glass in Yorktown Heights, just off the Taconic Parkway in Westchester, New York, the structure houses the prestigious IBM research team on speech recognition and the rest of IBM's U.S. research division. Around this table, the ad, the lips, the sweater prompt just as much relaxed amusement and speculation as at Kurzweil Applied Intelligence, and no more sense of being close to its source. "Sex," someone blurts out amid general laughter, "it's the first time IBM ever admitted it had sex!"

It is "tea" time in Yorktown Heights. About half the speech recognition team has gathered for Coke and cookies and tea in the late afternoon. It is a casual and confident group, and coffee cake is passed freely around, along with jokes and complaints about IBM's bureaucracy. When would the 3080 mainframe be made available to the team as promised; when would the signal processor snafu be unscrambled; and when would the right chips from Essonnes in France become available?

No ties, pinstripes, or business suits are to be seen here either, except on the one woman in the group. A motley spread of accents—from British Oxbridge to Middle European, from City College to Delhi University—would defy the most advanced speech recognition system in the world. Luckily, however, the man in front of the room presses a button and the ad springs into motion, declaring that this team's machine *is* the world's most advanced. After the ad ends, the man recites a series of sentences full of mind-bending homonyms, similar to the wright, write, right, rite array, and the machine efficiently sorts them out and again displays the correct spelling on the screen.

This feat seems more remarkable because the man's voice shrouds every word in a thick tangle of phonemes from both Central Europe and the streets of New York. The speaker is Fred Jelinek, for nearly fifteen years IBM's chief of research in speech recognition. He knows that the homonym demonstration is the forte of his machine, with its emphasis on context and probabilities rather than on looking for clues in the sounds themselves. An intense, jovial figure with curly gray hair and an impish sense of humor, he is using this teatime—a daily rite at the center—to rehearse for a performance the following Monday at Ann Arbor, Michigan. Then he will leave for Europe to address a plenary conference on information theory the next week in Berlin. . . .

It has been a long way to Berlin, a long way since the uniformed figure in black, emblazoned with swastikas, ousted him from his room as a ten-year-old boy in bed with pneumonia. It has been a round-about loop since the grim reapers of the Holocaust sent one after another of his friends away to death, until he was left alone with his family cowering in a basement in Prague, and then they came for his father, too. A doctor, his father died of typhoid attempting to keep inmates alive in a concentration camp. On one occasion, sideswiped by a paddywagon full of Nazis and prisoners as he walked across a bridge, the boy escaped only by stripping away his Jewish star and telling a series of artful lies. As a Jew, he was banished from school for several years under a law that made it illegal for him even to be taught at home. From then on, he was not only fatherless and friend-less but also hopelessly behind in all his classes.

Gnawing always on the small boy's mind was the bitter knowledge that his father had actually dispatched all his possessions to London in a plan to emigrate before the war. But he then decided to remain in Czechoslovakia, confident that his friends in the community would protect him from the Nazis. After the war when the Communists came, his mother resolved not to make the same mistake twice. They fled quickly to join family in the United States.

Then sixteen, Jelinek was once again years behind his classmates in school, years behind in the English language, in social skills, and at first in his courses. His family penniless, he delivered clothes for a cleaner, stripped varnish for a furniture store, pasted decals on ugly lamps at a lamp factory. An ungainly boy, he was often fired; he dropped lamps or fell asleep. But eventually he got through high school and into night school at City college, and in the course of time, as an unexpected brilliance in mathematics began to emerge, he won a scholarship to MIT.

So now it is October 1986 and in Berlin they want information theory. Jelinek is the man to give it to them. When he had arrived at MIT some twenty-five years before, the campus was largely in the thrall of Claude Shannon, the pioneer of information theory.

Shannon's focus was on how to encode and send a message through a noisy channel. Jelinek's interest came to focus on speech recognition, which was the opposite of Shannon's concern. In effect, speech recognition required detecting and decoding a message from a noisy channel, the human larynx and vocal cords. But just as Mead and others found Shannon's work deeply fruitful in the field of computation—the Twenty Questions model of information entropy, for instance—Jelinek found it fruitful in speech recognition. By reversing Shannon's mathematics, he found that in principle, if not in practice, Shannon's information theory could be used as effectively in identifying words in a flux of sound as in sending them.

Shannon used an ingenious multi-dimensional geometry to point the way to solving such problems. Any sample of sound can be perfectly analyzed by Fourier processors and filters into a set of sample amplitudes. Then a group or vector of the samples can be mapped as points in multi-dimensional space, where the number of dimensions is equal to the number of sample amplitudes, and the energy of the sound is equal to the square of the sample amplitudes. Using a complex mathematics resembling Euclidean geometry, Shannon showed that a message would rise to the "surface" of the "multi-dimensional hyperspace" and could be infallibly distinguished from a shroud of noise if the entropy of the transmission was within the capacity of the channel.

Like most of the computations of the microcosm, the concept of the surface of a multi-dimensional hyperspace transcended the sensory world. The efficacy of such a bizarre contrivance in solving a practical problem of communications is a powerful example of what physicist Eugene Wigner called "The Unreasonable Effectiveness of Mathematics in the Natural Sciences." Enthralled by such unreasonable effects, Jelinek studied information theory at MIT under Ernest Guillemin, one of Shannon's teachers, and under Robert Maria Fano, another brilliant refugee from the Nazis and one of Shannon's most eminent students, and had written his thesis for a committee including Shannon himself.

After leaving MIT, Jelinek made a pilgrimage back to Prague. There he met one of the few Jewish acquaintances of his childhood who survived the war, Milos Forman, the film director. With Forman, he also met his own future wife, a beautiful actress and director named Milena. Scheming for the next three years to bring Milena into the West, he finally succeeded by getting former MIT president Jerome Wiesner, President Kennedy's science adviser, to intercede with Cyrus Eaton, who made a personal appeal to Nikita Khrushchev himself.

In the United States, Milena got a fellowship to study linguistics with Noam Chomsky, the pioneer of deep structural linguistics at MIT. Chomsky represented the opposite school of linguistic interpretation from Shannon. Rather than focusing on the mathematics of the sounds, he delved deep into the structure of the language and from there into the structure of the brain. The history of computerized speech recognition swings between these polar approaches, Shannon's and Chomsky's.

Shannon's school relies on the unreasonable effectiveness of mathematics, probability theory, signal processing, and pattern classifying; Chomsky's school depends on the reasonable notion of identifying words by studying the syntax and semantics of language and its use by human organs—vocal cords, aural cochlea, and brains. Attending Chomsky's classes with Milena, Jelinek began a long and deeply ambivalent involvement with the Chomsky school. Beginning as a devotee, he ended as a nemesis of linguistics.

Both intrigued by the analysis of language and frustrated by its perplexities, he devoted years to resolving the problems of identifying linguistic structures. He even went to Cornell to study under an eminent linguist. But no sooner had Jelinek arrived in Ithaca than the linguist abandoned the field to write operas. Back in Cambridge, Chomsky embroiled himself in operatic political polemics that diverted attention from his disappointments in the field he pioneered. In the end, facing the iron test of actually decoding speech, Jelinek found the techniques of Shannon and the statistics of probability far more powerful tools than Chomsky's more specific universal rules of usage. As Shannon saw it, mathematics dominated grammar even in the domain of language itself.

The crucial test came at IBM, where Jelinek arrived in 1972 to work on a speech recognition project. Soon put in charge, Jelinek has managed this project ever since.

By October 1986 in Berlin, he could report a string of impressive achievements in speech recognition. Vindicating Shannon's information theory in this domain, Jelinek had reduced the problem of recognition exclusively to the statistics of probability. To Jelinek, any use of experts, heuristics, rules, or inference engines merely pollutes the pure power of mathematics. To make his point, Jelinek spoke with dismissive relish about the initial "success" in voice recognition, achieved in a project at Carnegie Mellon in the early 1970s.

In a precursor to the Jelinek-Kurzweil duel of the 1980s, the Advanced Research Projects Agency of the Department of Defense (now known as DARPA) had launched a five-year race among five teams to create a 1,000-word speech recognizer. Among the five were groups from MIT, Stanford Research Institute, Bolt, Beranek, & Newman, and Carnegie Mellon. The winner was a Carnegie Mellon team led by an Indian immigrant named Raj Reddy. This team had created a system called Hearsay, which identified words by using a series of "experts." One judged the sound, another the word, a third the grammar, a fourth the meaning, and a fifth the context. These experts would all make their choices and list them on a "blackboard" where a final expert would determine the correct word. It was a true exercise in artificial intelligence, based on the Chomsky school, contrived by distilling and integrating phonemic, lexical, syntactic, linguistic, and semantic expertise from leading professors in each field. The only problem was that it didn't work. The experts constantly stumbled over each other and ended up in a snarl that only sometimes could be decoded.

Nonetheless, Reddy's Carnegie Mellon team won the DARPA prize. By chance, in one of Reddy's classes was a graduate student in statistics and information theory named James Baker. As the deadline approached, Baker created a 1,000-word system called Dragon which actually performed speech recognition. Reddy made a few adjustments, changed the name to HARPY, and entered it into the contest, shoving Hearsay aside. HARPY worked slowly—fifty times real time —and it was restricted in its sentence structures, but it got the answers right. DARPA correctly concluded that with the rapid pace of advance in computer hardware the system could eventually be made to work in real time. Carnegie Mellon thus won the most famous victory for artificial intelligence largely by eschewing five years of efforts to apply AI techniques and adopting a program that combined rudimentary statistics with ingenious and ambitious engineering.

From Jelinek's perspective, the few artificial intelligence rules and heuristics included in HARPY only "screwed up" Dragon and failed to contribute at all to its effectiveness. The power of the system derived entirely from its use of statistical theory to determine the probability of sequences of words given various acoustic inputs. No rules or linguists need apply. The machine would simply determine which word conforming with the acoustical evidence was most likely to follow a previous word based on a large volume of statistics of word sequences.

Any sentence unfolded as a pattern of acoustic energy varying over time in a Markov chain of probabilities. In the Markov chain, the probability of a subsequent word in the chain is determined by the previous word and a statistical analysis of the words that followed it in a database of usage. As Jelinek insists, there is no artificial intelligence or linguistics to it, only the supreme power of probability theory.

Baker's triumph vindicated Jelinek's approach, which he had been developing at IBM. Soon Baker arrived at Yorktown Heights, and all would have gone well if he had not brought with him his wife Janet and demanded that she be given equal pay and prestige. IBM was not used to package deals, or dictation, and the Bakers soon departed. Today Dragon Systems, consisting mostly of Jim and Janet, is a small firm using the same mathematical techniques as IBM. Dragon is the world leader in providing software packages that can give speech recognition capabilities to 386-based personal computers.

Meanwhile, at IBM, Jelinek moved from one triumph to another. The first program was executed on an IBM mainframe operating in batch mode on a vocabulary and syntax provided by patent attorneys. The system's algorithms achieved 97 percent accuracy in recognizing one thousand words. If the words were spoken continuously, it worked at 350 times real time; with isolated words (separated by short pauses), it operated at 50 times real time. Again the fast pace of advance in hardware led Jelinek to believe that he could achieve a real-time system within a few years.

The optimism was short-lived. It turned out that the success of the first system was attributable chiefly to the rigid syntax of patent attorneys and to the approach of waiting to the end of a sentence to operate a small batch process on the sound signals. With more flexible grammar, more varied word usage, and more rapid choices, the system collapsed into the same kind of paralysis as the old Hearsay program. Tensions mounted at Yorktown Heights as Jelinek and the team struggled to come up with better software.

On Jelinek's staff, however, was a young British engineer named Ken Davies who suggested that better hardware would redeem the project. More speed would allow them to preserve the purity of information theory, racing through the Markov chains ever faster, without indulging in the crippling complexities of the expert system. The key to more speed was more parallelism. Davis urged them to use a new array processor from Floating Point Systems that could do vector quantizations of the acoustical signals, transforming the words as they were spoken into twenty dimensional clusters of data, suitable for the Shannon algorithms of multi-dimensional geometry. By 1983, IBM exhibited a new system that worked in real time. It used a $500,000 IBM mainframe and three Floating Point array processors, each $200,000, a $1 million mass of computer hardware that would fill an office to recognize five thousand words.

Jelinek continued to improve the statistical algorithms and to compress the information in ever more parsimonious ways. IBM continued to feed its mainframes vast amounts of documentary material to refine and extend the probability statistics on an ever larger vocabulary. But the success celebrated in the ad in 1987 was achieved chiefly by the development of new hardware. As Davies puts it, "The best things are never planned." Working in Switzerland to develop an ideal modem for sending computer signals over telephone lines, an IBM Fellow named Gottfried Ungerboeck developed a digital signal processor that could perform 10 million instructions per second. This speed compared favorably with the Floating Point array systems. Ungerboeck persuaded an engineer working at IBM's plant at Essonnes outside of Paris to make a prototype of this system on silicon in the form of a gate array.

This semi-custom device, designed for a modem that was never built, became the Hermes chip at the heart of the speech recognizer celebrated by those fabulous new IBM lips. Using three boards centered on the Hermes, plus huge amounts of memory and glue logic, the Jelinek team created a voicewriter that could run in an IBM AT personal computer and an outside box and would cost some $20,000 to build. With a major drive to integrate the signal processors and the glue logic on custom silicon chips, the machine could be reduced in time to fit in the AT itself and cost under $10,000. It would be a product. In 1989, with the cooperation of IBM's custom semiconductor division, the entire recognition processor could be reduced to one chip and produced at first for a few hundred, later for a few score dollars.

It would be an impressive success for the approach of Fred Jelinek and an intriguing speech for the Berliners. To Jelinek, the statistical system realized a democratic ideal. Because the data organizes itself according to its own probable sequences of words, with the word patterns emerging as vector quantizations—clusters of mathematical information—the program represents spontaneous order. The expertise wells up from the information and organizes itself like a soccer team, with no expert rules from the top.

It is the pattern of Jelinek's tea table itself. He does not dictate and IBM's hierarchy did not dictate. Each part of his team did its job with relatively little interference. Jelinek knows little about the hardware in the system and the hardware experts know relatively little about the compression and clustering algorithms at the foundation of the software. Neither group pays much attention to the arduous details of the endless database analysis of written materials. Each group does its own job its own way. Jelinek sees it as a moral triumph as well as a practical fulfillment of information theory and the purity and power of Shannon's mathematics. It should be a triumph for IBM as well.

At IBM, however, things are rarely so simple. Occasionally a project at IBM acquires the high priority of the personal computer effort, developed and launched in three years by an intrapreneurial team in Boca Raton, Florida. But in general IBM's internal politics and bureaucracy make it impossible to get a major new product to market in less than four years after development is completed. Even as late as 1989, the Jelinek effort was still chiefly a research project, still unable to achieve high priority in the company. Far from getting a top-priority drive to build cheap and fast custom versions of the glue and processing boards, the Jelinek team was having trouble acquiring the commodity chips that ran its current hardware. Jelinek would make speeches and win prizes. But the initial winner in speech recognition was the unlikely entry from Raymond Kurzweil.

It seemed unlikely to Jelinek because it adopted many of the concepts of artificial intelligence that had climaxed in the paralysis of Raj Reddy's "Hearsay." Kurzweil's machine had experts galore, rules, heuristics, even structural linguistics. It had a new "blackboard" and a new supreme expert to resolve disputes. It used statistics only as part of a seven-expert system. To Jelinek, it was a reactionary device.

Nonetheless, Jelinek was not distressed. In a way, IBM resembled an Eastern European economy, entirely dependent for market research and pricing on entrepreneurs in the West. If Kurzweil could demonstrate a market for Voiceworks, Jelinek could more easily win the attention and favor of headquarters than if Kurzweil failed. In a similar way, it was likely that if IBM succeeded, it would increase demand for the Kurzweil machine and accelerate its acceptance by other computer companies. The bitter rivals, in a way, were allies on the microcosmic frontier.

The Kurzweil machine did have some advantages over IBM's. In various forms Voiceworks had existed as a product for three years. It was built of custom and semi-custom devices that reduced its cost far below the level of the IBM prototype. It used less than one third as much memory. And it did not wait to the end of sentences to validate or "firm up" its word choices. The Kurzweil machine operates fully in real time. Kurzweil never looks back.

Still, although Jelinek respects the ingenious shortcuts by which Kurzweil and Dragon got to market with speech recognizers, the IBM researcher remains serenely confident that his approach—the most purely digital approach to an analog challenge—will win in the end. The IBM machine depends on blazingly fast processing of huge amounts of memory defining the probabilities of an immense number of word sequences accumulated over years of computation. Living in the midst of the most advanced digital computing laboratory in the world, the IBM researchers know that such computer power will be available at a reasonable price on a single chip relatively soon and that memory technology improves at its compound rate of 30 percent annually. In just a few years, the emergence of affordable machines performing gigaflops, on gigabytes of memory, containing statistics from yet more millions of terabytes of documents will make robust and reliable continuous speech recognition a reality. And only IBM will have that immense database of probabilities.

Carver Mead, however, thinks that IBM will merely bind itself in endless Markov chains—that it will take more than the brute force of statistics reliably to recognize words. Let us visit a smaller laboratory on the frontiers of a different approach.

CHAPTER
22

The Neural Computer

The eminent professor of computer science, like many in his trade, seemed to be showing off his connections in industry. Projecting the design of a massively parallel processor on the screen, he said, "Now I've been up in Silicon Valley, talking to the guy who made this thing and. . . ."

Why was this class laughing? They were in on the secrets of a new movement in computing and they viewed their professor and his connections with humorous admiration. No matter how brilliant the creations of entrepreneurs, the ultimate seeds of industry are usually planted in universities, in classes of playfully conspiratorial students and teachers.

In the mid-1980s, even as Kurzweil launched his climactic products, such a new movement of ideas was gaining momentum in classrooms and laboratories, and was sprouting up in new firms near the most seminal campuses. In the course of time, this movement could transform the technology and economics of the quantum era.

Throughout the history of technology, innovations have tended to spring from such new sources, from scruffy and obscure professors and graduate students, entrepreneurs and engineers, who missed the bus to an earlier bonanza. Usually callow and obsessed like the young Shockley, Kurzweil, or Mead, the revolutionaries have defied their elders and imposed their ideas on a resisting world by the force of willful intellect and enterprise.

In the midst of what was perhaps the most promising ferment in the late 1980s, however, could be found a familiar figure. Now in his fifties and at the very pinnacle of his reputation as the prophet both of VLSI and of silicon compilation, Carver Mead was embarking on an almost entirely new field of study and innovation: a new course, a new book, a new set of inventions that paralleled Kurzweil's products in both speech recognition and music synthesis, even a new company called Synaptics. And his students were laughing along happily with him in his adventure in innovation.

It is always difficult to define the moment when a set of academic ideas begets an industrial genesis. But the balmy air of the Caltech campus was fragrant with possibilities in early 1981, when Richard Feynman called Mead and suggested that they join with John Hopfield in a new course on "The Physics of Computation." Feynman was the most charismatic scientist of his era. Hopfield was a physicist with a mathematical bent who had pioneered the theory of neural networks, digitally simulating the behavior of the human brain and launching a new field of study on campuses across the country. Mead was a polymathic theorist, who offered a rare ability to embody and test his ideas in novel chip and computer designs.

Wielding formidable egos and ambitions, the three Caltech luminaries shared only a reputation for genius and a suspicion that conventional computer theory—the set of ideas at the heart of the quantum era—was breaking down. They sought new concepts and metaphors to give new coherence and direction to the industry.

Feynman proposed to discuss quantum devices, using quantum resonance effects and other phenomena as computing elements or primitives. Bad health and other distractions (such as a best-selling autobiography and leadership in the Shuttle Disaster inquiry) prevented Feynman himself from pursuing the idea very far. But quantum devices have become a focus of lively investigation at Caltech, MIT, Texas Instruments, Bell Labs, and other institutions in the United States and Japan.

Returning like Mead to the inspiration of Ludwig Boltzmann, Hopfield had introduced an ingenious and mathematically fruitful physical model of neural networks as energy systems seeking equilibrium. Appealing both to physicists and mathematicians, Hopfield's concept has spurred kindred researches across the country and around the world.

Mead, however, slipped steadily away from the physics agenda. Teaching the Feynman course for three years between 1982 and 1984, he found himself instead focusing on the *biology* of computation. As he had worked little in biology since his association with Delbrück, he had to learn it anew. He consulted Caltech biologists; he raided their classes for students who could help him learn the field. He worked closely with a graduate student named Misha Mahowald. Returning to Feynman's class in April 1986, Mead was ready to explain his new approach in a lecture entitled "Biological Structures as an Inspiration for Computation":

"I would maintain that biology has always been the inspiration for computation. In fact, as near as I can tell, it's the only computational metaphor we have." He pointed out that Alan Turing's definitive paper of the 1930s, which introduced the basic concepts of serial processing, found its inspiration in the way mathematicians prove theorems. So did the computer model of John von Neumann.

"Now, at Caltech, I have to remind you that even mathematicians are biological entities." Copying the thought processes of mathematicians, engineers produced the digital computer and founded an industry. Nonetheless, Mead believed that this foundation is beginning to crack. For new inspiration, he sought "much simpler biological systems than mathematicians, since the way mathematicians work, nobody can understand them anymore, including mathematicians; in fact, they've even proved theorems about that." Indeed, Gödel's famous proof had shown that no mathematical scheme could entirely escape self-contradiction. If mathematics was widely regarded to be in crisis, so—whether we knew it or not—was the digital computer.

Using the techniques of digital mathematics, computers could never achieve perception. Perception was less a logical than an analogical process. The perceiver did not abstract from reality; he merged with it. He did not use numbers or other symbolic logic; he used eyes and ears. Mathematicians take eyes and ears for granted, but computer scientists cannot.

As an alternative biological model of computation, Mead suggested the human brain and showed a slide of the human neural system. It was this massively parallel computing device that provided the occasion for the classroom guffaws. "Now I've been up in Silicon Valley, talking to the guy who made this thing," he said, "and *he* told me.

. . ." But whether it was God or Gordon Moore whom Mead had been consulting, there was nothing original in itself about treating the brain as a computational device. The entire movement of artificial intelligence and its academic counterpart, cognitive science, had sprung from a similar approach.

In the course of increasing the cost-effectiveness of computing by a factor of 10 million since World War II, however, scientists and engineers had adopted a materialist fallacy. They began imagining that the brain itself was reducible to a physiological machine. They saw digital computers as a possible model of human intelligence. By making ever larger and faster machines, they supposed that they might begin to excel the human mind. They declared their devices "faster than thought," their memories more capacious than brains, their wires and buses more efficient than nerves. As long as the problem was essentially mathematical and defined in advance, they were right.

Nonetheless, the dreams and goals of artificial intelligence remained far beyond reach. Marvin Minsky, Kurzweil's teacher and perhaps the leading theorist in the field, finally concluded that 90 percent of human intelligence is acquired during the first year of life. It turned out that the human eye could perform more image processing than all the supercomputers in the world put together. It became clear that in a crucial way the supercomputer itself was obsolete and that by focusing on the material totems of computation—ever faster central processing units (CPUs) and ever more dense and speedy dynamic random access memory chips (DRAMs)—the computer industry would fail to fulfill its own promise.

The AI movement dedicated itself to transcending these limitations. The problem was that AI theorists remained preoccupied with symbolic logic, continued to see the brain as a "physical symbol system." Still, this effort bore considerable fruit. Simulating cognitive functions, AI exponents developed thousands of software programs and scores of computer prototypes. In laboratories across the country and in new computer firms, engineers created general-purpose parallel machines ostensibly patterned on the human mind.

Mead believes that parallel processing is inherently application-specific. Thus he doubts that any one architecture will come anywhere near repeating the success of the Von Neumann model. But as a longtime advocate of parallelism, he saw the new machines as an important advance. In particular, he was impressed by Daniel Hillis's Thinking Machine—created in Kendall Square next to MIT—which used some 64,000 processors in parallel. In addition, N-Cube, a firm started by Mead's former student, Steven Colley, was leading the industry in selling parallel minisupercomputers that performed up to five hundred megaflops. Mead's colleague Charles Seitz had developed an ingenious hypercube model for Ametek. For many tasks these machines promised a several thousandfold increase in efficiency over a serial processing architecture. But, Mead believed, their ultimate promise was limited by a misunderstanding of the nature of thought.

As in the mathematical models of Turing and Von Neumann, thought was still conceived as essentially digital and syllogistic. Turning from the mind of the mathematician as a biological model, many AI theorists adopted the even more bizarre model of the mind of the positivist logical philosopher. Creating intelligent machines came to mean manipulating symbols to create "general problem solvers," in which every cog turns according to the rules of logic. But as Anya Hurlburt and Tomaso Poggio of MIT pointed out, "A general problem solver could never solve the problem of driving a car." Most of the brain cells of the human being are devoted not to cognition, on the mathematical model, but to perception. In order for computers to become fully useful and accessible, they would have to gain the power to perceive as well as compute.

To achieve this goal, Mead reversed the usual approach of artificial intelligence. Rather than treating the computer as a possible model for a better brain, he analyzed the brain as a model for better computers. Taking the human model truly seriously, he concluded that those computers would have to be massively parallel and application-specific, consisting mostly of components specialized for particular tasks. Furthermore—and this was the revolutionary idea—the new machine would have to be heavily analog.

These ideas provoked less friendly classroom guffaws when presented by a Mead admirer at a conference at MIT late in 1986. The MIT professors were advocating optical computers that used photonic transistors directly translated from digital silicon. They laughed out loud at Mead's idea that the absence of digital optics in the brain —the world's only fully successful image processor—cast serious doubt on the prospects for optical image processing with digital transistors.

At MIT and throughout the computing establishment, it went without saying that a return to analog was reactionary. Regardless of what went on in the brain—a rather slow and low-powered system in any case—establishment scholars saw digital computers as the supreme achievement of their science. To them, the proposal for an analog computer based on natural models was exactly comparable to a proposal to revive the Pony Express on the grounds that God did not make airplanes.

No matter how much the computer could excel human beings in multiplying or storing numbers, however, it was vastly inferior in processing images and sounds. In fact, it takes the most advanced supercomputer hours or even days to accomplish feats of seeing and hearing that are routinely performed in real time by human eyes and ears. Most human perception, moreover, is completely beyond the powers of a computer.

To Mead, the industry's obsession with increasing the speed of CPUs by another MIPS or megahertz, or reducing the size of lithographic geometries by another fifth of a micron, or expanding the capacity of memory chips by another factor of four, could yield only small gains compared to the achievement of genuine perceptual powers. Perception, he believed, was fundamental to learning and to thought. Regardless of how fast a machine can manipulate symbols, unless it can interact perceptively with its environment, it will have to wait around for humans to identify the content of the symbols and verify the validity of the results.

No thought process or computation can be faster or more reliable than the flow of facts that fuel it. As AI pioneer Terry Winograd has written with Fernando Flores: "Computers neither consider nor generate facts. They manipulate symbolic representations that some person generated in the belief that they corresponded to facts."

Perhaps the most important consequence of this limitation is the software bottleneck. Von Neumann computers will never be able to transcend the stalemate in software design. In the late 1980s, for example, a two-year delay at Microsoft in finishing a new operating system for Intel's 386 microprocessor left much of the personal computer industry with 4-MIPS machines, commanding a potential 4-gigabyte memory address space, lurching along like Intel engineers in their Ferraris during rush hour on Route 101.

In order to escape this cul-de-sac of high-powered speed—with systems that constantly hurry up to wait—the computer would have to learn from its environment and adapt to it. Learning entails perception. By imitating the perceptual and adaptive powers of biological systems, the industry could be relieved of some of the burden of constant software revision.

Operating thousands of times more slowly than transistors, neurons are the ultimate slow and low device. But somehow they can perceive and learn many million times faster. To achieve these quantum speeds, computer science would once again have to eschew high-powered speed and analyze the paradox of slow and low.

Mead's early mentor Delbrück had presented the alternative agenda, "the as yet totally undeveloped but absolutely central science: transducer physiology, the study of the conversion of the outside signal to its first 'interesting' output."

For all the instrumental usefulness of digital machines, so it seemed to Mead, they could not really think, because they divorced themselves from this "absolutely central science" of sensory physiology, the deeper dimensions of input processing. As Hurburt and Poggio put it, "Vision is more than a sense; it is an intelligence." By restricting vision to the role of a separate input faculty for computers, conventional AI cripples both vision and computing. Vision is reduced to a pixel transmission device and the computer is reduced to a logical processor. Neither can achieve the integration of perceptual and logical powers that characterizes human thought.

By rendering all inputs as numerical and discrete objects, digital machines neglect the interplay between subject and object and they ignore the endless webs of interconnection among objects. All facts presume a framework of interpretation beyond the facts and irreducible to them. Once an object has been rendered in digital language—defined, structured, measured, and sequestered—most of the real thinking has already been done.

The world consists of protean webs of interrelated objects, richly mediated by the experience of the observer. The digital computer rips them from this subjective womb, cuts the organic whole into inorganic parts, reduces them to mathematical symbols, and then shuffles them at megahertz speeds as if they were discrete entities. In the process, meanings and analogies bleed away everywhere.

The biological metaphor suggests an effort to stop all that invisible bleeding. Human thought is not the manipulation of dead symbols. Crucial to the thinking process is the conceiving of symbols, which begins with perception, an analog mode. The brain has no inherent knowledge of Boolean algebra, but it commands an endlessly rich array of physical processes. The fundamental problem of the digital computer is that, unlike the human brain, it begins by casting aside all its analogical potential. By moving instantly to Boolean logic, it eschews the computational resources inherent in the physics of the system, those precious "Analogies" that "know all the secrets of Nature."

As Misha Mahowald put it: "Take the electron. It took me six years of a Caltech education to understand the Schrödinger equations about the wave behavior of electrons. But the electron already *knows* the Schrödinger equation. It doesn't even think it's hard. It computes Schrödinger equations day and night." Similarly, the cochlea in the inner ear constantly performs what amount to fast Fourier transforms of sound pressure waves in real time. Using such physical processes as a mirror for the world, the human brain sees and senses by analogy. Essentially rejecting this rich fabric of analogies, the digital computer has no way to relate in depth to the real world.

As long as the computer is being used to manipulate numbers or other symbols, this limitation doesn't matter. But after thirty years of progress in number crunching and symbol shuffling, the computer is now beginning to come up against some of the hard problems of Minsky's first year of life, and grown-up computer scientists are howling like babies. Whether a robot trying to find a wrench, a continuous voice recognizer searching for a word in a spectrogram, or a strategic defense system looking for a missile in the clutter of sky, the most crucial goals of current research and development depend not on cognition but on perception. In finding the significance of images, whether auditory or visual, the digital computer is a preposterous kludge.

By directly converting sound waves into electrical sine waves, the cochlea essentially performs a Fourier analysis that a digital computer can achieve only after a complex analog-to-digital conversion and a Fourier transform in a digital signal processor, followed by a digital-to-analog conversion. This process takes several expensive microchips full of analog and digital circuitry comprising millions of transistors. The cochlea does it all in one analog step. It does not have to develop a complex algorithm for Fourier analysis; it already knows Fourier analysis. It doesn't even think it's hard.

In an image processor, the scene begins as an analog event: photons streaming into the lens in parallel as combination of waves and particles constantly varying in intensity. But the computer immediately uses an array of analog-to-digital converters to change these light intensities into a stream of bits relating to pixel positions on the screen. This process effectively captures the two-dimensional image. But the problem of perception is not getting images; the camera is fine at capturing images and the computer can store them as a pattern of dots. The problem is to find anything important in the image, to identify meaningful events, to seek out significant change. The problem is to see, which is a process of thinking.

Current systems for machine vision used in robots are based on television cameras that take in images at a pace of thirty frames per second, convert them into digital form, and then compare them with a template containing a desired pattern—for example, a wrench. Beyond the ability to compare one frame to an ideal pattern or to a previous frame stored in memory, the systems are completely dumb. To the computer, the picture is merely a mass of digital bits; the machine has no way to draw back and gain perspective on larger shapes or movements. The computer must operate at tremendous speed just to present the images, let alone understand them. If the scene is complicated by an unexpected shape, a flash of light, an interplay of moving objects, the machine will fail entirely.

Suggestive of the problem is the phenomenon known as aliasing: the reversal of direction in propellers, wagon wheels, and other quickly rotating objects. In sampling a scene in movement, the system sacrifices tremendous amounts of data. Then, by complex digital image processing—by averaging intensities between two frames and by other artificial techniques using hundreds of chips—the machine proceeds to reconstruct some of the information it threw away. But most of the local analog variation is never retrieved.

Human vision also begins with two-dimensional light patterns transmitted from a retina with 100 million photoreceptors, each covering a particular point in the view. If the brain was a camera, you could peel off Polaroids behind the retina. If the brain was a digital computer, it would have to process these pixel inputs at a pace of hundreds of millions a second, entailing billions of multiplications and additions. The math could probably be performed by a million PCs operating in parallel. But still the full image would slip away because too much information is lost in moving from three dimensions to two. As Kurzweil suggested, billions of PCs would actually be needed. But the brain is not a camera or digital computer; it has found a better way, the analog way.

For several years Mead has been trying to emulate this better way. He has been seeking a method for processing the image the way the eye does, in analog form. Almost everyone in the industry knew this was futile. Analog devices are too large and crude and slow; photoreceptors are too cumbersome to couple to the processing devices; there are no adequate analog memories to store the images; the analog devices would hog power and seethe with heat. To make a useful system that could operate in real time would require not merely small silicon chips connected with wires but entire wafers interconnected on the silicon itself.

Several companies, including Eugene Amdahl's famous Trilogy, had already wasted billions of dollars, using far smaller and more efficient digital technologies, without producing useful wafer-scale devices. Heat differences across a wafer warped the silicon, and the defects inevitable in any large wafer forced so much redundancy that the entire system became uneconomic.

Yet that day in Feynman's class room in 1986, Mead presented to the class a radical innovation. Drastically expanding the microcosm, he launched a new era of analog wafer-scale electronics and a new neural technology for processing images. Until Mead's breakthrough, microcosmic technology was restricted to tiny chips the size of fingernails. Beyond these millimeter dimensions, engineers had to leave the microcosm, using expensive wires to communicate with other chips. After printing hundreds of dice (unpackaged chips) all conveniently together on a wafer, engineers had laboriously to scribe them in preparation for cutting them apart with diamond saws, follow five more exacting steps before encapsulating them in their plastic packages, all so they could be reattached by wires to other similarly packaged dice. For this assembly operation, the dice are usually flown in wafer form to Bandung or Manila, Seoul or Malacca, halfway around the world from where the wafer was fabricated.

Integrating the circuits on the wafer itself—wafer-scale integration—obviates this entire expensive and cumbersome flight from the microcosm to islands in Asia. Wafers are between five and eight inches in diameter and can hold the equivalent of hundreds of chips. On such wafers, Mead estimates that within a decade as many as 10 billion transistors can be implanted and interconnected without ever leaving the microcosm. Such a wafer-scale device could hold the equivalent of scores of supercomputer CPUs or gigabytes of memory. It would make possible, once again, with exponentially heightened capabilities, what Intel memorably called in 1972 "a new era of integrated electronics." Most of all, it would enable the creation of analog machines that could actually perceive and learn.

Mead's system was not merely a hypothesis; he offered a prototype that actually worked. Based on the actual electronic design of the retina, the Mead device could follow a rotating fan without aliasing and could adapt to changing intensities of light. It was a dramatic first step toward the creation of an effective real-time image processor on monolithic silicon.

To the argument that such a system would require exotic designs and materials, Mead offered his device in plain bulk CMOS silicon. To the argument that it would use too much power, he ran the CMOS transistors—both the N- and the P-type devices—at subthreshold voltages like the micropower systems in digital watches. Because the structure was massively parallel, it offered natural redundancy; like the human brain, it could operate even if a portion of the cells were damaged.

It was this redundancy and low power that made possible the implementation of the devices in wafer-scale dimensions. Creating a technology with a high latitude for error, he had used design ingenuity to overcome the remorseless constraints, restrictive manufacturing tolerances, and rigorous quality controls of submicron chipmaking. Moving toward wafer-scale and using the most routine CMOS technology, he had expanded the potential domains of microcosmic design by a factor of hundreds.

Even the photoreceptors were accommodated without any complicated addition to the CMOS process. Mead's solution to the photoreceptor puzzle was typical of his ingenuity. Every CMOS designer faces a fundamental problem. Between every cell's two transistors—the complementary N-type and P-type devices—is a potential bipolar transistor. This is called the latchup, or parasitic device. Every CMOS process has to figure out some way to neutralize it. But like Bob Widlar before him—and thousands of flora and fauna before Widlar—Mead decided not to kill the parasite but put it to work.

Rather than neutralizing the latchup transistor, Mead enlarged and enhanced it. Responding to light on its p-n junction, the latchup transistor became a superb photoreceptor. With an amplification factor of 1,000, it was far better than the usual photodiodes that would have had to be added to the machine.

Taking analog electronics out of its intermediate position between the macrocosm and the microcosm, Mead had created for the first time a fully microcosmic analog technology. Creating whole wafers full of thousands of analog devices using far less power than digital switches, Mead actually brought analog more fully into the quantum economy than the most advanced digital integrated circuits.

Yet this slow and low, massively parallel analog system, cobbled together on a few dollars' worth of bulk CMOS, could outperform supercomputers costing $20 million dollars or more in identifying motion. With suitable refinements, it could offer an architecture for solving some of the most intractable problems of image processing, from SDI to machine vision. At a time when some experts were lamenting an alleged "graying of technology," Mead had shown the great power of design innovation.

The retina chip, however, was only one of the products of Carver Mead's lab. In a first step toward solving the problem of continuous speech recognition, he had used similar analog processors to create an auditory system based on the cochlea and further layers of neural "wetware" beyond the human ear. Mead believes that efforts by Kurzweil and IBM can never go much beyond discrete words without a more complex model of hearing than these machines possess.

Starting with an evolutionary metaphor, he contends that the key to human survival was identifying the location of sound: the directional source of the crackling footsteps or other signs of predatory beasts. Because this faculty necessarily progressed to the highest level of refinement—"the animals that had it ate the ones that didn't"— Mead suggests that the directional sense is the foundation of the hearing process. Yet models such as Kurzweil's and IBM's, essentially based on the cochlea alone, fail to capture the multi-dimensional cues of the localization function in the ear. The ear does indeed perform a massive Fourier analysis, but this is only the beginning of the hearing function.

Biologists believe that localization is performed in three ways. For horizontal cues, the brain measures the time disparity between the signals in the two ears and it computes the interaural spectral shift caused by the head blocking out high frequencies in sounds from its other side. This computation is performed by an organ picturesquely named the "medial superior olive." Vertical cues derive from echoes as sound bounces off the big and little flaps on the ear. Biologists have modeled all these processes, and Mead took inspiration from them to create an aural computer.

Like the retinal chip, the aural device consists of ordinary CMOS circuits that will cost only a few dollars to manufacture. Yet it is a massively parallel analog machine that performs computations that confound a supercomputer. It represents a tentative first step toward reproducing the kind of hugely complex physical primitives which the human ear uses as mirrors for sound.

From the eye and the ear, Mead proceeded to the problem of music synthesis that Kurzweil had largely mastered with his 250 synthesizer. Once again the biological model of computation, with its density of processors and localized memory, yielded massive benefits over the usual Von Neumann architectures.

The most powerful existing music synthesizers, based on sound samples from instruments, require huge amounts of memory and wire. Indeed, even with ingenious compression algorithms, memory comprises close to half the costs of Kurzweil's machines. Rather than storing and synthesizing sounds, Mead decided instead to synthesize the instruments themselves. What differentiates a violin from a loud-speaker playing the same spectrum of frequencies is the specific reso-nances of the violin body. By creating mathematical models of the physics of the devices, Mead thought it would be possible to achieve greater verisimilitude with less silicon than most current synthesizers and capture a more complex interplay of musical effects.

Transforming wave forms of different intensities into unified im-ages, the synthesizer problem resembles the challenge of the retina. Any sound can be reduced to a fundamental frequency and a set of "partial" or sympathetic frequencies. These frequencies act within an "envelope" of amplitudes affecting volume and timbre over time (the attack and decay of the sound). Synthesizers can generate the frequen-cies and partials separately, multiply them by their volume envelopes, and add them all together to get the desired result. Called "additive synthesis," it is the essence of all direct creation of musical sounds.

The problem is that the process requires such immense amounts of computation that it cannot be done in real time without supercom-puters. Composers can create music with full richness and accuracy at the computer. But it may take days to produce twenty minutes of sound. Live performance is impossible.

To overcome this problem, synthesizers use many ingenious tactics, such as FM synthesis, invented by Richard Chowning of Stanford, which has yielded more money for Stanford than any other patent. But like the memory and algorithms of compression and extrapolation used in the Kurzweil system, FM synthesis simulates sounds rather than reproduces instruments.

For a new inspiration, Mead and his students, led by John Wawrzy-nek, returned to the early analog synthesizers. These machines used an analog oscillator for each frequency and "partial," and patch cords to interconnect them for a particular sound. Slow and inaccurate, the analog synthesizer was far inferior to modern digital equipment and had been relegated to oblivion by scientists in the forefront of the field. Mead saw, however, that the machine was performing Fourier synthesis with linear differential equations. Moving into the microcosm, he knew he could produce scores of such processors on one chip and interconnect them not with wire patches but with programmable flipflops. Duplicating entire analog synthesizers on single chips—produceable for a few dollars apiece—he could show how entire orchestras could be contrived with a few score chips.

In this case, he would not be using analog devices themselves. Music does not allow the imprecision of existing analog technology. But his approach was analogical in nature. Rather than converting sound waves into digital forms, storing them in memory, and then extrapolating from them, he created a physical analog of the instrument. Using programmable processors, he could readily connect his digital devices to create analogs of cylinders, vibrating strings, marimba spars, chimes, and other generic technologies. The cylinders, for example, could simulate flutes, recorders, or organ pipes, while the strings could evoke pianos and string instruments.

Reproducing the harmonic content of notes is relatively easy and results in pure sounds that are readily identified as electronic. The hard part is simulating the attacks, the noisy non-linear effects of blowing into a mouthpiece or striking a string. But by combining pure sounds with attack resonators and noise effects, Mead and his students succeeded in imitating the complex timbres of real instruments.

Containing sixty-four processors with reconfigurable interconnections, each chip can capture one musical instrument voice, whether a flute or a single piano wire. Played through a piano keyboard, each of these chips can generate musical sounds in real time, when a Digital Equipment VAX minicomputer would take some thirty minutes for a second of sound.

First unveiled in 1984 at the International Conference of Computer Music at the Pompidou Center in Paris, the chimes alone caused a sensation. Overcoming a key problem of synthesis—the interplay of sound waves over time—the first vibrations continued for many seconds and mingled realistically with later notes; harsh, noisy attacks mellowed into pure tones. Extending the repertory, the goal of Caltech student musicians and engineers is to produce a cheap but superior piano, with each key activating one processor chip.

These chips, each equivalent to six hundred minicomputers, are currently fabricated in 3-micron geometries far short of the state of the art. Potentially manufacturable in volume for a few dollars apiece, the devices illustrate the powerful effects of design innovation using parallel architectures and analog models. Many analysts bewail the future of the U.S. semiconductor industry because of somewhat lower yields in manufacturing commodity memory chips. But Carver Mead has demonstrated that design innovation can bring results thousands of times more significant than manufacturing refinements.

To achieve some of these advances, Mead has joined with Federico Faggin, of silicon gate and microprocessor fame, to help launch a new company called Synaptics. With enthusiastic support from venture capitalists, the firm is using neural network technology to address problems of machine vision, continuous speech recognition, and intelligent memory that seem increasingly difficult and expensive on digital systems. Their first chips were celebrated—along with superconductivity—in *Business Week*'s annual review as one of the five most important innovations of 1987. Within five to ten years, Faggin predicts, "We will be able to build neural systems chips the size of a whole wafer and containing over one billion components." With such devices, which can be made without major process advances, analog electronics will leap ahead of digital in the descent into the microcosm.

A billion components on one piece of silicon—or even the 10 billion predicted by Mead—will radically increase the perceptual powers of the computer. But both Mead and Faggin warn against the notion of artificial brains. The brain after all contains some 10 to the 15th synapses, each equivalent to ten silicon components, for a complexity at least seven orders of magnitude greater than any feasible computer. But even these measures bespeak the materialist superstition. The human mind operates by principles beyond current scientific knowledge. The purpose of neural technology is not to usurp the mind but to serve it with the first truly intelligent machines.

Carver Mead, however, is primarily a teacher. His greatest contribution in the past was not developing a new theory of VLSI design but reducing it to intelligible form in his book with Lynn Conway. The book introduced an entire generation of students to his concepts and changed them from an academic theory to a widespread industrial practice. No technology can finally prevail unless it attracts large numbers of students. Ralph Gomory, IBM's chief scientist, has said that he canceled his Josephson Junction project in superconductive transistors not so much because of inherent weaknesses in the technology as because of the tiny team of engineers qualified to work on it. "It was 70 of us against 70,000 silicon guys," he said.

The movement toward silicon compilation essentially began with far fewer than seventy students. It transformed the industry when scores of thousands of students had studied the Mead-Conway text and used it to design chips. Mead hopes that his new book, *Analog VLSI and Neural Systems,* will have an even greater impact. It offers a simple and practical program for translating biological insights into working electronic devices. By producing scores of analog chips that perform at levels far beyond conventional technology, Mead's students have already proved that the method works. Working with Mahowald and others to contrive a common language for electronic and biological processes, he succeeded in overcoming the separation of realms and languages that Norbert Wiener denounced in *Cybernetics*. A biologist with no experience in engineering, Mahowald has designed several chips that outperform supercomputers in specific tasks.

Like the Mead-Conway volume, *Analog VLSI and Neural Systems* was taught at MIT in manuscript to forty students a year before publication. Analog computing, apparently, no longer seems so reactionary in Cambridge. Like the man suddenly startled to discover he had been speaking prose, the MIT professors had been doing it all their lives.

In a speech reprinted in the November 1986 issue of Caltech's *Alumni* magazine, William J. Perry, president of Hambrecht & Quist Technology Partners, predicted a hundredfold rise in the cost-effectiveness of computing over the next decade. Formerly Undersecretary of Defense for Research and Engineering and now one of the nation's leading venture capitalists, Perry would seem to have a superb vantage point for appraising the possibilities of the technology. His prophecy seemed optimistic to most of his audience of research directors from Fortune 500 companies. Yet his estimates seemed drastically pessimistic from the vantage point of Carver Mead's laboratory hidden away in the university's service building just 100 yards from Perry's podium.

With silicon compiler technology rapidly spreading through America's engineering community, Carver Mead predicts a thousandfold rise in the number of designs to be produced by the U.S. semiconductor industry. With increasing inspiration from the biological model, each of these application-specific designs is likely to achieve levels of efficiency hundreds of times the usual commodity chips. Mead suggests that the next decade will see a rise in computer efficiencies tens of thousands of times greater than Perry believes and that these gains will change the world. His own inventions confirm his vision.

Perhaps processing speed was not the crux of the problem; perhaps memories did not have to be dumb; perhaps you could throw away the rigid computer clock. With such insights, the industry began making tentative first steps into Minsky's first year of intelligent life. Initially, the results would be unimpressive; the computer still would not be able to understand words or see images or recognize continuous speech. The computer industry had thrived doing well what humans do badly; AI would have to cope by doing badly what humans do well. But the new spirit of humility opened new vistas of possibility.

With the keyboard inputs collapsing into the microcosm and replaced by tiny sensors and speech recognizers, the computer at last could be reduced as small as needed. Merging with the world through analog devices, it could bring more and more of human life into the realms of the microcosm. In the process, the computer would even usurp that most powerful and conspicuous of electronic devices, the television set.

CHAPTER

23

The Death of Television

In pressing toward the frontiers of the human mind, Mead's ventures in analog computing opened new horizons for the twenty-first century. But in the usual rhythm of the microcosm, the same advances in the silicon art used in Mead's analog devices quickened progress still more on the digital side. Gains in the sensory powers of computers—giving them more immediate and accurate data to process—would unleash still more powerful feats of digital processing.

In the early years, the computer itself had been entirely analog. But the inevitable flaws of analog, which compounded in long computations, made a deadly contrast with the flawlessly extendable methods of digital. You might make an adder with two wires that would take a thousand digital wires and switches to duplicate. But digital did not deteriorate with complexity while analog became a swamp. In the end, the superior accuracy of digital had allowed digital to take over almost the entire computer.

The result was a hypertrophy of digital powers and a relative withering of analog connections to the world. Like Robert Widlar in the early era, Mead, Kurzweil, and others set out in the 1980s to redress the balance. Giving the computer new links and senses, they could provide the analog grounding for still more powerful digital computers in the decades ahead.

Meanwhile, year after year, the armies of digital engineers moved ever deeper into the microcosm, overthrowing all the remaining bastions of conventional or macrocosmic analog. Even the most obvious of analog processes—the reproduction of sound waves—became increasingly digital. The compact audio digital disk emerged as the most rapidly rising consumer product since World War II. Digital control systems appeared in automobile engines, in microwave ovens, in refrigerators, in telephones. Digital engineers looked hungrily at every remaining analog socket or system. Digital tape and digital thermometers entered the market. Only in one ubiquitous machine was the digital tide held back.

That machine was the imperial TV, running in the average American household some four hours every day. Although digital features were added from time to time, the television set remained doggedly an analog device. TV had to be analog because broadcast signals use the scarce resources of the radio frequency spectrum (for copper cables) and digital video takes between seven and ten times more spectrum space than analog. With scores of channels of high-quality digital television, there would be no room left in the spectrum for cellular phones or CB radios or any other mobile technologies.

Television thus was forced to accept the efficient inaccuracies of analog rather than the prodigal precision of digital. But no one seemed to see a problem. Television seemed impregnable. Most experts assumed that TV—with its increasingly formidable analog electronics and digital enhancements—could forever resist the movement into the digital microcosm. To extend the future of broadcast television into the twenty-first century, the Japanese invested as much as $2 billion in developing new analog codes and digital cosmetics for analog video. They called it high-definition television (HDTV).

As a technology, however, television is dead. It does not matter what kinds of cosmetics are applied to its essential analog features. The life and death of a technology does not depend on its pervasiveness or its popularity. It depends on its prospects. How it fits with the directions of complementary technologies and how much improvement it offers. How fast it is advancing in comparison with potential substitutes. By those standards, television is as dead as the vacuum tube or any other declining technology. In fact, TV may be the last vacuum tube.

If TV is dead, HDTV is stillborn. How could this be so? With its large screen of nearly filmlike clarity and proportions, high-definition television seems a superb product that could survive deep into the next century. Influential observers such as engineer-Congressman Don Ritter called this new form of television "one of the most important inventions of the twentieth century." Others called it the "crown jewels" of new technology. Still others assert that HDTV will determine whether the United States has a future in electronics.

This is the way the relatives always talk in the presence of dead technologies. This is the way politicians always talk in the presence of influential constituents and petitioners. Dead technologies are easy to describe and dramatize. They do not change much. They employ lots of worthy people and influential lobbyists. But at the very moment HDTV became the cynosure of the media and the politicians, it was being blown away by the power of the computer.

In the future, says Nicholas Negroponte, head of MIT's Media Lab, "Your TV set will probably have 50 megabytes of random access memory and run at 40 to 50 MIPS. It will be basically a Cray computer." Within seven years, progress in parallel processing and silicon compiling will bring down the price of such computing power for video processing to well under $1,000.

The various components of such a computer are already being prepared in the workshops of such U.S. firms as Intel, Apple, IBM, Texas Instruments, Kurzweil Applied Intelligence, Sun, Next, and scores of others. Most of these companies do not see themselves in the television business. But if they pursue the promise of the technology with entrepreneurial verve and vision, they will have a greater impact on the future of TV than Sony or Matsushita.

It is hard to imagine a more bloated and vulnerable technology than the analog TV. In an age when computers will be responsive to voice, touch, joysticks, keyboards, mice, and other devices, television is inherently passive, a couch potato medium. To get minimal effects of control—windowing, zooming, and other manipulation—the analog signals must be digitized. Otherwise analog television is mostly a take-it-or-leave-it system.

This problem cannot be overcome in any system that is compatible with the current television standard, which originated for black and white images in 1940 and was adapted for color in 1953. Containing the color information in a high-frequency subcarrier on top of the television transmission signal, this technology doggedly thwarts interactive manipulation.

As an analog system, a TV—no matter how much improved—

must perform complex electronics of aerial capture, filtering, processing, and synchronizing signals, all just to eliminate the artifacts,
smears, streaks, jitters, color skews, weather spikes, multi-channel interference, and other noise and static of the broadcast transmission. It
also must now do digital conversion and processing to further improve and control the picture. All this processing is applied to films
and programs that already have undergone as many as six generations
of analog production and post-production steps—building up errors
all the way—before the signal is even yielded up to the clutter of the
air.

All this heroic processing and reprocessing, analog and digital, is
aimed to achieve one amazing result: a signal far less clear and accurate
than the signal a computer will have at the outset before it does any
processing at all. The digital video information in a computer can be
raised to any arbitrary level of resolution. The reason computers are
not used today for showing films and other full-motion realistic video
is the enormous cost of the memory and processing speed required to
transmit millions of pixels to a screen thirty times a second. High-
definition television, for example, uses about a million pixels, and each
pixel entails some 12 bits of information. At a rate of 30 frames per
second, this resolution means a total of 360 megabits per second,
requiring memories delivering information to D to A converters at a
pace of 3 nanoseconds per bit to feed the screen with the hue, bright-
ness, and definition of moving images.

With fast memories using parallel access, these rates are perfectly
possible, but not cheap enough for a consumer product. Sun workstations with similar video capabilities were priced near $7,000 in 1989.
Personal computer monitors long could attain this resolution only if,
like the basic Macintosh, they lacked color.

Nonetheless, the writing is on the screen. To eclipse television technology, the digital computer needs only a few more years of learning
curve pricing and a means to deliver digital video signals to the computer. The learning curve is a sure thing. The digital channel awaits
the widespread extension of fiber-optic cables to the home. Scores of
thousands of times more capacious than copper cables, digitally
switched fiber optics will permit a computer to send or receive high-
definition video to or from any address accessible to phone lines anywhere in the world.

This technology exists today at AT&T's Bell Laboratories, at Bell-
core Laboratories of the regional telephone companies, and at many

experimental sites. It will move rapidly through the United States as the phone companies, electric utilities, and others are released from archaic regulations and given a chance to profit from it.

Whether transmitted over fiber-optic cables from anywhere in the world or dispatched from writeable optical disks in the machine itself, the digital data of computers begin at a level of accuracy and reliability which the television set may never attain at all. Moreover, the computer will command a global cornucopia of programming and nearly infinite libraries of data, education, and entertainment, with full interactivity at the behest of the customer, while television will be heavily tied to the lowest common denominator broadcasts of centralized media.

Foreshadowing the death of television is the rush to digital among related technologies. All the most promising storage media—magnetic, electronic, or photonic—are inherently digital. All interactive technologies are inherently digital. All fiber-optic channels are inherently digital. Just as important, all new audio technologies—a crucial complement to vivid video—are increasingly digital.

The conflict between the computer and the television set is not merely a technological issue. If tomorrow's video technology is merely an improved version of the same through-the-air broadcast technology used since the birth of television, the Japanese will continue their domination of consumer electronics. The Japanese have been focusing on advanced television technologies for twenty years and now dominate most of these markets. They are at least a decade ahead in HDTV.

Most analysis focuses on this alleged threat from Japan. But the real danger posed by continued dominance of broadcast television is continued control by politicians everywhere of what the people can see. Currently, for example, they restrict the use of U.S. programs in Europe and the Third World. They fill the air with government propaganda. In the United States, government bodies assign franchises and other monopolies, and regulate the networks and local stations. With a limited number of government-regulated channels, television will necessarily resort to violent and vulgar programming to capture the largest possible markets of passive and impressionable viewers. Defying the customized interests and narrowcast offerings of other media, broadcast television will continue to stultify its audiences by approaching them at the lowest terms of mass appeal rather than summoning their creativity and individuality. Today, politicians around the world are propping up this obsolete system in the name of progress and aiding Japanese dominance in the name of competing with Japan.

For thirty years, however, the power of the microcosm has allowed computer technology to advance thousands, or even tens of thousands, times faster than television technology. The fourfold improvement in resolution offered by Japanese HDTV is impressive only by the standards of the television industry, which makes one step forward every twenty-five years or so and calls it a revolution. Landmarks to date are monochrome television in 1923, color TV in 1950, and the videocassette recorder in 1972. Meanwhile, between 1961 and 1989, the speed of a computer operation increased 230,000 fold. Recent innovations promise a dramatic new acceleration of computer progress during the very period that HDTV will enter the fray.

Far from a revolution, the HDTV system is an extension of current top-down, take-it-or-leave-it television, using program input from a relative handful of government licensed broadcasters (eight channels for all of Japan, for example). It is designed to receive analog signals from satellites, over the air, or down conventional cables. It is inherently passive and receptive.

As management consultant Peter Drucker says, a competitor should never try to catch up. The goal should instead be to move ahead, with a new technology at least ten times better than the old.

Unlike HDTV, the new video computers will easily pass the Drucker tenfold test. They are telecomputers and they soon will have virtually no current television technology in them—no high-voltage vacuum tube CRT with an electron gun spraying serial streams of charge millions of times a second, no high-powered high-frequency antennas, transponders, and gigahertz converters, no processing of analog waves, interference patterns, smears, streaks, and weather spikes at all. The telecomputer of the future will be a solid state machine, tied to fully digital low-powered fiber-optic networks. It will be capable of connecting to any digital database of any sort anywhere in the world. It will be cheaper, more energy-efficient, and far more powerful than any television-based technology.

The United States has five times as many computers in use as Japan and, even per capita, an installed base of computer power nearly three times larger than Japan's. In the late 1980s, there were already some 40 million personal computers up and running in this country—about one for every five TV sets—but the gap was rapidly shrinking.

Computer technology is America's strength and opportunity in consumer electronics. Already available in the United States are video technologies far excelling high-definition television. One is called DVI, for digital video interactive. Now the property of Intel Corporation, it promises to combine Intel minds and memories with Intel images.

When Jack Welch of General Electric sold RCA to the French, he spun off RCA's Sarnoff Labs to Stanford Research Institute. At the Institute, the DVI project attracted the attention of Gordon Moore. When venture capitalists tried to capture the product for far below its worth, Moore told Intel to buy it outright. The dominant company in the last age of computing, Intel by this one step moved toward dominance in the next era as well.

As a result, the most important product at Intel is not its highly hyped 486 and i860 microprocessors but its DVI chipset. Now oriented toward compact disks, this invention is applicable to any digital medium, and it represents a technological advance in video imaging that far surpasses any feature of HDTV.

A critical problem in transforming analog images into digital data is computer storage and transmission of the voluminous bits. With a capacity of some 600 megabytes of data (the equivalent of about 600 big books), compact disks currently find their chief use in high-definition audio, which requires about a hundred times less data than video. For showing video images over a digital system, CDs are far too small and slow. (The last generation of laser video disks were mostly analog). With existing technology, a compact disk could hold less than a minute of full-motion digital video, and would take some two hours to show it. The system is far from a full entertainment medium.

The Sarnoff Labs DVI breakthrough transforms the technology. Working under the direction of Sarnoff Labs' Arthur Kaiman and DVI inventor Lawrence Ryan, Sarnoff engineers startled Microsoft's CD-ROM conference in March 1987 by announcing and demonstrating three breakthroughs: (1) a digital compression system that can effect a 100 to 1 reduction of audio-visual data, allowing storage of 72 minutes of sound and full-motion video on one CD; (2) a super-fast playback technology that can produce sound and full-motion video in real time; and (3) a fully interactive graphics controller that offers a dazzling array of video effects for games and educational uses. Since these initial announcements, first Sarnoff and now Intel have been rapidly improving the technology and hundreds of companies have been developing software for it.

DVI is a dramatic expression of the new balance of power in this new industrial era. For rapid compression of the images, a Von Neumann system proved too slow. Instead, Intel uses a massively parallel system combining sixty-four microprocessors (transputers from Inmos). Intel is also transporting the system to an Intel hypercube parallel computer. Arriving in parallel, video images map well to parallel computer systems. Intel also offers a digitizing board for edit-level video that can achieve compression adequate for pre-production purposes. When a videotape, game, or other program is completed, it can be compressed on the parallel supercomputer. With the rapid improvement under way in digital electronics, however, Intel expects eventually to offer systems for rapid compression on small computers.

The execution of the DVI chipset was also a supreme fulfillment of the Mead paradigm. The two video-processing chips—one to manipulate the images interactively and the other to play them back in real time—are also heavily parallel in architecture. Complex VLSI devices, they were designed in months using a Silicon Compilers Genesil program.

With DVI technology, not only will a feature-length movie soon be held on one CD, but any frame can be randomly accessed and interactively manipulated. For example, in a real estate presentation, the viewer sitting at a screen can enter a 360-degree picture and "walk through" a room; he can change the wallpaper or move and reupholster the furniture or even open shades and look out the window—all on the screen. Or a medical student could perform a simulated surgery; a student pilot could fly a plane over a fully realistic countryside. A tourist could visit a Third World country and view any chosen set of scenes or events without having to drink the water. An art student could walk through the Louvre at any chosen pace, watching short biographies of selected artists or visiting the scenes of particular paintings, all without leaving his home. The viewer of a suitably televised football game—perhaps at his local high school—could record it and then watch the event from any chosen vantage point on or off the field.

Using any digital medium, especially fiber-optic cables and advanced optical disks, DVI will eventually provide smart graphics capability that far transcends anything in the Japanese HDTV system at a much lower potential cost.

What makes DVI such a breakthrough, though, is not just its capabilities, which can be reproduced by multi-million-dollar supercomputers, but its low potential cost. With Intel's microchip expertise, the entire DVI system will soon be reduced to just one circuit board and in volume production could soon be sold for far less than a current high-end TV set. Combined with Intel's 50 MIPS microprocessor, the i860, DVI technology means that Negroponte's graphics super-computer, projected to be decades away, will soon exist at Intel—at about one ten-thousandth of the cost of a Cray.

For all its powers, the Intel system offers a mere glimpse of the coming age of telecomputers. Increasingly responsive to the voice, movement, or touch of the viewer through new analog technologies, computers soon will so enrich the power of communications that people will be saved the tedium, expense, and energy waste of conventional travel, whether for work, entertainment, or education. Individuals will truly live in the world and will create their own worlds as individuals in the spirit of microcosmic emancipation.

This dream of a fully digital telecomputer, based on silicon and glass, depends on the continued wiring of America with the necessary cables. The telephone laboratories are in the forefront in developing the network, signal-processing, and digital-switching technologies that will be necessary to deliver full interactive video to American homes. Unfortunately, however, many politically powerful media interests, from newspaper publishers to local monopoly cable franchisees, want to hobble the Bell companies on the grounds that they are dangerous monopolies. By restricting the profits that the phone companies could earn by wiring up the nation with glass, many cable firms want to hold back the digital tide and perpetuate their monopolies. For them, Japan's HDTV system is preferable.

Free access to the limited resource of the radio frequency spectrum already represents a powerful subsidy for broadcast interests competing with computers and fiber optics. Far from expanding these subsidies by granting extra bandwidth for through-the-air analog broadcast technologies such as HDTV, however, the FCC should actually begin a phased withdrawal of current TV spectrum space. Since television has the far better alternative of fiber and cable, the FCC should continue doggedly to deny TV broadcasters scarce spectrum resources. Spectrum space is sorely needed by such mobile systems as cellular phones, laptop computers, paging systems, CB radios, air and ground traffic controllers, and a wide array of other portable uses for police, medical, military, trucking, Federal Express, and other vital public needs. The ultimate vision of the Dick Tracy wristwatch phone, using spectrum on every hand, moves closer year by year.

With its commanding leads in many of the most important com-

puting and telecommunications technologies, U.S. industry has every
hope of dominating the telecomputer business of the future. The
danger in confronting the Japanese challenge is not that the United
States lacks government involvement. The danger is that current governmental policy is dominated by broadcast interests with a huge
investment in Japanese-dominated analog television.

One of the most important applications of learning curve theory is
the growth share matrix pioneered by Bain & Company and the
Boston Consulting Group. It rates businesses by their prospects for
future growth and market share, classifying them as stars, cash cows,
wildcat question marks, and dogs, depending on the growth rates of
their markets and their share of them.

The matrix applies to nations as well as to firms. For the United
States, television is a dog. American firms have a small share of a
sluggish market. TVs are ubiquitous and everybody makes them. That
is why U.S. companies left the business. But dogs have a peculiar
quality. When you see them in Washington, they are always decked
out like poodles. Politicians cannot resist them. In fact, government
is always the dog's best friend.

To the politician HDTV is the crown jewel of technology, the most
important invention of the last twenty years. It will decide whether
the United States has a future in electronics. But the fact is that
television is dead. The only way HDTV will triumph is if Congress
saves it. By committing the U.S. government still more deeply to its
subsidies for obsolescent broadcast media, a U.S. program for HDTV
will probably help assure the Japanese huge new markets in home
entertainment, because the Japanese are some ten years ahead in
HDTV.

If instead Congress commits itself to the vision of a fiber-wired
America, the United States can win because the United States in many
ways is years ahead of the Japanese in computing. The problem is that
current governmental policy props up the past in the name of progress
—fights microcosmic change in the name of combating the Japanese.
For the rest of this century, the great political conflict in most industrial nations will pit the forces of the microcosm against the forces in
revolt against its powerful waves of creative destruction.

PART

The Quantum Economy

V

CHAPTER
24

The New American Challenge

In the new economy, human invention increasingly makes physical resources obsolete. . . . Even as we explore the most advanced reaches of science, we're returning to the age-old wisdom of our culture. . . . In the beginning was the spirit, and it was from this spirit that the material abundance of creation issued forth.
—RONALD REAGAN, at Moscow State University

The laws of the microcosm are new and contrary to ordinary experience. Even those leaders of the new technologies who operate deep within the microcosm often seem to forget or ignore its most powerful truths. In economics and political theory, in sociology and philosophy, these laws are mostly unknown. But however slowly theory catches up to practice, the microcosm will increasingly dominate international reality, subduing all economic and political organizations to its logic.

Futurists and other sages have long agreed that the microcosm would shape the future of nations and industries. They thought it would demand great collective efforts and masses of capital. Both the use and manufacture of computers, they foresaw, would entail huge economies of scale, fostered by the state.

In the 1960s, many observers believed they could discern the pattern already in the United States, then as now the world leader in the use of new technology. In 1967, Jean-Jacques Servan-Schreiber alerted the world to *The American Challenge (Le Défi Américain)*. The challenge as he saw it was the military industrial complex: the combination of American multi-national corporations, government research and development laboratories, Pentagon money and management on the new frontiers of high technology.

The new technologies, the French editor asserted, inherently dictated the kind of capital-intensive and highly collective effort that had put a man on the moon and an IBM mainframe in every data-processing center. Governments not only would need to plan and subsidize the new industries but also would have to protect and support the throngs of expected victims in existing businesses.

Servan-Schreiber experienced an unusual and presumably gratifying success. Not only was his book a best-seller around the world, but nearly all his prophecies were taken to heart and all his prescriptions were adopted. The governments of Europe joined in launching a series of large projects, led by multi-national cartels, to exploit the new technologies. Eureka, Esprit, Prestel, Teletext, Informatique, Antiope, Airbus Industrie, Alvey, JESSI, the Silicon Structures Project, and other major government initiatives focused on answering the American challenge with a colossal *défi Européen*.

European tax systems were severely skewed to favor the formation of capital and the concentration of investment. New training programs proliferated to ease and inform the transition from schools into the new pattern of employment and to adjust to the expected loss of jobs. Immigrants across the continent were paid to return home, supposedly to open new opportunities for citizens. The European welfare states and regulatory agencies expanded massively to protect their workers and entrepreneurs from the dire effects of technological advance.

Though some of these policies were predictably disastrous, the vision that led to them seemed plausible to nearly all expert observers. Indeed, the critical insight—that the dominant technology of the age will demand economic and political accommodation—is certainly correct. In the 1960s most analysts shared Servan-Schreiber's vision of the new technology.

During this period, the economics of data storage and transfer favored large machines. Most of the industry still operated in the macrocosm of electromechanics. Made of different materials by different means, devices for switching, storing, and transmitting data were separate and required expensive interfaces.

As a fixed cost of a computer, such electromechanical peripherals enhanced the economies of scale for centralized processors, with large attached storage facilities. For example, the vacuum tube drivers, sense amplifiers, and other support for early magnetic core memories at IBM weighed several thousand pounds and were hundreds of times heavier than the memory core planes themselves. Since the cost of the peripherals was essentially independent of the size of the memory, small amounts of storage cost hugely more per bit than large amounts.

All these conditions of a still essentially electromechanical computer industry favored the mystique of IBM and its Von Neumann mainframes. People spoke in hushed tones about "giant computers" and a famous market survey estimated a total world demand for about fifty of them. The computer sat on a pedestal in the central processing room, guarded by a credentialed guild of data-processing gurus.

It was these conditions that led Servan-Schreiber and the other experts to see the computer as a leviathan instrument of Big Brother and to prophesy the emergence of huge organizations to manage its powers. Predictions of a totalitarian 1984 gained credibility from the image of Big Blue in suits of gray. Not only would computers gravitate to the hands of the state and other large bureaucracies but the computer would give the state new powers to manage the economy and manipulate individuals.

The world Servan-Schreiber predicted, however, never appeared. The Servan-Schreiber prophecy was based on the Von Neumann paradigm. The Mead vision would capsize its every key assumption. In the new regime, the Von Neumann components—storage, interconnect, and processor—all would be made of the same materials by the same process. The entire computer would become increasingly solid state. Working memory would be on silicon. Even the controllers to mass-store disk media would be solid state integrated circuits and monolithic analog-to-digital and digital-to-analog converters.

Unlike the electromechanical peripherals of magnetic cores, these silicon peripherals would cost radically less the greater the volume of production. In other words, the more computers were built and sold, the cheaper the memory access would be. Ultimately, controllers would plunge down the learning curve and cost a few dollars, an amount suitable for personal computing. The previous premium for separation of functions and for scale of operation and storage gave way to a still greater premium for integration and miniaturization in a world of widely distributed processing power. The computer on a chip became both vastly cheaper and more powerful than the computer on a pedestal.

Some of the key new technologies originated in the laboratories of huge companies such as IBM, RCA, and AT&T. But the microcosm reduced the inventions to a size and portability that made them uncontrollable by large institutions. These mind-scale machines could escape in the minds of engineers and be reproduced in weeks on workstation screens and radically improved in months in new companies.

The firms spearheading the move to the microcosm, therefore, were not the megacorporations of industrial policy dreams but the rebels of TI and Fairchild, TEL and Intel, and dozens of others making systems from their chips. The resulting array of personal computers, engineering workstations, database servers, desktop publishers, and silicon printers emerging in the real year of 1984 unleashed forces of liberation that would foil every tyrannous portent of *1984* and other industrial fictions.

Servan-Schreiber's industrial policies left European technology aswim in an alphabet soup of acronyms while European entrepreneurs negotiated with bureaucrats and union bosses for the right to fire a saboteur or close down an obsolete plant. Trying to enter the microcosm clutching macrocosmic institutions and technologies, the parts of Europe where Servan-Schreiber's views were adopted suffered the worst slump of the post—World War II era. In nearly twenty years after the publication of his book, no net new jobs were created on the continent. During that period Europe fell ever farther behind in the very information technologies that were targeted by national industrial policies. Their small share of the microchip and computer industries, for example, fell some 20 percent between 1967 and 1987.

Meanwhile the United States followed totally different policies. Rather than extending national regulation and control over the "central nervous system" of finance, transport, and communications, the United States deregulated finance, telecommunications, energy, trucking, and air transport. Rather than increasing taxes on individuals and subsidizing corporations as Europeans did, the United States drastically lowered tax rates across the board, to an eventual top rate under 30 percent for individuals and closely held businesses, and under 35 percent for normal corporations. Rather than concentrating investment in industrial conglomerates, the United States liberated pension funds and other institutions to invest in venture capital for risky new companies.

These policies, beginning in the late 1970s, fostered a massive upsurge of entrepreneurship and innovation. Not only did venture capital outlays rise by a factor of 200 but new public issues of shares on the stock market rose some tenfold; small business starts nearly tripled, from 270,000 in 1978 to some 750,000 in 1988. The result, contrary to predictions of rising unemployment, was a total of some 15 million new jobs, with a record of some 67 percent of adults in the labor force in 1988 (compared to 58 percent in Europe). Real per capita disposable income rose 20 percent. Surging 30 percent between 1981 and 1989, even industrial production rose three times faster in the United States than in the European realms of industrial policy.

Ignoring all this history, many experts now urge the United States to reverse policy and embrace the Servan-Schreiber code. They declare that the new technologies demand drastically more government guidance and aid rather than less. They point to Japan's Fifth Generation Computer Project and to similar European schemes; they decry the imbalance in U.S. trade accounts; they panic at the Japanese lead in DRAMs and in HDTV; and they predict a steady decline in the international position of the United States on the frontiers of technology.

Once again these voices are wrong, victims of a common nostalgia alien to the meaning and prospects of the quantum era. They are wrong first of all on the facts, misreading the recent history of the computer and semiconductor industries. More importantly, they are wrong on the future, deeply misunderstanding the meaning of the microcosm.

American companies hold some 70 percent of the worldwide computer market and in many ways are increasing their lead in the technology. With accelerating technical progress, the value added in the industry has shifted rapidly from hardware to the software that makes the hardware useful. Summing up the situation is Intel's 32-bit microprocessor, the 386. Intel invents it, IBM and scores of other firms adopt it, and all must wait for some three years for Microsoft, a relatively new firm headed by thirty-three-year-old Bill Gates, to produce the software to exploit its full powers.

Gates's firm is now worth several billion dollars and the value of its software to the users is far more. Yet the most complex and expensive capital equipment used in this enterprise, other than Bill Gates's mind, are desktop computers. This is typical of the industry.

Software is chiefly a product of individuals working alone or in small teams with minimal capital. Between 1975 and 1985, some 14,000 new software firms rose up in the United States, lifting the U.S. share of the world software market from under two thirds to more than three quarters. Since 1985, U.S. software output has grown faster even than Japan's. Software comprises a four times greater share of computer industry revenues in the United States than in Japan, and the United States produces more than four times more marketed software.

Mostly created since 1981, the new U.S. software firms transformed the computer from an esoteric technology into a desktop appliance found in 20 percent or more of American homes and ubiquitous in corporate workplaces. In the per capita application of PCs, the United States in the late 1980s led Japan by approximately three to one.

Analysts focusing on "Fifth Generation computer projects," on mainframe systems, and on supercomputers, swoon worshipfully before HDTV screens and dismiss these PCs as toys. So did the experts at IBM a few years ago. But at the same time that the United States moved massively into microcomputers, small systems surged far ahead in price-performance. In cost per MIPS (millions of instructions per second, a very rough but roughly serviceable measure), the new PCs are an amazing ninety times more cost-effective than mainframes. With the ever-growing ability to interconnect these machines in networks and use them in parallel configurations that yield mainframe performance, microcomputers are rapidly gaining market share against the large machines.

For many years, IBM expected the new network technologies to favor sales of its large computers. It was believed that networks would have to be hierarchical, with a mainframe host at the top of the pyramid. By the late 1980s, however, IBM had to give up this hope. Even the first generations of peer networks, linking intelligent machines and specialized processors, proved to be 28 percent cheaper than networks of mainframes and terminals offering the same performance. With the coming of a new standard for network interconnections, the hierarchical network will eventually become obsolete.

The future of computing lies not in the Cray 3 or the IBM 3090 supercomputers or all the IBM imitations in Japan but in the Intel 386, 486, and the Motorola 68000 family of chips. It lies in their array of networking co-processors, and in reduced instruction set (RISC) processors optimized for parallel use. These devices are creating entire new businesses in computer-aided engineering workstations, in specialized semiconductors, in embedded microprocessors, in superminicomputers, in graphics computers, and even in minisupercomputers. Yet large machines still dominate data processing in Japan and most of Europe and the Japanese hold under 10 percent of the U.S. microcomputer market. During the late 1980s, even their chip designers were still doing the vast bulk of their computer-aided engineering on mainframe terminals.

Driven by the proliferation of software packages, the more rapid diffusion of computer technology to individuals in American homes, schools, and workplaces is as important a source of future competitiveness for the U.S. economy as the Japanese lead in robotics and manufacturing is for Japan.

In fact, it was the U.S. software lead, not a hardware lag, that caused the high-technology trade gap. Most of the imported hardware was components and equipment needed to respond to the unexpected impact of a booming U.S. software industry.

Software tends to be a culturally conditioned product that enhances the need for imported hardware but is difficult to export. Impelled by demand created by the new software packages, computer sales lurched up by one third a year between 1981 and 1984, creating a need for memory chips that the design-oriented U.S. chip firms could not begin to fulfill.

In a key strategic move, IBM chose not to invest in special factories to meet its own huge DRAM needs. Ambushed by immense but unpredictable demand for its personal computer, IBM in 1982 resolved to let the Japanese fill its memory sockets while IBM pursued later DRAM generations. When the price of the chips plummeted well below the cost of production, IBM's decision—followed also by all the other U.S. computer firms—became a spectacular strategic coup. During the peak quarters of 1983 and 1984, IBM was ordering more memory chips a month than U.S. domestic manufacturers could turn out in a year. Since memory chips comprise as much as a third of the manufacturing costs of small computers, IBM and the other U.S. computer firms gained a subsidy from Japanese memory producers far exceeding the trade gap in chips.

Essentially, the U.S. let the world's DRAM producers, targeting these obsolescent "technology drivers," subsidize our computer firms with cheap memory chips. Meanwhile American entrepreneurs drove the technology off in new directions. The foreign companies achieved a hugely expensive and mostly inflexible capacity to produce commodity memories at a loss. Meanwhile the DRAM producer with the smallest chips and by far the cheapest process turned out to be Micron.

In time the industrial politicians will learn a key lesson. By targeting a technology, you become a target yourself. In this case, the European bureaucrats were saved from their folly chiefly by the inability of their subsidized firms to produce DRAMs in volume. But the Japanese conglomerates lost close to $3 billion on the product between 1985 and 1987.

In the face of the upsurge of demand for computers, U.S. makers of peripherals, monitors, keyboards, disk drives, printers, power supplies, and other apparatus also faced a problem of inadequate production capacity. The result was an avalanche of imports from Far Eastern plants, often owned or controlled by Americans. Between 60 and 70 percent of the hardware components in a mid-1980s personal computer were imported from Asia. A key instrument of U.S. competitiveness, the PC was also the driving force behind the trade gap.

In a surprise, however, even the famous IBM personal computer clones were increasingly manufactured in the United States. From Tandy and Dell to an array of mail-order startups, American firms used highly integrated components—mostly designed by Chips & Technologies—to produce personal computers in the United States that were both cheaper and better made than IBM's because they used far fewer chips.

Nonetheless, preoccupied with the balance of trade as an index of competitiveness, many analysts were full of fear and trembling. Americans import far more computer products than any other country and run a large trade deficit with Asia in both systems and components. Trade deficit fetishists prophesied doom.

The trade deficit with Asia in computers and semiconductors, almost entirely the result of a booming computer market, mostly stems from overseas manufacturing by U.S. firms and comprises less than 4 percent of U.S. computer industry revenues. U.S. companies still hold more than two thirds of the total world market even for peripherals. In 1988, U.S. producers of hard disk drives, largely manufactured by wholly owned branches and subsidiaries in Asia, took between 70 and 80 percent of the world market.

In the broader electronics industry, U.S. companies produced more than $195 billion of electronic equipment in 1988, some 80 percent more than Japan. Even excluding some $50 billion of defense electronics, the United States substantially outproduced Japan. Indeed, during most of the 1980s, U.S. electronics production firms may well have grown faster than Japanese output, which was vastly inflated in the data by an 80 percent rise in the value of the yen. Including the fourfold U.S. edge both in software and in telecom equipment revenues, the United States continues to hold the lead in the keystone products of the age.

The computer industry is not a special case. It exemplifies the new technologies of the quantum era. But the lesson is deeper than a mere failure of insight by economists and other experts. The lesson is that the United States made gains in competitiveness during the 1980s chiefly by ignoring nationalist fears and fetishes and statist industrial plans. Instead, the United States pursued strategies suited to the global microcosm.

Rather than trying to become self-sufficient in computing by some massive national project in defiance of the market, the United States paid for imports by adding enormous new value to them. Shopping in the farthest corners of the globe to find the best suppliers of components to implement their systems and software, U.S. entrepreneurs ended up making America the center of the industry and the prime source of world economic growth.

Ignoring a few failed sallies into industrial policy, the U.S. approach to information technologies has been a supreme success. But the Cassandras have an answer. They say the future will be different. Computer technologies are now allegedly maturing. As growth slows and large firms marshal their forces to achieve market share, America's dependence on Japan will gradually cripple and destroy the U.S. industry.

This view has even less merit than the claims of U.S. failure. It entails an acute misreading of the technology. Computer industry growth is not slowing, the technology is not "maturing" (they mean ossifying), and large companies are not gaining ground (although new firms are growing large). The pace of progress in computers is now on the verge of a drastic acceleration. Summoning these advances is a convergence of three major developments in the industry that also play to American strengths in the microcosm. Indeed, pioneering in silicon compilation, artificial intelligence, and parallel processing, U.S. entrepreneurial startups are already in the lead in shaping the key technologies of the coming era.

Contrary to the analysis of the critics, the industry is not becoming more capital-intensive. Measured by the capital costs per device function—the investment needed to deliver value to the customer—the industry is becoming ever cheaper to enter. The silicon compiler and related technologies move power from large corporations to individual designers and entrepreneurs.

The true economic implications of silicon compilers go beyond lowering capital costs for current designs. As Carver Mead asserts, silicon compilation will allow chip designers to take advantage of ever-increasing component densities on chips—densities impossible with pre-compiler design methods. This evolution also will favor the American industry.

Until the end of the century, chip densities are likely to continue rising. Even without contemplating the analog advances of Carver Mead, former Stanford chip guru James Meindl (now provost at Rensselaer Polytechnic) and Howard Bogert, then of Dataquest, both have estimated a billion digital transistors on a chip by the year 2000. When even millions of transistors can be put on one chip and be produced by the millions, the unit material costs drop to virtually nothing. The value is in the design. If the design is merely a routine memory, the price of the chip will eventually drop to near the cost of the material it contains. That material is mostly sand. On this foundation did the national strategists of the world build their industrial policies.

Embodying an entire mainframe system, with working memory on board, a billion transistor chips will climax the computer's collapse into the microcosm. But as Howard Bogert shows, memory will be the least efficient way to use such chips. With a billion transistors, one chip or wafer-scale device could hold a thousand VAX CPUs working in parallel or twenty Cray 2 supercomputer CPUs. But with memory, it could hold only one fortieth of a Cray 2.

As Mead has long predicted, the superchips of the future will stress logic rather than memory. As application-specific massively parallel processors made possible by silicon compilation, the new chips could achieve levels of vision, voice recognition, and other artificial intelligence that would yield radically new powers. As the computer collapses into the microcosm, the off-chip memory, increasingly, will be on optical disks. The value of the chip will depend on the usefulness of the design.

The distinction between hardware and software continues to blur. As Raymond Kurzweil has observed, "When you buy Lotus 1-2-3 on a floppy for $300, you know you are paying for information; but when you buy a computer, you think it is different; that's hardware. But the bulk of the cost of a computer comes from the chips and all but about 2 percent of their cost also derives from the costs of the information the hardware embodies."

As we begin to write massive amounts of highly specific logic directly onto ever denser chips, the distinction between hardware and software will all but vanish. With logic processors increasingly cheaper and more efficient than memory and connections as a use for silicon, software will increasingly assume the form of custom logic rather than code inscribed on memories.

As software hardens into crystals of silicon, the balance of power shifts in America's favor. As hardware designs increasingly embody software concepts, the Americans can parlay their software lead into an increasing dominance in the entire industry. As in Mead's prophecy, software will subsume hardware. Algorithms will become architecture. General-purpose systems will give way to special-purpose devices.

Special-purpose systems cannot be designed, built, or sold by the techniques appropriate to the era of commodity chips. Driving the technology in the quantum era will not be Goliath fabs that can produce millions of units of one design but flexible design and manufacturing systems that can produce a relatively few units of thousands of designs. Designers will need to be increasingly close to the marketeers, and both will have to understand the customer's needs.

This is not just Mead's prophecy; it is a fact. With the industry in the midst of a tenfold increase in the number of chip designs to be produced annually, future factories will have to produce hundreds or even thousands of designs a year. Relieving some of this pressure on the low end will be Actel and other desktop silicon-printing technologies. But the increasing number of designs ranges from the smallest to the densest devices and transforms the challenge of semiconductor fabrication.

With chips increasingly made for specific applications, silicon efficiency will hugely increase and silicon usage will tend to decline. The dominant technologies will probably eschew photomasks, which will be difficult to manage and store for thousands of different chips. In place of these physical embodiments of the design, subject to dirt and defects, foundries will store designs digitally in the microcosm and use various means of writing directly on the wafer. Whether with electron beams, with lasers, with ion beams, or with some new way of atom-by-atom fabrication such as molecular-beam epitaxy, the creation of new chips will become even more fully a quantum process.

Economies of scale will probably decline for all but the most widely demanded devices. As markets grow, many competitors will command a world-scale factory. Like chemical producers today, they will move beyond the point where increasing volume can substantially lower costs. Value will again be pushed to the designs and to the individuals who create them.

In a sense the Cassandras are right: the U.S. semiconductor industry will disappear. It will become a different kind of business, dominated by design and systems. Semiconductor companies no longer are chiefly making components. In the future, components will be almost entirely part of the chip. Even while losing market share in the remaining commodity chips, the United States can continue to lead the move into the microcosm. Most important, U.S. companies can still dominate the computer industry and its software, which is what the contest was all about.

While American economists ululated about declining competitiveness, we won the first phase of the competition. Whatever happens in the future, we should learn the lessons of this success, not forget them in a hypochondria of decline. We won because we dispersed power among thousands of entrepreneurs rather than concentrated it in conglomerates and bureaucracies. We won because we went for growth opportunities rather than for trade surpluses. We won because we didn't try to do everything at once. We won because we were not afraid of the international division of labor. We had a global orientation rather than a national industrial policy.

It is the power of the microcosm that ultimately confounded the policies and predictions of Servan-Schreiber and vindicated the policies and prophecies of Carver Mead. During the very period that the Europeans placed their reliance increasingly on the state's ability to gather capital and experts, the balance of power in the world shifted massively in favor of the individual. For Servan-Schreiber was right. The organization of world industry would indeed be shaped by the nature of the new technology.

The new technologies of the microcosm—artificial intelligence, silicon compilation, and parallel processing—all favor entrepreneurs and small companies. All three allow entrepreneurs to use the power of knowledge to economize on capital and enhance its efficiency: mixing sand and ideas to generate new wealth and power for men and women anywhere in the world.

Indeed, the power of the microcosm is so great that it has provoked a powerful revolt.

CHAPTER
25

Revolt against the Microcosm

Microcosmic technology restored the United States to global leadership. But it also wreaked creative destruction in many American companies. It fueled wave after wave of entrepreneurial startups that encroached on profits in the business plans of AMD, National, Fairchild, Mostek, TI, and even Intel. It threatened the position of scores of computer firms, from Control Data, Sperry-Univac, AT&T, and Datapoint, to Apollo, Convergent, Data General, and Commodore. It made obsolete thousands of products, engineering skills, patents, fabrication lines, and expensive capital equipment. It embarrassed bureaucrats from the Kremlin to the Champs-Elysées. No respecter of persons, credentials, finances, investments, megacorporations, or even governments, sooner or later the power of the microcosm was bound to provoke rebellion.

The revolt against the microcosm took many forms, from appeals for protectionism and industrial policy to litigation against startups and calls for collaborative research. The signal triumph of the revolt was the semiconductor trade agreement between the United States and Japan in August 1986. The results fully illustrated that men and nations that fight the microcosm fall before its powers.

Because of excessive investment in Goliath fabs, the Japanese had lost nearly $3 billion on DRAMs, more than twice the U.S. losses, when the agreement was signed. But the revolt against the microcosm by U.S. industrial lobbyists and politicians saved the Japanese from their folly.

Under pressure from a few U.S. firms, the U.S. government pushed MITI to manage Japanese production. As usual with government intervention in global markets, the agreement was both ill-timed and misapplied. Between August 1986 and August 1987, while demand for DRAMs rose some 30 percent, MITI administered a 32 percent reduction in the output of the then dominant 256K chip. Japanese firms cut back DRAM investment and shifted efforts from current generation chips to the next generation.

These measures did very little for U.S. chip firms, most of which had long ago left the DRAM market. But the U.S. industrial policy did wonders for the Japanese semiconductor industry, restoring its profits and margins and allowing Japanese firms to move up market at the expense of U.S. computer producers. Until the agreement, Japanese companies were fighting fiercely among themselves and with the Koreans for share of the DRAM market, pushing prices down for the benefit of U.S. computer firms. After the agreement, this rivalry slowed sharply and was replaced by administered pricing and production quotas. Not only did prices soar, but for a time many smaller U.S. firms could not get DRAMs in bulk at any price. Personal computer prices actually rose. Thus the trade agreement reduced demand for the microprocessors and custom chips actually made by U.S. companies.

Except for IBM, which by then had become the world's leading DRAM producer, U.S. computer firms suddenly had to pay between two and three times as much for DRAMs as their Japanese competitors who made them in house. Helping the established semiconductor firms and IBM in their revolt against the microcosm, the U.S. government had launched a menacing attack against the single greatest U.S. industrial strength: the innovative computer firms—from Sun to Zenith—and their specialized chip and software suppliers, which were collectively gaining market share against both IBM and the Japanese.

In itself, however, the trade agreement was less significant as a force than as a symbol of an ideological sea change in American high-technology firms and among their advocates in the media and academy and lobbyists in Washington. As the microcosm wreaked its transforming impact through the mid-1980s, a new coalition of Silicon Valley entrepreneurs and leftist professors emerged to war against what Robert Reich of Harvard termed "chronic entrepreneurship," "cowboy capitalism," and other colorful epithets.

This new group rejected the claims of Servan-Schreiber that the barriers to entry in electronics were too high to be overcome without major government aid. Instead, they argued that the barriers were so low that the U.S. industry was splintering into vulnerable fragments. Some analysts even maintained that the big companies needed government aid to protect them from the world's entrepreneurs.

A protest rang out from the leaders of Silicon Valley itself that far from an asset, the startup culture had become American industry's gravest weakness in competing with Japan. Intel chairman Gordon Moore denounced the financiers of new firms as the "vulture capitalists" of Sand Hill Road. Offering a theory of immaculate conception for Intel's silicon gate, Andrew Grove said that Intel's founding contrasted with the rise of new startups taking key technologies from their parents. He added that his company was threatened as much by Wall Street as by Japan. Not only did American vultures and headhunters prey on his best personnel, but the stock market appraised the fabs and fantasies of Intel defectors at many times the value of the real chips and achievements of Intel itself.

In a memorable encounter at a Dataquest conference in 1987, Jerry Sanders of AMD confronted Pierre Lamond, once a leading figure at National and now a leading venture capitalist. "You took my best guys to Cypress," Sanders said, "and they made $3 million last year on $14 million in sales. Fast static RAMs. Big deal. Static RAMs are not exactly a major innovation. [They were introduced at Intel in 1972.] But that's okay, so far so good. The problem is what to do next. No one denies that small firms can benefit from being focused. But they all want to get big and that means unfocused. After the first fifty million, believe me, it gets much harder. You can't fill large fabs with niche products. So far these companies have just ripped off the big firms without contributing anything significant. I don't think any of them will ever become a major factor."

Lamond tried to laugh off the complaints of Sanders and the others as merely the maundering of old men who want the world to stop turning as soon as they get on top. Such is the understandable response of all today's young entrepreneurs to the charges of their elders. But amid the late-1980s carnage in microchips, with the Japanese sweeping ahead in some measures of market share, the easy answers beg the question.

The leaders of the flagship firms of Silicon Valley were bringing a serious message. In the future, success would depend on cooperation and cheap capital, planning and political savvy, lifetime employment and government research. To Silicon Valley's newer entrepreneurs, it was as if Julius Erving had solemnly accosted Michael Jordan and informed him that his future in basketball was over unless he learned not to jump.

Weakened, in the words of Reich, by this "chronic entrepreneurialism," catastrophe was foreseen for many semiconductor firms. "The Japanese are systematically destroying the U.S. industry, piece by piece," Grove said. "We are in a demolition derby, and some of us are going to die."

For the Silicon Valley patriarchs, the remedy was intervention by the U.S. government to reshape and subsidize the industry. Their ideas bore fruit in Sematech, a chip-manufacturing research and development consortium to be funded half by industry, half by government, and headed by Robert Noyce. At best, Sematech could only undo some of the damage already done to U.S. manufacturers by non-commercial Pentagon priorities, personnel demands, export controls, immigration restrictions, and other government costs. But the idea of industrial policy seemed to vindicate the leading scholars of the left. With high entrance fees and other requirements hard for smaller firms to meet, Sematech seemed in part a tax-paid collaboration of the established firms and IBM in revolt against the microcosm.

Reich, Lester Thurow of MIT, and Chalmers Johnson of Berkeley have long maintained that whether mobilizing for war or for peace, government nearly always plays the leading role in evoking, developing, and financing new technology. Entrepreneurs can best serve as niche suppliers or subcontractors to major firms.

In the face of the demonstrable success of Silicon Valley, surging forward with ever greater force as government purchases dwindled, these arguments long seemed unconvincing. In the late 1980s, however, a new school of more sophisticated analysis has risen to buttress the argument. Knowledgeable about technology and industrial structure, some of these academic critics, such as John Zysman and Michael Borrus of Berkeley, emerged as influential consultants to Congress and the federal bureaucracy. The most complete and compelling case, however, came from a young scholar at MIT named Charles Ferguson, who arose—even before the publication of his magisterial thesis —as a vocal Cassandra of American microelectronics.

A former consultant at both IBM and LSI Logic, Ferguson showed intimate knowledge of the computer establishment and the semiconductor industry alike. His case against American entrepreneurs rang with conviction and authority. His conclusion was stark. Echoing Grove, Sporck, and Sanders, he declared that without urgent and incisive government action the American semiconductor producers will be unable to compete in global markets. The U.S. merchant semiconductor firms will perish. Since semiconductors are at the heart of all advanced computer, telecom, and defense technologies, the failure of U.S. microelectronics will cripple the American economy and threaten the national defense.

With an eye cynical but shrewd, Ferguson describes the life cycle of a typical startup in America's information industry: It begins with a breakaway from a major company, often an entire team, including engineers, managers, and marketing personnel, all offered prospects of personal wealth far beyond what a large firm could provide. They win an investment of up to $20 million from Sand Hill Road. They lease equipment, office space, and even sometimes a manufacturing facility. They hire additional engineers and workers from existing firms.

With a frenzied effort and commitment, they launch a new device —a CAD system, a non-volatile memory, an application-specific chip, a hard disk drive—mostly based on work at a major firm and competitive with its product line. With the large firm weakened by their defection, it cannot respond quickly. The new device of the startup outperforms all others in the marketplace and sparks massive orders which the startup cannot fulfill.

Never fear, though, help will be on the way. A horde of imitators, also funded by venture capital (fifty-four disk drive firms, forty-seven gate array companies, twenty-five workstation firms, and eleven non-volatile memory makers, for example), rush for the market. The initial startup struggles to diversify and escape the stampede. Expenses soar. Mezzanine financing is needed from the venture funds. They are too deeply committed to deny it; several millions more are provided. The firm teeters but clings to its lead. Finally, if all goes well, the company at last concocts a small quarterly profit, and with the hungry enthusiasm of venture capitalists, underwriters, lawyers, accountants, consultants, it sprints toward the tape: its initial public offering.

Nonetheless, in most cases there are already key defections. In part fueled by them, the competition—whether other startups or major firms—has become more powerful. Yet the new company forges ahead, expanding its manufacturing capabilities and its marketing effort. The total investment rises toward $200 million. But the mood has decisively changed.

The stock price is slumping and the initial frenzy of work has flagged. Believe it or not, disgruntled shareholders have launched a class action suit, alleging they were misled by the business prospectus and demanding several million in damages (they, or at least their lawyers, will often get it). To add injury to insult, Hitachi and Toshiba are threatening to enter the market. The venture funds have found a new fashion; Wall Street is bored by semiconductors or CAD; the top engineers are restive as they watch their option values decline.

Almost overnight, the firm will begin to age, and the paper millionaires who run it find themselves suddenly poorer. They are deep in debt and the IRS is attacking some of their tax shelters, used years earlier. They are slipping behind on their deadlines for the next product generation, the next payment on their platinum card, and the mortgage on their Los Altos home. They are desperate for further funds. None are available in the United States. They turn abroad.

Another way of putting it is to say that they decide to sell out. They resolve to put their company—chiefly the new technologies it owns—on the market. In the mid-1980s, several American startups did indeed sell out to Asia when the going got tough.

The American system says, "So what?" Ferguson says, this is a disaster for the American system. In selling where the buyers are, Japan or Korea, the entrepreneurs are not selling just their own assets; they are selling the public assets of America.

Embodied in the firm is not only the work and ingenuity of the owners but also the value of their accumulated experience at a major firm, the value of subsidized research conducted at U.S. universities, and the value of the damage inflicted on the parent company from which the entrepreneurs initially broke away. In addition, there is the value of the licenses, patents, standards, and architectures involved in the product and the value of a vantage point in the U.S. marketplace. Finally, according to Ferguson, there is the value of implicit subsidies in the U.S. tax code favoring small and growing firms that can show rapidly rising capital gains, R&D expenditures, tax-loss carryforwards, and investment tax credits.

Although some of Ferguson's claims were plausible, the truth is that large corporations exploit tax benefits more massively and resourcefully than small ones do. Morever, by 1987, with capital losses mostly undeductible and capital gains taxed like ordinary income, the capital gains "subsidy" became a sharp penalty in proportion to inflation and risk. In the face of this tax attack on equity and risk taking, many firms of all sizes fled into takeovers and buyouts converting stock into tax-deductible debt. But individual investors in risky technologies suffered severely.

Reich and Ferguson are not content merely to criticize U.S. entrepreneurs. They have a far larger argument to make. To Reich and Ferguson, the entrepreneur's behavior shows the crippling limits of laissez-faire economics. Personal optimization is suboptimal for the country and industry. Entrepreneurial self-interest leads, as by an invisible hand, both to the wealth of persons and to the poverty of nations.

According to Ferguson, a basic law of information technology dictates that in design, assembly, training, testing, networking, or marketing, costs rise with complexity and complexity rises with the number of objects to be managed. For example, the number of connections between nodes of a network increases as the square of the number of nodes.

This rule applies remorselessly, whether to the proliferation of transistors on high-density chips, of components in complex computer systems, of lines in software code, of terminals in a data-processing network, of human interactions on a factory floor. The result is that complexity swamps all systems made of large numbers of simple components, such as the American semiconductor industry produced in its heyday. Bringing together the once separate worlds of computers, factory automation, and telecommunications, these complex systems require technical, financial, and even political resources that cannot be mustered by small companies. Themselves comprising an industrial system of uncoordinated and narrow-based components, the entrepreneurial system thus will suddenly be swamped by a tsunamic wave of complexity which they will not be able to understand or manage.

The only answer, Ferguson maintains, is hierarchical organization: competitive rules, modular interfaces, and industry standards imposed from the top. This theme is familiar to all followers of design pioneers Carver Mead and Lynn Conway, and other industry observers, who have also urged top-down design approaches as the answer to the complexity problem of electronic design. Ferguson, though, is not speaking of designing a chip or a computer but of designing an industry or even an economy. Modularity and hierarchy must be forced from the top. They require rules and standards to guide the producers of future modules and connectible systems. They require control by large firms and governments.

The large costs of developing standards inflict continuing penalties on the industry leader. It must make the technical effort to determine the standards, and the political effort to impose them. These efforts expand markets and lower prices for the entire industry. But in a fragmented marketplace, the costs cannot be recovered. Unless disciplines are imposed on the industry, either by government or by oligopoly power, the companies that pay for these standards and architectures become sitting ducks for rivals that can avoid these costs.

In information technology, Uncle Sam has been the sitting duck. Nearly all standards and architectures in hardware, software, and network interfaces have originated in the United States, largely with IBM and AT&T or under their sponsorship. These standards collectively cost scores of billions of dollars to develop and to propagate. But because of anti-trust laws, lax enforcement of intellectual property rights, and MITI's prohibition of auctions for licenses in Japan, this huge capital endowment of innovations has become available at far below its cost. Overall, between 1956 and 1978, when most of the key breakthroughs occurred, Japan is estimated to have paid a total of some $9 billion for U.S. technologies that required—depending on assignment of costs—between $500 billion and $1 trillion to develop.

The real cost may have been even higher, however, because by setting critical design targets, standards shift the focus of competition from the invention of products to their manufacture on a global scale. Because manufacturing is itself a complex system, power again flows upward. A globally competitive new plant for high-volume commodity production costs some $250 million and dictates high-capital intensity. Requiring both global communications and local integration of an array of technologies—from magnetics and materials science to photonics and robotics—manufacturing imposes huge economies of technical scope.

As Ferguson concludes, the technologically optimal form for the global information industry might well comprise IBM, AT&T, and a few firms in Japan. Oligopoly is optimal, not only for the competitive advantage of particular nations but for the welfare of the world. Yet the United States remains afflicted with an extraordinarily fragmented information arena: a chaos of companies, architectures, operating systems, network protocols, semiconductor processes, design systems, software languages, all controlled by entrepreneurs avid for personal wealth, or by sophisticated employees poised to flee to the highest bidder, whether a venture capitalist or a foreign conglomerate. Except at IBM and possibly AT&T, American electronics defines or defends no larger interests.

As Ferguson and the Silicon Valley patriarchs insist, the law of complexity pushes decisions ever higher up the hierarchy and dictates ever larger organizations. The system is sick and chronic entrepreneurialism is the disease. The remedy, Ferguson, Reich, and others agree, is a heavy onslaught of governmental subsidies and guidance, designed to favor established firms and counteract the existing bias toward new companies. Although most of the critics deny it, the essential recommendation is to copy Japanese political and economic institutions in much the way the Japanese have copied our technologies.

As its advocates point out, the key rule in industrial policy, which distinguishes it from mindless nationalization of the economy, is always to favor and subsidize success rather than protect failure—to help Americans lower prices for others rather than to force others to raise prices for Americans. Some of the industrial policy suggestions —large increases in spending on science and engineering education, fierce protection of intellectual property rights, persistent effort to pry U.S. goods into protected foreign markets, a long-term Pentagon concern to aid U.S. technology—follow the essential rule of favoring success.

Some suggestions—for example, parity in tax treatment of capital gains and other income—have already been much overdone in current tax law. But for all the apparent sophistication and plausibility of the new critique of Silicon Valley, the analysis is just a further revolt against the microcosm and thus poses serious dangers to American technological leadership. A better theory is available. It begins with a better understanding of Japan.

For the last thirty-five years, Japan, an island nation poor in foods and fuels, has focused its energies on production for export. Exalting manufacturing over finance and litigation, favoring engineering education over the softer arts and sciences, this country half as populous as America now has 15 percent more engineers and one eleventh the lawyers. It has seven times as many manufacturing small businesses. It has maintained a rigid patriarchal family structure that has created a dutiful workforce and led to crime and divorce rates one eighth of America's. It has kept students in school 40 percent more hours every year; it has attracted the best performers to Saturday night *juku* classes in science and computing. It has maintained a work week 20 percent

longer. It has banished adversary unions. It has maintained apparent
personal savings rates five times higher than the United States, resi-
dential accommodations less than half as extensive, and manufacturing investment rates at least twice as high. Until the mid-1980s, its
government spending and effective tax rates were far lower and its
policy persistently favored manufacturing high technology over all
other areas of concentration.

The United States meanwhile defended the free world, maintained
wartime tax rates until the mid-1960s and then allowed inflation to
impose them again in the 1970s. In the process, the IRS confiscated
most of the nation's savings that were not invested in housing and
collectibles and inflicted a trauma on savers not forgotten for a decade.
The United States launched a campaign against family roles and rules,
achieved unparalleled rates of crime and drug addiction, nourished a
media and intellectual counterculture hostile to business and technology, and taught more students sex education and psychology than
physics or calculus. At age seventeen, the top 5 percent of U.S. math
students are inferior on standardized tests to the average Japanese of
that age. The United States ignited a national fever of litigation and a
national obsession with claims of "rights" and a national disdain for
disciplines and duties. Since 1965, American men have shortened
their average workweek by 15 percent, and women have shortened
theirs by 20 percent. The United States cultivated contempt for blue-
collar work, fostered class war between unions and management, ha-
rassed successful businesses with anti-trust suits, liability claims, and a
three-thousandfold increase in regulations.

This contrast is limned in overly broad strokes, but it captures the
essence of the situation. The Japanese have trained a workforce and
nourished a culture suitable for large-scale manufacturing, making
them the most efficient mass manufacturers in the world. The result
was Japanese dominance in a series of manufacturing industries, beginning with shipbuilding and steel, continuing with motorcycles,
automobiles, and consumer electronics. Perhaps most important, in
the mid-1980s the Japanese prevailed in machine tools and robotics.
Anecdotal evidence from firms such as IBM and Allen-Bradley suggests some improvement since then, but the Japanese still lead in
many applications.

Reich, Thurow, Ferguson, and the other analysts now attribute any
U.S. problems in information technology not to this obvious Japanese strength but to excessive American entrepreneurialism. There are
two fatal flaws in their argument. The first is that information tech-

nology is the one key area the Japanese have failed to dominate. The second is that in every one of the industries in which the Japanese did prevail, they generated *more* companies and *more* intense domestic competition than the United States.

The Japanese created three times as many shipyards, four times as many steel firms, five times as many motorcycle manufacturers, four times as many automobile firms, three times as many makers of consumer electronics, and six times as many robotics companies as the United States. It was established American oligopolies in consumer electronics such as RCA, Zenith, and General Electric that gave up the color television industry to the more entrepreneurial Japanese. It was vertically integrated American automobile firms that succumbed to the more fragmented and entrepreneurial Japanese automobile industry.

At various times during the last three decades Japan has boasted fifty-three integrated steel firms, fifty motorbike producers, twelve automobile firms, and forty-two makers of hand-held calculators. In recent years, the Japanese have launched some two hundred and eighty robotics firms, eight makers of videocassette recorders, thirteen makers of facsimile equipment, and twenty copier manufacturers. The Japanese rate of business failures has been about twice America's.

Indeed, the only key industries in which the United States launched dramatically more companies than Japan were computers and semiconductors. In chips, it was the United States that led, 280 to 20, and in computers and software, the United States led by thousands. It is no coincidence that the United States has done far better in these fields than in virtually any other industries targeted by Japan.

Contrary to the usual impression, the United States at the outset was probably farther ahead of Japan in automobiles, television sets, and steel than in semiconductors. Except for the immediate years following World War II, the Japanese have never lagged far behind the United States in electronics. Excelling America in some areas of chip production ever since Sony began making transistor radios in the late 1950s, Japan actually exceeded the United States in unit output of cheap (non-military) transistors in 1959.

The Japanese have never imported more than 25 percent of their semiconductors. Their huge unit sales in consumer electronics entailed early mastery of mass production of high-quality linear and discrete circuits, and later impelled them to master CMOS for low-power devices. In 1969, the Japanese created a small Dynamic RAM prototype. In 1972, less than a year after Intel, NEC began mass production of DRAMs. Moreover, by the late 1970s, their computer technology was by crude hardware benchmarks comparable to the IBM machines on which it was based.

Nonetheless, the cultural and political advantages that allowed them to dominate one after another major industry, from shipbuilding and automobiles to TVs and digital watches to VCRs and robotics, failed to give them leadership in information technologies. What needs to be explained is this American success.

The answer from the left is the Pentagon. Defense industries provided the major initial markets both for digital computers and for integrated circuits. The advocates of industrial policy claim that the United States benefited from military spending and research during the years of American dominance and then later frittered this advantage away by chronic entrepreneurialism.

This is the opposite of the truth. During the post–World War II and Korean War era, when defense necessarily dominated the entire technology arena, the growth and progress of U.S. electronics were relatively slow. As the decade approached its end, Japan came close to parity in semiconductors and the Russians launched Sputnik. The United States was even losing ground in key military technologies. Dominated by the Pentagon and afflicted by 90 percent tax rates, three recessions, and the Korean War, the fifties were a disastrous period for American technology and economic advance.

During most of the 1950s, the United States had one of the slowest-growing economies in the entire world. The semiconductor industry did not burst into its upward spiral until the 1960s, when tax rates were slashed and the military share of industry sales dropped some 80 percent.

What actually ignited the U.S. resurgence is the newly maligned startup system itself. It released energies that were often stagnant in large firms. Its high rewards for innovation spurred extreme efforts among otherwise slackly motivated American personnel. The wide-open U.S. university system attracted immigrants who would otherwise have languished overseas or worked for rivals. The equally open entrepreneurial culture then kept these foreigners in America, where immigrants made key contributions to every major company and aided virtually every technical breakthrough. The promiscuous culture of Silicon Valley, with workers and engineers constantly moving from one firm to another, fostered a wildfire diffusion of technology that compensated for the lack of national coordination. Venture capitalists began ranging the world—from Israel to Japan—in search of new talent to finance and bring public in U.S. markets.

The venture capital system also compensated for America's relatively high cost and low availability of capital. By converting scarce savings to equity and targeting the most risky and promising technologies, the venture capitalists greatly enhanced the quality of U.S. capital formation. For all its flaws, the Silicon Valley system was nearly perfectly targeted to overcome American limitations in both personnel and funds.

During the early 1980s, however, the pattern of excess described by Ferguson in fact appeared. The reason was the extinction of the venture capital industry between 1970 and 1977 with a top capital gains tax rate between 35 and 49 percent, plus state and local levies. Since most nominal gains were mere inflation, the government confiscated more than 100 percent of all real gains in the economy during this period. Then, between 1978 and 1981, the capital gains tax plummeted to a federal level of 20 percent, inflation collapsed, and venture funds rose some two hundred fold. The venture capitalists and entrepreneurs were inundated. Hence the excesses that so revolt the critics of Silicon Valley.

Yet the damage was less severe than Ferguson would suggest. Many of the technologies sold abroad were already obsolescent or otherwise less useful than he suggests. U.S. entrepreneurs may well have exploited the Koreans more than the Koreans exploited them. In any case, many of the startups of the early 1980s have thrived and are contributing to American leadership in key technologies, from high-performance CMOS and custom chips to hard disks and parallel computers.

Meanwhile, the United States did not forgo the services of large, global, fully integrated firms, with essential lifetime employment and long-time horizons in research and development. IBM, AT&T, and TI were world leaders in semiconductors from the beginning, and as the years passed they were joined by Digital Equipment Corporation, Hewlett Packard, and Delco of General Motors.

When analysts claim that the Japanese surpassed the American semiconductor industry in 1987 and 1988, for example, they make no adjustment for a 55 percent rise in the value of the yen and the government chip agreement's artificial price hikes for Japan's DRAMs. These punishing government policies—devaluing the dollar and pumping up Japan's chip profits—disguised an actual increase in the American lead in the real value of chip output. Moreover, these statistics exclude "captive producers," all of which are defined as American. This means they ignore two of the largest U.S. chip producers, IBM and AT&T, and one of the most resourceful ones, Hewlett Packard.

In 1988, IBM produced an estimated $3.7 billion worth of semiconductors, making it—with any reasonable adjustment for the soaring yen—the world's leading manufacturer of chips. Other captive producers made over $2 billion more. Including the contribution of the captives, U.S. microchip output exceeded Japan's by one fifth in 1987 and comprised some 50 percent of world production. Adjusting for the 55 percent yen upsurge in 1987 and 1988—enhancing the dollar value of Japanese home production—the United States continued to keep pace with Japan in microchip output through the end of the decade.

Although IBM's costs are believed to be substantially higher than Japanese costs, in 1987 IBM became the first company to massproduce one-megabit DRAMs on eight-inch wafers, and their access time of 80 nanoseconds was among the best in the industry on standard devices. IBM's four-megabit design is probably the most manufacturable in the industry, and indeed was being used in IBM computers in 1988. Working together with Perkin-Elmer, IBM developed a new photolithography system combining the throughput of a projection aligner with the resolution of a stepper down to geometries under .4 microns. While remaining cautious in expanding capacity, Texas Instruments has also kept up with the Japanese in leading edge semiconductor technologies and actually demonstrated its four-megabit prototype before any of its rivals.

Hewlett Packard's plant in Corvallis, Oregon, meanwhile, was producing thousands of different custom chips from designs originating in forty-six divisions, with yields comparable to the best Japanese facilities. It is possible to celebrate the superb performance of the Japanese companies and to recognize the real vulnerabilities of the U.S. industry without forgetting the continuing strengths that underlie the still large U.S. lead in information technology.

CHAPTER
26

The Law of the Microcosm

In the worldwide rivalry in information technology, the greatest American advantage is that the U.S. system, for all its flaws, accords best with the inner logic of the microcosm. When Ferguson says that complexity increases by the square of the number of nodes, he makes a case not for centralized authority but for its impossibility. Beyond a certain point comes the combinatorial explosion: large software programs tend to break down faster than they are repaired. Large bureaucracies tend to stifle their own purposes.

Endemic in most industrial organizations, private and public, this problem becomes deadly in the microcosm. The pace of change creates too many decisions, too many nodes, to be managed effectively in a centralized system.

Rather than revolting against the microcosm, the advocates of industrial policy should listen more closely to the technology. For what the technology is saying is that the laws of the microcosm are so powerful and fundamental that they restructure nearly everything else around them. As Mead discovered, complexity grows exponentially only off the chip. In the microcosm, on particular slivers of silicon, efficacy grows far faster than complexity. Therefore, power must move down, not up.

This rule applies most powerfully to the users of the technology. In volume, anything on a chip is cheap. But as you move out of the microcosm, prices rise exponentially. A connection on a chip costs a few millionths of a cent, while the cost of a lead from a plastic package is about a cent, a wire on a printed circuit board is 10 cents, backplane links between boards are on the order of a dollar apiece, and links between computers, whether on copper cables or through fiber-optic threads or off satellites, cost between a few thousand and millions of dollars to consummate. The result is that the efficiency of computing drops drastically as efforts are made to control and centralize the overall system.

Provided that complexity is concentrated on single chips rather than spread across massive networks, the power of the chip grows much faster than the power of a host processor running a vast system of many computer terminals. The power of the individual commanding a single workstation—or small network of specialized terminals—increases far faster than the power of an overall bureaucratic system.

The chip designers, computer architects, and process engineers using these workstations—more potent by far than mainframes of a decade ago—are far less dependent on bureaucracy for capital and support than their predecessors. The more intellectual functionality placed on single chips and the fewer expensive interconnections, the more the power that can be cheaply available to individuals. The organization of enterprise follows the organization of the chip. The power of entrepreneurs using distributed information technology grows far faster than the power of large institutions attempting to bring information technology to heel.

Rather than pushing decisions up through the hierarchy, the power of microelectronics pulls them remorselessly down to the individual. This is the law of the microcosm. This is the secret of the new American challenge in the global economy.

So it was that the new law of the microcosm emerged, leaving Orwell, Von Neumann, and even Charles Ferguson in its wake. With the microprocessor and related chip technologies, the computing industry replaced its previous economies of scale with new economies of microscale.

Nonetheless, the new rules are still not well understood. For example, in the last five years, IBM, Digital Equipment Corporation, Hewlett Packard, most of the remnants of the once awesome BUNCH (Burroughs, Univac, NCR, Control Data, and Honeywell), as well as the entire Japanese computer industry, time and time again have foundered unwittingly on these new facts of life in micro-electronics. While new multi-billion-dollar industries emerged in networked workstations led by Sun and Apollo, and in minisuper-computers, led by Convex, Alliant, N-Cube, Multiflow, and many others, the leading firms still upheld the Von Neumann model. Computing should be centralized and serial rather than parallel and distributed. Even DEC allowed a whole $3 billion industry of workstations to erupt in the midst of its market rather than give up the concept of the time-shared VAX-run network.

But the force of change was inexorable. Mainframes remained valuable for rapid and repeated disk access and transaction processing for banks and airlines. But ninety times more cost-effective, small computers were barging in everywhere else. Distributed peer networks of specialized machines and servers were becoming ever more efficient than hierarchical networks run by mainframe or minicomputer hosts. In many special applications, such as graphics and signal processing, specialized embedded microprocessors would outperform large computer systems.

The law of the microcosm has even invaded the famous Cray super-computer, the pride of the Von Neumann line, still functioning powerfully in many laboratories around the globe. Judging by its sleek surfaces and stunning gigaflop specifications (billions of floating point operations per second), the Cray appears to be high technology. But then there is the hidden scandal of the "mat." Remove the back panel and you will see a madman's pasta of tangled wires. The capacity of these wires—electrons can travel just nine inches a nanosecond—is the basic limit of the technology.

The Mead mandate to economize on wire while proliferating switches has led inexorably to parallel architectures, in which computing jobs were distributed among increasingly large numbers of processors interconnected on single chips or boards and closely coupled to fast solid state memories. Such machines were often able to outperform Von Neumann supercomputers in various specialized applications. A symbolic victory came when HiTech, a parallel machine made of $100,000 worth of application-specific chips, challenged and generally outperformed the $15 million Cray Blitz supercomputer in the world chess championships for computers.

The very physics of computing dictates that complexity and interconnections—and thus computational power—be pushed down from the system into single chips where they cost a few dollars and are available to entrepreneurs. This rule even constrains the future of

IBM. Ralph Gomory, IBM's chief scientist, predicted in 1987 that within a decade the central processing units of supercomputers will have to be concentrated within a space of three cubic inches. The supercomputer core of the 1990s will be suitable for a laptop.

Following microcosmic principles in pursuing this goal, IBM in December 1987 joined the venture capitalists and sponsored the new supercomputer company headed by former Cray star Steven Chen, designer of the best-selling Cray XMP. The Chen firm will use massive parallelism to achieve a hundredfold rise in supercomputer power.

Such huge gains were only the beginning. But there will be one key limit: stay on one chip or a small set of chips. As soon as you leave the microcosm, you will run into dire problems: decreases in speed and reliability, increases in heat, size, materials cost, design complexity, and manufacturing expense.

Above all, the law of the microcosm means that the computer will remain chiefly a personal appliance, not a governmental or bureaucratic system. Integration will be downward onto the chip, not upward from the chip. Small companies, entrepreneurs, individual inventors and creators, handicapped persons, all will benefit hugely. But large centralized organizations will lose relative efficiency and power.

This is nowhere more evident than in telecom systems, once a paradigm of the oligopoly so cherished by Ferguson and his allies. Still America's most comprehensively organized business institution, the telecom network is being inexorably redesigned in imitation of the microcosmic devices that pervade it. Though afflicted by obtuse regulators fearful of monopoly, the telecom establishment is actually under serious attack from microelectronic entrepreneurs.

Once a pyramid controlled from the top, where the switching power was concentrated, the telephone system is becoming what Peter Huber of the Manhattan Institute calls a "geodesic" network. Under the same pressure Carver Mead describes in the computer industry, the telecommunications world has undertaken a massive effort to use cheap switches to economize on wire. Packet switching systems, private branch exchanges, local area networks, smart buildings, and a variety of intelligent switching nodes all are ways of funneling increasing communications traffic onto ever fewer wires.

As Huber explains, the pyramidal structure of the Bell system was optimal only as long as switches were expensive and wires were cheap. When switches became far cheaper than wires, a horizontal ring emerged as the optimal network. By 1988, the effects of this trend were increasingly evident. The public telephone network commanded some 115 million lines, while private branch exchanges held over 30 million lines. But the public system was essentially stagnant, while PBXs were multiplying at a pace of nearly 20 percent a year. PBX's, though, now are being eroded by still more decentralized systems. Increasingly equipped with circuit cards offering modem, facsimile, and even PBX powers, the 40 million personal computers in the United States can potentially function as telecom switches.

Personal computers are not terminals, in the usual sense, but seminals, sprouting an ever-expanding network of microcosmic devices. AT&T and the regional Bell companies will continue to play a central role in telecommunications (as they are released from archaic national and state webs of regulation). But the fastest growth will continue to occur on the entrepreneurial fringes of the network. And inverted by the powers of the microcosm, the fringes will become increasingly central as time passes.

Nevertheless, the question remains, will the producers of the technology be as bound in their business tactics by the laws of the microcosm as their customers are? In the face of repeated Japanese victories in mass manufacturing of some essential components, mostly achieved through government-favored conglomerates, is it credible that entrepreneurs can be as powerful a force in the production of the technology as they are in its use?

The answer is yes. The industry must listen to its own technology, and the technology dictates that power gravitate to intellect and innovation over even the most admirable efficiency in mass production. The evolution of the industry will remorselessly imitate the evolution of the chip. As chip densities rise and suddenly inexpensive custom logic replaces memory, specialized computers-on-a-chip will become the prevailing product category. Even more than today, chip designers will be systems builders and systems builders will be specialists.

The world of the collapsing computer is a world everywhere impregnated with intelligence in the form of scores of thousands of unique designs. Even today, the "cowboy capitalists" of the U.S. industry reap far more than their fair share of value added and profits. As the power of custom designs grows and the relative importance of the general-purpose computer, patched together with general-purpose chips, declines, innovators and entrepreneurs will be more favored still. And these innovators and entrepreneurs will enjoy the same advantages of increasingly cheap and increasingly powerful design and production equipment that has been bequeathed to entrepreneurs throughout the economy.

As the world economy becomes more global and interdependent, the voices of autarky will be able to summon a crisis a month for the rest of the century. The Europeans and Japanese are taking over our silicon production, our photomask firms, our wafer suppliers. But to say that foreign conglomerates will dominate the world information industry because they have the most efficient chip factories or the purest silicon is like saying the Canadians will dominate world literature because they have the tallest trees. Or Kodak will dominate the movie industry because it is the most efficient producer of film.

In appraising the predictions of the prophets of oligopoly, there is a powerful test. Most of the recent spate of predictions were first made in 1985. Since 1985, we have seen several years of change and turmoil in the industry. Over the five years between 1983 and 1987, a record ninety-nine new semiconductor firms emerged. If the prophets were correct, these companies would be in trouble. But the vast majority are already profitable: their total revenues exceeded $2 billion by 1988, and their success ratio far excels the performance of the startups in the supposed heyday of the industry between 1966 and 1976. By most standards, the average productivity of the new firms is far higher than the output of established firms.

If the doomsayers were correct, few of these companies could afford to build semiconductor factories. But between 1985 and 1988, some two hundred minifabs have been constructed for making leading edge custom semiconductors. Some of these facilities offer the most efficient advanced submicron processes in the industry. Contrary to widespread contentions that the industry is growing ever more capital-intensive, in fact the efficiency of the new production gear far exceeds its incremental costs.

The appropriate measure of production in information technologies is not output of "chips" or even "computers." Nor is the appropriate standard output in dollars, for the cost per unit is plummeting while the real value of the output soars. The correct measure is functionality, or utility to the customer.

In cheaply delivering functions for the customer, the new semiconductor production gear every year advances decisively over its predecessors. Such dynamic startups as Chips & Technologies and Weitek have avoided direct fabrication of chips, farming out their production to Japanese and American firms with excess fab space. But other startups, such as Cypress, Integrated Device Technologies, and Performance, have created some of the most advanced fabs in the industry and lead most of the established companies in a variety of advanced CMOS products. One of the fastest-growing and most profitable firms in the industry, T. J. Rodgers's Cypress in 1988 barged massively not only through the $50 billion barrier cited by Sanders but through the $100 million level as well. With no federal aid, Performance exceeded all the goals of the 2 billion-dollar Pentagon VHSIC program for submicron geometries, and produced CMOS static RAMs in volume with speeds excelling all the large firms in the industry. Xicor has mastered non-volatile memory technologies that have so far defied the best efforts of Intel and the Japanese. Micron Technology is not only producing the industry's smallest DRAMs; it is also making some of the fastest large CMOS static RAMs. Intel, the virtual inventor of the DRAM, in 1988 agreed to sell Micron-made chips under the Intel label and obtained warrants to purchase $11.6 million worth of Micron stock.

The advocates of industrial policy are enthralled by supercomputers as the key to competitiveness. Yet the leading U.S. supercomputer firm is Cray Research, a relatively small firm, deeply dependent on the new generation of semiconductor companies for key technologies. Gigabit Logic provides its superfast gallium arsenide logic devices; Performance supplies CMOS chips interfacing with the gallium arsenide devices; Micron furnishes 256K fast static RAM built to Cray specifications. "I don't know what we'd do without the startups," said Seymour Cray, a leading investor in Performance and fervent admirer of Micron.

The first successful billion-component chip will be designed, simulated, and tested on a massively parallel desktop supercomputer that will yield functionality far beyond its cost. Whatever the chip will do, those twenty Crays of computing power will make it incomparably more potent than current microprocessors which take scores of designers to create and are built in a $150 million plant. It will be manufactured on a laser direct write system or an X-ray stepper that will also far outproduce its predecessors in the functions it will offer per dollar.

Doubtless all the business magazines will still be speaking of the incredible cost of building and equipping a new semiconductor plant. But in functional output per dollar of investment, the fab of the year 2000 will be immensely more efficient than the "Goliath fab" making DRAMs today. By the law of the microcosm, the industry will continue to become less capital-intensive, more intensive in the individuals' mastery of the promise of information technology.

Contrary to the usual assumption, even the need for standards favors entrepreneurs. Standards often cannot be established until entrepreneurs ratify them. Based on an operating system from a startup called Microsoft, the IBM PC, for example, did not become a prevailing architecture until thousands of entrepreneurial software and peripheral companies rose up to implement it. Similarly, Ethernet, launched at Xerox PARC by Robert Metcalfe, gained authority despite the opposition of IBM, when scores of companies, from SEEQ to Sun to DEC to Metcalfe's 3-Com itself, adopted and established it. Just as public utilities open the way for thousands of small firms to use electricity and other services, information industry standards emancipate entrepreneurs to make vital contributions to the economy.

Finally, if the doomsayers were correct, the United States would still be losing market share in electronics. But since 1985, U.S. firms have been steadily gaining market share in most parts of the computer industry and holding their own in semiconductors (adjusted for the rising yen).

The revolt against the microcosm remains strong among academic intellectuals. Many critics of capitalism resent its defiance of academic or social standards of meritocracy. William Gates of Microsoft left Harvard for good his sophomore year. The system often gives economic dominance to people who came to our shores as immigrants with little knowledge of English, people unappealingly small, fat, or callow, all the nerds and wonks disdained at the senior prom or the Ivy League cotillion. Some academics maintain a mind-set still haunted by the ghost of Marxism. But the entrepreneur remains the driving force of economic growth in the quantum era.

Rather than pushing decisions up through the hierarchy, the power of microelectronics pulls them remorselessly down to the individual. It applies not only to business organization but to the very power of the state as well.

CHAPTER
27

The Eclipse of Geopolitics

As the chip reorganizes industy and commerce, so also will it reorganize the powers of states and nations. The laws of the microcosm subvert any attempt to capture, intimidate, confine, or overwhelm the exertions of mind by the tyranny of matter.

The great liberator in the quantum economy is the mobility of mind. A million and eventually a billion switches on a chip means that one free mind plus a workstation can outperform any array of regimented minds. Using computational subroutines, one mind can do work that in the past required dozens of subordinate minds devoted to routine tasks. Using that master of repetition, the computer, humans can offload all subhuman work and rote.

Potentially eliminating most rote jobs, the micocosm makes obsolete the hierarchies and capital fixtures that once controlled, channeled, paid, and patronized rote workers. Mind flees the corporate traps of the pre-quantum economy.

The mobility and ascendancy of mind among all the forms of capital deeply undermines the power of the state. Quantum technology devalues what the state is good at controlling: material resources, geographic ties, physical wealth. Quantum technology exalts the one domain the state can never finally reach or even read: mind. Thus the move from the industrial era to the quantum era takes the world from a technology of control to a technology of freedom.

The most evident effect of the change is a sharp decline in the value of natural resources. The first Industrial Revolution vastly increased the value of materials. All the dirt, rock, and gunk that had been ignored for centuries suddenly acquired worth in the age of mass manufacturing. The new industrial revolution is a revolution of mind over matter, and it is rapidly returning what used to be called "precious natural resources" to their previous natural condition as dirt, rocks, and gunk.

The use of steel, coal, oil, and other materials is plummeting as a share of value added in the economy. As a symbol of the shift, consider two smelting processes. Smelting iron, you banish silicon in the slag as dirt; "smelting" silicon, you get rid of the iron as conductive waste. Following Moore's parable of MOS, a silicon chip is less than 2 percent raw materials. A few pounds of fiber-optic cable, also made essentially of sand, carries as much information as a ton of copper.

The coming decades may bring about a further transformation of the economics of energy, as superconductive elements promise to increase by compounding factors of between 5 and 100 the efficiency of every wire, receiver-transmitter, electric motor, solar collector, power generator, and magnetic battery. By allowing sharp reductions in the weight of a wide array of mechanical systems, this technology may further accelerate the decline in the value of raw materials in the economy. The drop in their price is not a caprice of the market but a fundamental event, reflecting a permanent technological change.

The decline in the value of raw materials has changed the foundations of geopolitics. Raw materials have long constituted a leading reason and reward for military aggression. In the past, ownership of particular regions imparted great political and economic power. The balance of power in Europe depended in part upon who controlled the coal and steel in the Ruhr Basin. The Ruhr Basin is now a European sink of government subsidies.

We live in an epoch when desert-bound Israel can use computerized farming to supply 80 percent of the cut flowers in some European markets and compete in selling avocados in Florida; when barren Japan can claim to be number one in economic growth; and when tiny islands like Singapore and Hong Kong can far outproduce Argentina or Indonesia.

To comprehend the change, consider a steel mill, the exemplary industry of the previous epoch: A huge manufacturing plant entrenched near iron and coal mines, anchored by a grid of railways and canals, served by an army of regimented workers, all attended by an urban infrastructure of physical systems and services. At every step the steel mill can be regulated, taxed, and controlled by government. Compare this massive array of measurable inputs and outputs to a man at a computer workstation, with access to databases around the world, designing microchips of a complexity exceeding the entire steel facility, to be manufactured from pattern generation tapes. Even the tape, the one physical manifestation of his product, has become optional. Without any fixed physical manifestation at all, the computer design can flow through the global ganglion into another computer attached to a production line anywhere in the world.

Quantum products resemble books more than steel ingots. The value is in content, not substance. A book costs some 80 cents to print; the bulk of its value is created by its author, publisher, distributor, and retailer. No one imagines that the printer commands crucial manufacturing knowledge that will allow him to dominate literature. VCRs are similar. This $6 billion industry is enabling the creation of a $10 billion industry in VCR software that has revitalized the U.S. film industry.

The value of a videocassette is not chiefly in the tape or even the player; it is in the information, the movie. Even the marketers of videocassettes gain a larger share of their value than the manufacturer. Manufacturing remains vitally important. But the analysts who claim that manufacturing is the prime source of new value have succumbed to a totemistic worship of material things in an age of ideas.

Even observers who comprehend the nature of information technology, however, often fail to understand its radical effect on international economics. The decline in the value of raw materials entails an equal decline in the value of geography. In an age when men can inscribe new worlds on grains of sand, particular territories are fast losing economic significance.

Not only are the natural resources under the ground rapidly declining in value, but the companies and capital above the ground can rapidly leave. Commanding a worldwide network of transport and communications, the businessman sends wealth flashing down fiber-optic cables and caroming off satellites at the speed of thought rather than of things.

Capital is no longer manacled to machines and places, nations and jurisdictions. Capital markets are now global and on line twenty-four hours a day. People—scientists, workers, and entrepreneurs—can leave at the speed of a 747, or even a Concorde. Companies can move in weeks. Ambitious men need no longer stand still to be fleeced or exploited by bureaucrats. Geography has become economically trivial.

Except perhaps for tourism and historic sentiment, control of particular territories confers virtually no business advantage. Walt Disney could even make a dump like Anaheim, California—or Orlando, Florida—into a leading tourist attraction. It is people and ideas that matter, not places and things.

Just as the value of the knowledge outstrips the value of the material in most modern goods, so the knowledge of the value moves faster than its vessel: money moves far faster than goods. This change has transformed the dynamics of global markets. As most economists see international exchange, goods come first, and capital movements follow to compensate for imbalances of trade. But in the quantum era, capital markets are far more efficient than goods markets, which are hobbled by transportation costs, cultural differences, and protectionism.

Therefore capital movements now often come first and goods follow. For example, during the middle and late 1980s, high Japanese savings rates and declining rates of economic growth prompted a tenfold rise in overseas investment, largely in the United States. This capital flight from Japan was a capital surplus for the United States, and reflected confidence in the leadership of the United States in the global quantum economy. Usually depicted as a trade deficit, the Japanese goods in fact flooded into the United States to balance the influx of investment. This inflow of capital allowed U.S. citizens to purchase large amounts of highly desirable industrial equipment, components, and other merchandise from Japan, and allowed U.S. industrial production to outpace Japan's, all without reducing our purchases of domestic goods.

In the global quantum economy, there is no more reason for a balance of merchandise trade between nations than between any two American states. National politicians constantly talk of international trade. But in the microcosm, capitalist countries no longer trade. Across increasingly meaningless lines on the map, entrepreneurs rush huge and turbulent streams of capital, manufacturing components, product subassemblies, in-process inventories, research and development projects, royalties, advertising treatments, software programs, pattern generator tapes, technology licenses, circuit board schematics, and managerial ideas. Many of the most important transactions consist of electronic or photonic pulses between branches, subsidiaries,

contractors, and licensees of particular companies, and defy every cal-
culation of national debt and exchange.

The central conflict in the global economy pits the forces of statism
against entrepreneurs using the new microcosmic technology to integrate world commerce. Defying national boundaries, entrepreneurs
have knitted together a dense global fabric of manufacturing suppliers
with near just-in-time delivery commitments. They have ferreted out
the world's best human resources without regard to nationality, setting up factories and laboratories in foreign countries, and they have
brought in the immigrant engineers and researchers who are revital-
izing the best American firms. They have summoned an explosive
expansion of international investment ties that leaves the entire history of national economies in its wake.

During the 1980s, the U.S. government was mostly on the side of
the entrepreneurs. But bureaucracies around the world still struggled
to maintain their powers. Even in the United States, they choke trade
with quotas, suborn business with legislative bribes, spurn and insult
foreign investors, hobble high-tech exporters, thrill bigots everywhere
by smashing Japanese radios on the marble steps of Congress, amaze
other Asian exporters with ever more bizarre "dumping" charges,
stifle immigration and put our precious resource of foreign doctoral
students into a national security ghetto, mug key capitalist allies and
suppliers such as Korea and Taiwan, moon over socialist killers and
parasites from Nicaragua to Mozambique, burden producers with
new taxes, and wage war in general against the key sources of growth
in world capitalism.

In their most deadly assault on U.S. high technology, the bureaucrats destroyed the value of their dollars in order to favor increasingly
uncompetitive heavy industry. The result was to strengthen Japanese
electronics firms vastly by nearly doubling the value of their capital,
profits, and market share without in any way helping innovative U.S.
companies. The majority of high-tech products offer unique functions
rapidly moving down the learning curve with heavy investment in
both R&D and capital equipment. A one-time drop in price is worthless when it comes at the expense of skewing investment away from
American innovation and strengthening foreign rivals.

As the driving force in the rise of the world's central strategic technology, the U.S. upsurge in computers in the early 1980s was a major
factor in the rise of the dollar. The U.S.–led devaluation of the dollar,
at the behest of farmers and uncompetitive commodity producers,
showed a complete incomprehension by the U.S. government of the

true sources of competitiveness. A fruitless mercantilism, worshiping trade gap totems, threw away the fruits of American high-tech entrepreneurship and allowed the Japanese to buy at bargain rates what they could not create. A revolt against the microcosm allowed textiles and timber firms to export material goods against third world rivals while devaluing the creativity of American innovators on the frontiers of the global economy of the mind.

Nonetheless, barring war or other catastrophe, the bureaucrats will inevitably lose in the end. The global microcosm has permanently shifted the world balance of power in favor of the entrepreneurs. Using the planetary utility, they can avoid most of the exactions of the state. Without their fully voluntary cooperation a government cannot increase revenues, enhance military strength, provide for the public welfare, or gain economic clout.

Entrepreneurial freedom, however, means that trade cannot balance. The pattern of world trade chiefly reflects the planetary utility: the deep inner structure of the worldwide division of labor created by entrepreneurs. In particular, the industrial convergence between Asia and the United States on the frontiers of information technology has launched an intricate, highly specialized, and inextricably complex architecture of long-term contracts, complementary skills, and marketing relations that defy simple export-import accounting categories. In conventional terms, this is not trade but a form of horizontal and vertical integration across national boundaries.

Under the intensely nationalistic Andrew Grove, Intel Corporation, for example, became a fully international firm, deeply integrated with Asian competitors and collaborators. During the late 1980s, its fifth-largest customer and most bitter rival was NEC. All of its other leading customers in the U.S. computer industry, except possibly IBM, were utterly dependent on complementary devices produced in Japan. Mitsubishi Electric made an increasing share of Intel's EPROMs. South Korea's Samsung produced many of Intel's DRAMs. Zymos, an American firm mostly owned by the South Korean Daewoo, designed Intel's IBM clone chip sets; Daewoo, in turn, used Intel microprocessors in computers made in Korea and designed and marketed in the United States for several years by Leading Edge. And Intel packaged chips in the Philippines, bought capital equipment from Nikon, silicon wafers from Shin-Etsu, and photomasks from Dai Nippon. Even Intel's leading customer, IBM, turned to Shimizu of Japan to construct its world-leading 648,000-foot-square chipmaking factory in Fishkill, New York.

Intel during this period in the late 1980s was growing faster in real terms than any other major semiconductor company. But like thousands of other U.S. technology firms, it would collapse within weeks without imports from Japan and it would rapidly deteriorate without a constant flow of brilliant Asian immigrants.

Similarly, the fastest-growing computer firm during the late 1980s was Sun Microsystems. It bought chips from Toshiba and NEC, used Fujitsu to manufacture its next-generation microprocessor, licensed its workstation design to Toshiba, and used TEL to market its computers in Japan. Sun used chips from U.S. firms that were fabricated or assembled in the Philippines, Japan, Singapore, and Scotland, using equipment made in Germany, Switzerland, Holland, Japan, and the United Kingdom.

Most of the value added came from Sun designs, software, and marketing skills. The company embodies the entrepreneurial genius and steadily increasing competitiveness of the U.S. economy. But without Japanese components, Sun would set tomorrow.

In the same way, thousands of Asian firms would fail without U.S. designs, programs, markets, licenses, and personnel. U.S. firms and their Japanese partners and licensees, for example, hold more than half of the Japanese semiconductor capital equipment market. U.S. firms hold about 20 percent of the truly available Japanese chip market (excluding firms that themselves manufacture chips), about the same share that Japan holds in the United States.

Even the Japanese automobile industry would stagnate without American collaboration in sales, distribution, design, and even production. Increasingly, an American company with a majority of sales and a near-majority of workers in the United States, Honda made five out of the top ten of *Motor Trend* magazine's 1987 cars of the year. At a time when U.S. automakers increasingly were moving offshore, Honda continued to move to the United States, with a multi-billion-dollar investment program. Since an additional dollar of Honda's earnings was more likely to be spent in the United States than an additional dollar of Ford's earnings, the usual nationalist codes and assumptions were collapsing.

Around the world, economists continued to maintain that these fully profitable and desirable relationships put the United States dangerously in debt to other nations. Many pundits claimed that a wanton binge of consumption, mostly financed by foreigners, had destroyed U.S. competitiveness. But fully one third of the imports consisted of capital equipment imports. A dollar of the trade deficit was four times more likely to be spent on industrial equipment than a dollar of domestic GNP. What was happening was a surge of U.S. investment led by U.S. entrepreneurs pioneering in the microcosm.

The trade gap was a fully positive result of the U.S. entrepreneurial lead in the world economy. While the trade gap grew, the dollar rose in value, and U.S. rates of growth of investment, employment, and industrial production increased between two and three times faster than the rates of our trading partners, including for most of the mid-1980s Japan. Far from losing competitiveness, the United States steadily gained market share in the crucial new technologies of the quantum era.

The advocates of trade balance convey an image of autonomous national economies with trade and capital movements equilibrated at the borders by gyrating currency values. They may think they are for freedom, but in fact they are exchange rate mercantilists, revolting against the microcosm in the name of an economics of national control. Their great mercantilist triumph, driving the dollar to a level far below purchasing power parity, reflected the power of old established industries against microcosmic entrepreneurs expanding into the world.

The real target of the bureaucracy is not Japan or even trade. It is the entrepreneurial freedom conferred by quantum technology. Trade can never balance as long as entrepreneurs can dispatch capital wherever they wish in response to new opportunities or in flight from oppressive taxes and protectionist barriers. Therefore the drive to balance trade necessarily dictates an effort, whether direct or indirect, to control capital flows. So integrated is the microcosmic economy that the revolt against it cannot succeed without employing a degree of force and reach of control far beyond the present intent of the new mercantilists.

The nations that join this revolt will inevitably lose power in relation to the governments that liberate their people. During the 1990s, the move to the microcosm will continue to yield new wealth at an ever-increasing pace. As in semiconductors, telecom, and computers, the economies of the microcosm will overcome the economies of scale. A similar development is predictable in superconductors, which also partake of microcosmic powers.

Just as the first electrical motors were used simply to replace centralized steam engines, the first superconductive engines may well be installed by giant utilities. But the microcosm dictates once again the distribution of power, not its concentration. As the temperature of superconductivity rises, the economies of scale in coolants and other overheads decline. The economies of centralization will once again succumb to the economies of microscale and individual creativity. The costs of distributing power will drop in comparison to the economies of concentrating it. Like the automobile, the new technologies of power generation will endow and liberate individuals rather than institutions. Like microcosmic information, microcosmic energy favors peer networks over hierarchical systems.

This good news for individuals and entrepreneurs, however, is bad news for socialism. The state can dig iron or pump oil, mobilize manpower and manipulate currencies, tax and spend. The state can expropriate the means of production. But when it does, it will find mostly sand. For the men of production, the entrepreneurs, run for the daylight of liberty. One way or another, most of the time, the entrepreneurs take their money with them or send it on ahead. But always they take their minds, and knowledge is their crucial power.

Ideas are subjective events that always arise in individual minds and ultimately repose in them. The movement toward a quantum economy necessarily means a movement toward an economy of mind. Collective institutions will survive only to the extent that they can serve the individuals and families who comprise them.

This change in the balance of power between individuals and the state, human capital and material capital, also has transformed the nature of military rivalry. In the Pentagon today, military planners still speak of "landmasses" and sealanes, "strategic minerals" and "chokepoints," "soft underbellies" and crucial ports and passes. They try to protect "strategic technologies" and conceal new scientific knowledge. They ascribe great value to stretches of desert ruled by primitive tribes and regard particular barren points and peninsulas as a key in the lock of world supremacy. Control of specific countries is said to confer enormous power even if these nations cannot support themselves, let alone impart new strength to their conquerors.

Watching the Soviet Union gain control over one after another Third World backwater or turn once prosperous lands into penury by capturing them, many American observers believe that the Communists are winning the Cold War. But the Soviet strategy is obsolete. The Soviet Union could capture the entire Third World without in any way gaining military or economic power. In fact, each new conquest serves only as a new drain on the Soviet economy.

The military threats of the future come not from mass mobilizations for territorial expansion but from nihilist forces of terrorism and reaction, lashing out against the microcosmic wealth and creativity beyond their ken. The new nihilists will probably arise in many countries around the globe and will unleash military technologies with no geographic boundaries. To fight these threats, nations will have to employ ever more sophisticated information technologies. If terrorism increases, all democracies will face the challenge of using new methods of electronic surveillance, security, and control, together with new non-lethal weapons, without seriously infringing on the rights of law-abiding citizens.

The microcosm also offers a path of escape from the most horrifying form of state terror. The conventional notion is that the arms race threatens to spiral "out of control" unless it is limited by arms control negotiations. But this view is almost the opposite of the truth. In fact, the arms race is spiraling into the control of the microcosm. The chief threat to this salutary process is arms control negotiations that halt or inhibit technical progress.

For example, American missile warheads of the 1980s are only one seventh as destructive as the warheads of the 1960s, and Soviet warheads seem to have dropped in size by about one third. The total destructive power of the U.S. arsenal has declined over the last twenty years from 12,000 megatons to well under 4,000. This trend was achieved not through arms control but through nuclear testing and through the increased use of computer technologies to refine the accuracy of the weapons. As the aim improved from a radius of miles to a few hundred feet, the need for heavy explosives declined apace. The development of "smart" warheads will improve accuracies by a still greater measure and allow still more drastic future reductions in explosive power.

In general, the arms race began with hugely destructive and vulnerable weapons systems, kept in constant readiness and regulated by unreliable modes of command and control based on computers using vacuum tubes. With liquid-fueled missiles targeted chiefly on cities and bombers kept in the air, the defense postures of twenty-five years ago were far more dangerous, far more subject to accident or miscalculation, than the arsenals of the 1980s. In the absence of arms control, the defense posture of the 1990s will be safer still.

Substituting information for heavy materials, propellants, and explosives, this trend reflects the same forces that have devalued raw materials and territories in the global economy. The ultimate expression of this logic is strategic defense. With the far cheaper application of information technology, a Strategic Defense Initiative can largely neutralize the brute force of heavy missiles and nuclear explosives. SDI can greatly raise the cost of a successful nuclear attack. Thus it can reduce the incentives for any national program to acquire such weapons. Alone among feasible strategies, SDI can effectively inhibit the proliferation of nuclear weapons while allowing a continued slow reduction in military spending as a share of the GNP.

The law of the microcosm dictates a remorseless quest for cheaper ways of accomplishing both industrial and military goals. If information technology were not far more efficient than industrial technologies, it would not be rapidly increasing its share of the world economy.

The Soviet SS-18 is a 240-ton monster taller than the Washington Monument. Propelled by some 50,000 gallons of fuel, it bears between ten and thirty warheads each more powerful than the entire panoply of World War II and each capable of destroying an American city. Made of materials, fuels, and components that are not declining in price, this missile, fully armed, costs nearly half a billion dollars. But the SS-18 and all its progeny will fall like Goliath before a smart rock the size of a man's hand—a defensive device containing a computer the size of a fly that declines in cost or increases in efficiency at a pace of some 30 percent a year. The most powerful weapon in the history of the world, this monster's only long-run defense against U.S. information technology is Soviet disinformation, otherwise known as arms control.

Construction of an effective defense against missiles, however, depends on listening to the technology. SDI must fully conform to the laws of the microcosm or it cannot hope to work. Rather than using new systems to redeem old strategies, the microcosm dictates adoption of new approaches. As Raymond Kurzweil told then Defense Secretary Caspar Weinberger at a meeting with business leaders in 1987, SDI cannot work as a centralized system. It must follow Mead's laws. It must economize on wires, which are expensive, and multiply switches that are cheap. As much as possible, it must eschew complex software and integrate algorithms in architecture. It should use the natural redundancy of neural systems rather than risking the complexity explosions of error correction and process replication. Above all, SDI should rely chiefly on dispersed and largely autonomous interceptors rather than a vast integrated network like the NORAD air-defense system.

Encountering these realities, the SDI program has been slowly adapting to the laws of the microcosm. At first, researchers envisaged a completely centralized system. It was patterned on the battle management and command, control, and communications concepts of air defense, multiplied in complexity through four stages of closely coordinated space- and land-based systems and satellites.

Assuming the satellites escape laser attack, this centralized system would entail a large increase in supercomputer power. It requires a vast software program comprising many millions of lines of code to coordinate operations. Under the chaotic and unpredictable conditions of war, the software must appraise the progress of the attack, communicate the number, location, and trajectories of surviving warheads, and must manage communications to the four levels of interceptors.

Opponents of SDI have charged that this software mission is impossible. The problem of SDI in its centralized form, however, is not that it is impossible, but that it violates the intrinsic nature of the underlying technologies that make it possible. In response to the need for high levels of reliability in the system, the Pentagon began calling for increasing levels of redundancy and fault tolerance. In response to the widespread recognition that battle management software was the crucial constraint, analysts urged that the entire hardware development effort be integrated with the software development effort. In response to the huge need for gigaflops that the C3 apparatus entailed, the Pentagon launched a special parallel supercomputer project in Washington to increase cycles per second a hundredfold.

All in all, the SDI program was violating not only Mead's laws but also Drucker's mandate. It was failing to listen to the technology and it was solving problems and neglecting opportunities. Thus, it was wasting scarce resources on the most difficult challenges and ignoring the chance for a major redefinition of the program that would accord with the Mead and Drucker insights.

The message of the microcosm is that this effort is unnecessary. Consider first the hardware demands faced by the Pentagon's supercomputer project. This effort envisages creation of a command and control system a hundredfold more powerful than an existing Cray 2 supercomputer. James Meindl, however, estimates that by the year 2000, single chips will hold as many as a billion transistors. That means that a single chip could contain the central processing units of twenty Cray 2s. Manufactured in volume, these chips will drop in cost to below $100 apiece, and, depending on the efficiency of twenty-first-century manufacturing processes, could cost far less. Thus the hundredfold increase over the power of a Cray will probably be achievable with under ten chips.

Such new microchips would allow for creation of a highly parallel supercomputer many thousands of times more efficient than current Crays, with room for huge redundancy and fault tolerance. Running failsafe software developed over the next decade, these machines might be seen to allow efficient four-phased battle management that could indeed render the ballistics missile with nuclear payload largely impotent and obsolete.

This response, however, flouts the law of the microcosm. The kind of technologies envisaged by Meindl together with the analog pattern recognition capabilities forseen by Mead would obviate the entire problem of battle management and command and control. Instead of pursuing a centralized system or even a distributed system coordinated at a central point, the Pentagon should focus on the particular software requirements of widely dispersed and entirely autonomous installations on the ground, in space, or both.

Forget command and control. Take advantage of microcosmic technology: dispersed, local, massively parallel, application-specific devices that can recognize patterns, compute trajectories, and disarm a wide variety of possible weapons. As computers develop the sensory capabilities predicted for them by Mead, Kurzweil, and others, centralized battle management of SDI will be as pointless as linking every patrol leader in an infantry division to the Joint Chiefs' Situation Room.

Rather than a huge, expensive, and possibly vulnerable system usable only in response to a particular kind of missile attack, such a strategic defense program would produce a robust and flexible array of capabilities. These technologies of pattern recognition and weapons neutralization could serve both in strategic defense and in tactical warfare, against cruise missiles and sea-based launchers, against tanks and personnel, against attacks from the Soviet Union or from any Third World dictatorship.

Listening to the technology, strategists will eschew huge and specialized software programs, centralized computer systems, and complex and expensive defenses. Instead, the Pentagon should abet the industrial drive in the private sector toward a dispersal of cheap computer and sensory powers. Instead of competing with Cray, IBM, and a hundred other companies pursuing the grail of highly parallel computation, the Pentagon should harvest this privately developed hardware. Since hardware capabilities are improving faster than software, it should use hardware efficiency and speed to relieve pressures for software complexity. It should exploit the likelihood of as much as a millionfold increase in the cost-effectiveness of application-specific parallel processors performing feats of artificial intelligence and pattern recognition. Until land-based superconductivity becomes practical, the system might take advantage of the chill of space for superconductive technologies. Facing dispersed and flexible defenses, the software problem shifts to the offense.

These defensive systems should as much as possible use commercially available components. SDI will prove a Pyrrhic victory if it comes at the cost of competitiveness in commercial electronics. A leading advantage of a defensive strategy is that it conforms with the existing directions of the computer and semiconductor technology; it thus affords a way of strengthening the industry rather than merely depleting its personnel and other key resources.

The goal of using commercial components is to achieve redundancy by multiplying defensive modules that become more cost-effective every year. As President Reagan often insisted, the United States should not fear eventual Soviet acquisition of similar technologies; in fact, a Soviet SDI would obviate hair-trigger postures and further reduce the threat of war by accident or miscalculation.

Nonetheless, an open contest greatly favors Western abilities and would improve the national security far more than a futile attempt to conceal some more secret and specialized technology. Microcosmic defense dictates abandoning most of the export controls and secrecy regulations that have inhibited U.S. technical development in recent decades. As Edward Teller has pointed out, the United States has not competed successfully even with the Soviet Union in its secret programs such as missilery and explosives. But the Soviet Union cannot begin to compete with the United States in microcosmic technologies that are known by all.

Export controls become a form of arms control imposed chiefly on ourselves. We begin by embargoing advanced weapons technologies sent directly to Moscow; we end up seizing Apple computers on the docks in San Diego and barring urinalysis equipment because it contains embedded microprocessors available by the millions around the globe.

Determined to deprive the Soviet Union of the ability to create VLSI circuits, we end up delaying for months the shipment of wafer steppers from GCA to Hong Kong just as Nikon enters the marketplace. Thus we jeopardize the reputation of American companies as reliable suppliers to the world's fastest-growing markets in the Pacific and move the strategically vital semiconductor capital equipment industry increasingly to Japan and Europe.

We hold up licenses for 64K memory chips sent to Singapore while their price drops from more than $2 to 55 cents and the world supply rises to the billions. We restrict transmission of "sensitive but unclassified data" without any clear definition and cast a pall of uncertainty over every software or technology sale to a foreign purchaser who knows he will need regular service and updating of documentation. We bar immigrants from technical conferences devoted to the exploration of findings that only they can fully understand.

This approach reached a special pitch of absurdity when the United States excluded foreign nationals from a superconductivity conference in order to protect "our" technology, which had been discovered by two Swiss working for IBM in Zurich, was further exploited by several Chinese Americans working in Houston and Huntsville, and was pushed by Bell Labs only after an encounter with a Japanese researcher at a conference in New Jersey.

Subjecting the pullulating mecca of U.S. technology to the endless rigamarole of security clearances, licenses, citizenship laws, and nondisclosure agreements, snarling new technical papers in a web of bureaucratic delays, is entirely inimical to the dictates of the new science. In the long run, such restrictions on knowledge will be more devastating to U.S. defense than the work of thousands of spies.

Any successful strategy must be based on a realistic appraisal of costs and possibilities. The bad news is that the Soviet Union will inevitably steal almost everything we create. Stealing, bribing, filching, and confiscating are the moral and practical essence of socialism; it is all it can do. The more the arms race is reduced to a rivalry in cloaks and daggers, the better the Communists will perform. The only way we can keep technology from the Soviets is to keep it also from ourselves.

The good news is that the most crucial sources of technology are forever beyond the Soviet reach. Except for a few military codes and specialized devices, our national security depends not on protecting extant devices but on education, research, and enterprise for new devices. Technology is not a thing that can be clutched and concealed. It is chiefly a process, and the Soviet Union can acquire it only by abandoning socialism.

Targeting a few critical technologies, the Soviets have already achieved huge military power. Because mathematics is beyond the ken of bureaucrats, Soviet mathematics has maintained high levels of excellence. It has even led to the creation of valuable computer algorithms and architectures that compensate to some degree for a retarded information hardware industry. But the broad advance of knowledge is the effect of free inquiry. Secrecy not only slows the pace of discovery and creation and excludes some of our most loyal and ingenious immigrant workers; classification also tends to help the Soviets by solving the key problem of their spies: namely, identifying key technologies amid an otherwise baffling welter.

To preserve our national freedom, we must acquiesce in the erosion of nationalism. The spread of the technologies of microcosmic defense will breed a limited and largely benign impotence of nations. SDI will tend to frustrate the weapons of ultimate destruction. Other "smart technologies," such as hand-held "stinger" missiles, may allow guerrilla bands to fend off invading armies. (The real lesson of Afghanistan was that the poorly funded Mujhedhin with stingers could demoralize Soviet forces even more quickly than the North Vietnamese could frustrate the Americans.)

The promise of the new military technologies themselves, though, is less significant than the decline in the value of the objective. It is not that heavy missiles and massive armies can no longer achieve major military and territorial goals; it is that the goals are incomparably less valuable than the cost of reaching them. In fact, by taking over a territory and controlling it, a nation will destroy its economic value. Economic growth depends not on territory or control but on liberation and knowledge. The totalitarian conqueror destroys the value of anything he captures. In the quantum era, power derives not from secrecy and control but from knowledge and liberation.

By enhancing the risks and costs of war and sharply reducing its rewards, the microcosm forces nations to compete on its own terms, in modes of rivalry that emancipate citizens rather than enslaving them.

This is the age of the individual and family. Governments cannot take power by taking control or raising taxes, by mobilizing men or heaping up trade surpluses, by seizing territory or stealing technology. By imperialism, protectionism, and mercantilism, nations eventually wither and weaken into Third Worldly stagnation. In the modern world, slaves are useless; they enslave their owners to systems of poverty and decline. The chief source of the new wealth of nations is free immigrants; the nations of the baby dearth will compete for them to pay the pensions of their aging societies. Today, nations have to earn power by attracting immigrants and by liberating their people, their workers and their entrepreneurs.

The gains of the quantum era could yet be destroyed by some thug offering a final horrible holocaust to the Moloch of matter. But the logic of the technology, the logic of the microcosm, which is becoming the logic of history, runs the other way. History has capsized every prophecy of triumphant bureaucracy.

Rather than a New Industrial State, this era will disclose the new impotence of the state. Rather than the Revolt of the Masses under the leadership of demagogues, this era will see the revolt of the venturers against all forms of tyranny. Systems of national command and control will wither away. Systems of global emancipation will carry the day. The dismal science of the economics of aggregates—capital, labor, and land—will give way to a microeconomics of liberty. The beggar-thy-neighbor strategies of mercantilism—of trade as a weapon of state—will collapse before the strategies of global wealth creation under the leadership of entrepreneurs. The economics of scarcity and fear will surrender to the economics of hope and faith.

These are not mere prophecies. They are imperious facts of life. Any nation, corporation, or government that ignores them will begin to fade away. The new technologies—themselves largely the creation of Promethean individuals—completely transform the balance of power between the entrepreneur and the state. Inventive individuals have burst every link in the chain of constraints that once bound the entrepreneur and made him a servant of parliaments and kings. He is no longer entangled in territory, no longer manacled to land, capital, or nationality.

Even labor is losing its grip. Labor once was local and collective and tied the businessman to a specific place. A labor force commanding specific skills comprised a company's most critical asset and wielded great power over entrepreneurs. A skilled labor force still is the key asset of a company. But with automation performing most routinized labor, workers are no longer groups; their work is increasingly personal rather than collective.

To the extent that labor organizes to bind a firm to a specific community and attempts to extort benefits beyond its contribution, the firm can seek other labor around the globe. Speaking English, U.S. managers command the world's most widespread language; living in the United States, they enjoy a mecca of immigration. Like capital, labor is increasingly individual and international. Indeed, in the current era immigration is the most important source of new national power.

The United States already depends on immigrants for its most productive manpower. In fact, even black immigrant families are 50 percent more likely than native whites to earn more than $35,000 a year. By attracting labor and capital, a nation can increase its wealth and power far more effectively than by capturing them or piling up surpluses in mercantilist trade. A key to competition in the global economy will be who wins the hundreds of thousands of skilled workers and entrepreneurs now leaving Hong Kong.

These developments, however, are no special American monopoly. Indeed, much of the benefit will be lost if only the United States and a few Asian capitalist countries follow the crucial lessons of the new technology. For the central lesson is that information technology is not a zero-sum game to be won by some Fifth Generation monopolist. Information technology constantly redistributes its own powers as it is used. The final and most flexible source and vessel of these powers is the individual human mind. The power of information always ultimately gravitates to individuals.

This is not a world in which the gain of one nation can only come at the expense of another. All the world will benefit from the increasing impotence of imperialism, mercantilism, and statism. In this new economy of freedom, Americans must hope for the prosperity and freedom of Russians and Chinese. We must celebrate the successes of Koreans and Japanese. We must hail the increasing wealth and power of the Third World. Depending on an altruistic spirit, the microcosm requires not only a technological renaissance but also a moral renewal.

CHAPTER
28

Triumph over Materialism

The age of the microcosm promises material gains far exceeding all previous eras. But access to all these material riches will come on one key condition: we must achieve a moral and philosophical as well as scientific and technical escape from the materialist superstition.

A persistent canker of materialism survives at the very heart of the quantum economy. In computer science itself there persists the idea that mind is matter. Informing much of the agenda of artificial intelligence, this idea has committed a generation of computer scientists around the world to the most primitive form of the materialist superstition.

The idea of mind from matter resembles the notion of a perpetual motion machine. Centuries of dismal setbacks fail to deter its devotees. Max Delbrück compares the project to Baron Munchausen's effort to pull himself out of a swamp by his own hair. Yet all over the world today cognitive scientists are painfully yanking, twisting, and tugging on their scalps. The fact that they remain deep in the swamp only spurs new frenzies of effort.

In 1975, however, came deliverance for all the Munchausen materialists in cognitive science. Neurosurgeon Wilder Penfield reported discoveries that in the historic saga of the downfall of materialism may even rival the experiments of Michelson and Morley nearly a century before. While Michelson and Morley combed the cosmos seeking signs of material ether, Penfield pored through the material brain looking for signs of mind.

Penfield's researches were altogether as unique and unprecedented as the ether tests. He performed surgery on the brains of 1,132 epileptics over a thirty-year period. After removing part of the skull of the patient under anesthesia, he would bring him back to full consciousness. Then Penfield would explore the brain widely with an electronic probe. Since the brain lacks sensitivity, the patient, though fully alert, could feel no physical pain.

With the patient guiding the surgeon's explorations and reporting the results of stimuli at different cerebral sites, the doctor would often find and remove the cause of the epilepsy. In the process of pursuing this procedure with more than a thousand patients, Penfield gained a unique and voluminous body of data on the responses of different parts of the brain to an electrical stimulus. Combining these experimental findings with extensive studies of epileptic effects—also consisting of localized electrical bombardment of regions of the brain— Penfield made many original contributions to the mapping of brain functions. But as in the case of Michelson and Morley's ether tests, more important than all his findings was a momentous absence. Nowhere in the brain did he discover any evidence of mind: the consciously deciding, willing, imagining, and creative force in human thought.

Penfield sums up his conclusions:

> The electrode can present to the patient various crude sensations. It can cause him to turn head and eyes, or to move the limbs, or to vocalize and swallow. It may recall vivid re-experience of the past, or present to him an illusion that present experience is familiar, or that the things he sees are growing large and coming near. But he remains aloof. He passes judgement on it all. He says "things are growing larger" but he does not move for fear of being run over. If the electrode moves his right hand, he does not say "I wanted to move it." He may, however, reach over with the left hand and oppose his action.

Penfield found that the content of consciousness could be selectively altered by outside manipulation. But however much he probed, he could not enter consciousness itself. He could not find the mind or invade its autonomy.

Penfield concludes: "The patient's mind, which is considering the situation in such an aloof and critical manner, can only be something quite apart from neuronal reflex action. . . . Although the content of consciousness depends in large measure on neuronal activity, awareness itself does not."

Like Michelson and Morley before him, Penfield began his investigations with the goal of proving the materialist superstition. Like them, he failed utterly. "I like other scientists have struggled to prove that the brain accounts for the mind." Instead, he showed that the critical elements of mind—consciousness, will, commitment, decision, reason—are not locally manifest in the brain.

The intruding scientist could affect contents but not consciousness. The brain's activity always occurred within the dominating and enveloping radiance of an autonomous mind. Some epileptic seizures would break the brain's connection with mind. As a result, the person would become an automaton. Like a programmed computer, he could proceed with his existing activities—walk down a street or even drive a car. But he could not change his course or adapt to unexpected developments.

While Penfield's electrodes could cause specific events, he could not usurp awareness of them; the patient could always distinguish between authentic mental processes he had willed and processes evoked by the probe. Stampedes of electrons could not cause a conscious whim in the mind, but a whim of the consciousness could cause stampedes of organized electrical and chemical activity in the brain.

Penfield's book is entitled *The Mysteries of Mind,* and he did not purport to have resolved them before he died. But his conclusion accords with the reports of anguished armies of other investigators throughout the domains of cognitive science, neurophysiology, and artificial intelligence.

Computer pioneer Norbert Wiener made the essential point as early as 1948: "The mechanical brain does not secrete thought 'as the liver does bile,' as the earlier materialists claimed, nor does it put it out in the form of energy as the muscle puts out its activity. Information is information, not matter or energy. No materialism that does not admit this can survive at the present day."

The brain is a kind of computer. But it differs sharply from existing electronic computers. The kind of electrical probe used by Penfield would utterly scramble a microprocessor. Unlike a computer, the brain claimed a level of consciousness transcending local neuronal activity.

If the brain cannot function without a human mind, it should not be surprising that a computer cannot function without a human mind either. The idea of the computer as a mind is an idol of the materialist superstition. The human brain serves the mind. The AI vanguard wanted to show that the mind was merely brain in order to support the feasibility of building computer minds. But the industry will now have to give up its goal of usurping mind and focus more effectively on serving it, as the brain does.

In his book, Penfield gave a clearer view of the relationship between mind, brain, and computer: "Because it seems to me certain that it will always be quite impossible to explain the mind on the basis of neuronal action within the brain . . . and because a computer (which the brain is) must be programmed and operated by an agency capable of independent understanding, I am forced to choose the proposition that our being is to be explained on the basis of two fundamental elements." He sums up: "The mind seems to act independently of the brain in the same sense that a programmer acts independently of his computer, however much he may depend upon the action of that computer for certain purposes."

Penfield's researches were compelling and unprecedented. But it is impossible to prove a negative. That his elaborate probes and epileptic bombardments failed to find the mind in neuronal activity does not establish that the mind is not matter. In the long run there is a more powerful reason for rejecting the computer as support for materialism. The same microcosmic revolution that made possible the computer is destroying other materialist superstitions as well. The overthrow of matter by mind continues across a wide range of the sciences. But even today the rebirth of mind must combat some of the deepest superstitions of intellectual culture and society.

For all the claims of enlightenment and all the achievements of technology, most scientists and other intellectuals for more than two centuries have been dedicated to the overthrow and suppression of mind. Intellectual history pursued an agenda of self-destruction best described as materialist determinism. Asserting the apparent meaning of Newtonian physics, Pierre Laplace stated the proposition with the most explicit force: If he knew at one instant the positions and motions of every particle in the universe, he could theoretically compute its entire past and future. Even the mind is an epiphenomenon of the matter in the brain and is determined by its physical condition.

Computer science fell neatly into the agenda. By building a material brain, the computer scientist would confirm the possibility that intelligence could emerge from matter. Thoughts as well as things would succumb to the determinist scheme.

This belief was widely extenuated by assurances that it is true in theory alone. The needed data is necessarily unobtainable. The scientist of materialism denied every malign intent. His determinism served merely as a heuristic device. He was nowhere near a proof of materialism. But these reassurances were mostly ineffectual.

Man subsists not on his practical possibilities but on his ultimate philosophies, which he inexorably transforms into his religions. Materialist determinism became the implicit religion of science and the most profound enemy of all other religion and morality.

What mattered was not the practicality but the teleology and eschatology of determinist science: the goal of a thinking machine, the reduction of the brain to a material paradigm, the apparently continuing march of knowledge toward a fully mechanistic determinism. Even as the scientist promised to prolong life and enrich its pleasures, he seemed resolved to destroy its meaning.

As Steven Weinberg puts it, "The more the universe becomes comprehensible, the more it also seems pointless." Many physicists still entertain with interest Boltzmann's hypothesis that the visible universe is just an improbable fluctuation—a rare disequilibrium—amid a predetermined whirl down the entropic drain. The scientific revelation believed to underlie all the heroic achievements of modern man, that provides him his addictive wealth and comfort, his sustenance and support and his hope for health and happiness, is a revelation of eternal death.

Contributing to the vision of materialist determinism was the computer itself. Even human intelligence, the unique claim of the species on its road to ruin, was seen as likely to succumb to a mechanistic rival. In gaining dominion over the world, we lose control of our own machines.

Even the microcosm itself was deemed a vindication of the materialist movement. When switching speeds plummeted into the picosecond range (trillionths of a second), it seemed plausible to many analysts that for all its user friendliness and endearing idiosyncrasies —such as "common sense"—the brain was no longer "merely a computer," but was in fact far inferior to computers in speed and computing power. Theorists estimated that the human brain contains some 10 billion neurons (a number of switching devices within the reach of the supercomputing technology of the next decade) and each neuron switches at only one millionth of the speed of an advanced transistor.

The computer thus lent enormous new authority to the determinist vision. Not only could it be used to achieve enormous new advances across the range of the scientific enterprise but the computer itself constituted a paradigm for comprehending the very fabric of thought in materialist terms. The computer seemed to subordinate thoughts themselves to things, mind to matter, soul to solid state physics. It seemed to unmask the mind and demystify it, and thus to banish sacred and spiritual powers from the world.

Deeply believing—whether at some unconscious level or in a conscious paradigm of research—that human mastery of nature would end by destroying the mystery of human nature, the Western intelligentsia gullibly adopted a medley of materialist determinisms. From Marxism to behavioralism, from routine evolutionism to logical positivism, from deconstructionism to reflex psychology, scientists and scholars produced an unending steam of theories that reduced men to mechanism.

Artists always tend to follow the prevailing science, often with a lag as long as a century. Converging on the culture from every hall of academe, scientific determinisms prompted a bizarre stream of doom-laden or absurdist works of art and literature. The most prestigious novelists—from William Faulkner and George Orwell to John O'Hara and Robert Stone—left man in the twentieth century as arrant a victim of inexorable fate as he was in the great god-ruled tragedies of the ancient Greeks.

The counterpoint of determinism was existentialism. With history a saga of blind fate, heroism emerged chiefly as anarchic revolt by defiant individuals, supermen who leaped onto the stage abandoned by the gods and hurled thunderbolts of rebellion at the clockwork world. But by their very desperation and frenzy—and ultimate futility—the existentialist rebels seemed to vindicate the determinist philosophers.

With science believed to be abandoning man, and the dignity of his soul, men rose up to claim the dignity of science for squalid temporal schemes of the state and to ascribe soul to new machines. After contributing indispensably to the rise of science, religions began renouncing the great human enterprise of understanding and mastering nature, and thus failed to comprehend the real message of the microcosm or the scientific revelation of God.

Laboring desperately to reassemble the smithereens of a shattered grail of faith, the public clutched at a new pagan polytheism. Materialist intellectuals widely worshipped the god of the machine, artists reveled in the god of the senses, and churches retreated to a preindustrial vision of an ecological Eden. Even the most sophisticated apologists of religion expressed a nostalgia for Newton's cosmos as a causal clockwork and God as a watchmaker.

In presenting this idea of materialist determinism, therefore, science is playing with a fear afflicting the very heart of modern man. The fear is rarely declared to the world, but its outlines can be clearly descried in the culture of the time. Indeed, it is a fear that suffuses most of the art, literature, and philosophy of the West. The highest purpose of the leading universities, for example, seems to be to reduce philosophy to a mechanical positivism, to reduce history to statistical fluctuations and class exploitations, to deconstruct literature to a flux of words and writers' neuroses, and to banish heroes from human life.

The underlying vision emerges as a fable. The microcosm grants us incomparable riches, but only as part of a Faustian pact: a deal with the devil whereby we gain wondrous machines in exchange for our very souls. We give the scientist first our money and then the very meaning of our lives.

Plunging into the microcosm, from this point of view, we lose first our human dignity and then our free will. The microcosmic world is ultimately a tomb: the death site of both God and man. God dies with the eclipse of the Creator and the end of meaning. With the death of human dignity and uniqueness, man degenerates into a mere intelligent machine that differs from a computer chiefly by thinking more slowly and by more quickly and irremediably wearing out. As the capstone of the nihilist enterprise, science as a creation of free human minds also perishes in the microtomb.

Yet all this anguish and ardor—all the fatalist philosophies of nihilism and existentialism—are mere delusions arising from the great modern superstition of materialism. Refuted by an entire century of modern science and by the findings of computer theory itself, materialist determinism falls of its own weight.

The overthrow of matter reached its climax in the physical sciences when quantum theory capsized the rules that once governed and identified all solid objects. Physicists now agree that matter derives from waves, fields, and probabilities. To comprehend nature, we have to stop thinking of the world as basically material and begin imagining it as a manifestation of consciousness.

In other words, both at the highest levels of nature where meanings, purposes, structures, functions, and other conceptual entities prevail, so at the fundamental level of physical reality, thought dominates things. In the atomic domain, we do not find Newton's lifeless, opaque, inert, and impenetrable bits of matter. We find a high drama of richly intelligible activity, shaped by mathematical possibilities and probabilities, where electrons combine and disappear without occupying conventional time or space, and things obey the laws of mind rather than the laws of matter. Things combine and collapse, emerge and vanish, like tokens of thought more than like particles of mass.

The reverberations of this change sweep through social science as well. Gone is the view of a thermodynamic world economy, dominated by "natural resources" being turned to entropy and waste by human extraction and use. Once seen as a physical system tending toward exhaustion and decline, the world economy has clearly emerged as an intellectual system driven by knowledge. The key fact of knowledge is that it is anti-entropic: it accumulates and compounds as it is used. Technological and scientific enterprise, so it turns out in the age of the microcosm, generates gains in new learning and ideas that dwarf the loss of resources and dissipation of energy. The efforts that ended in writing E equals MC squared or in contriving the formula for a new photovoltaic cell or in inventing a design for a silicon compiler—or in spreading the message of a moral God—are not usefully analyzed in the images of entropy. To see the world primarily in terms of its waste products is possibly the most perverse vision in the history of science. The history of ideas cannot be comprehended as an entropic cycle of the production and disposal of paper products.

The materialist superstition succumbs to an increasing recognition that the means of production in capitalism are not chiefly land, labor, and machines, present in all systems, but emancipated human intelligence. Capitalism—supremely the mind-centered system—finds the driving force of its growth is innovation and discovery. In the age of the microcosm, value added shifts rapidly from the extraction, movement, manipulation, and exhaustion of mass to the creative accumulation of information and ideas.

At the heart of this progress is a bold human sacrifice. After centuries of struggle against the elements—human masses pushing and pulling on massive objects at the behest of armed rulers—man stopped hoarding gold and jewels, land and labor. He stopped trusting only the things he could touch and feel. He stopped believing only his eyes and ears. At last relinquishing the sensory world, he gained access to a higher power and truth. And all those other things are being added unto him.

Overthrowing the superstitions of materialism—of anthropomorphic physics, of triumphant entropy, of class conflict and exploitation, of national and tribal bigotry—modern man is injecting the universe with the germ of his intelligence, the spoor of his mind. Sloughing off every layer of macrocosmic apparatus, the computer will ultimately collapse to a pinhead that can respond to the human voice. In this form, human intelligence can be transmitted to any tool or appliance, to any part of our environment. Thus the triumph of the computer does not dehumanize the world; it makes our environment more subject to human will.

Giving up the superficial comforts of a human-scale world, mankind moves to mind scale. In the image of his Creator, he exalts the truly human—and godlike—dimension of his greatest gift from the Creator: his creativity. Giving up the material idols and totems in his ken, he is gaining at last his promised dominion over the world and its creatures.

The key lessons of the era spring from an understanding of hierarchy and freedom rather than of topology and determinism. Just as the large integrated circuits at the heart of computers cannot be designed from the bottom up, just as evolution could not have occurred from the bottom up alone, the human mind cannot be understood without a clear comprehension of the hierarchical structure of the universe.

The key error of materialism is to subordinate a higher level of creative activity to a lower one: the dramas of humanity to the determinations of lifeless and unintelligible monads of matter. Opening the atom to find a world of information, we discover that the inverted hierarchies of materialism are as false empirically as the ancient philosophers found them bankrupt logically. In the hierarchies of nature, the higher orders cannot be reduced to the sum of their parts. Even the concept of wholes and parts is misleading; every part is also a whole, and every whole is also a part in a higher order of meaning.

The long dream of physicists that by studying the structure of atoms we could comprehend the purpose of nature, or the dream of biologists that the study of evolution would reveal the meaning of life, have both proved false. In the same way, neither the physics nor the chemistry of DNA will explain the code of genetic information any more than the analysis of spectrograms will reveal the meaning of a word or phrase.

Far from a model of human intelligence that might excel its creator, the computer offers a very limited perspective on mind. The usual

assumption of materialism is that the brain, the hardware, comes first
and mind somehow emerges from it. But the computer offers a contrary example. The computer design itself is a software program and
determines the structure of the electronic circuitry that constitutes the
computer. Similarly, the brain consists of a rich panoply of neuro-
physiological processes that serve as analogs for physical experience
and perception. Yet neither system can function as an information
tool without a consciousness to interpret the information.

Knowing the location of every electronic charge and connection in
either brain or computer gives scarcely a clue of its purpose. It is the
human mind that brings meaning or semantic content to the syntax
of the machine, whether hardware, software, or wetware. The higher-
level languages of software lend significance to the dumb electrons
circulating through the system. But the language is still sterile until it
is translated by the human consciousness. Neither brain nor computer
could think without an embracing presence of mind. The higher-level
languages of software do not obviate the need for a controlling intelligence that transcends them.

Any lower level of existence finds meaning only in higher orders
outside itself. Physics finds meaning largely in chemistry, chemistry in
biology, biology in man, and man both in knowledge and culture and
in that higher level of meaning ordained by God. The study of science
is thrilling and rewarding on its own level, and imparts new powers
and technologies for action on higher levels. But movement up a
hierarchy leads to domains of increasing freedom and dignity and
spiritual fulfillment, not into a materialist or determinist trap.

In his efflorescent exposition of *The Age of Intelligent Machines,*
Raymond Kurzweil cites the phenomenon of dreams to show how
humans might imagine themselves free in a determined world. In
dreams, he says, we have an illusion of freedom and choice that van-
ishes when we awaken. But surely this analogy is wrong. The very
essence of a dream—or a nightmare—is a sense of being carried away
beyond choice or control. Whether in relief or anguish, we awaken to
freedom. In assuming consciousness, we move up the hierarchy from
involuntary association and drift to voluntary action and purpose.

The entire history of technology suggests a similar awakening to
new powers and horizons, new freedoms and opportunities. The new
age of intelligent machines will enhance and empower humanity,
making possible new ventures into outer space and new insights into
atomic structure. It will relieve man of much of his most onerous and

unsatisfying work. It will extend his lifespan and enrich his perceptual reach. It will enlarge his freedom and his global command. It will diminish despots and exploiters. It may even improve music and philosophy.

Overthrowing matter, humanity also escapes from the traps and compulsions of pleasure into a higher morality of spirit. The pleasure principle is the governing force of a physical determinism, whether in psychology, sociology, or economics. Physiologically defined, pleasure makes the materialist world go round. Optimizing for bodily pleasures, human beings can take their place in a determinist scheme. Believing in determinism, they can escape the burdens of a higher morality.

Pleasure seeking is entropy in social life. It leads to the image of dissipation as the effect of happiness. Moving in the direction of entropic forces—sliding down the gradients of our physiology—we live in a world of Boltzmann's laws rather than of his life and creations.

Overthrowing all forms of materialist determinism, however, the microcosm imposes its own more rigorous laws of freedom and responsibility. Requiring long planning and sacrifice and exacting disciplines, the microcosm epitomizes the longer horizons and altruistic rules of the life in the image of the Creator. The exaltation of mind and spirit leads to a higher order of experience and a richer access of power than any of the forms of physical dominance and exploitation that pervade the precincts of material pleasure. The first victim of the hedonistic vision is the disciplined quest for knowledge and truth.

The meaning of the microcosm is that far from a heuristic tool, the materialist assumption stultifies science itself. Any science becomes self-contradictory when it attributes to its object of study a nature that renders the science itself inexplicable or unintelligible. An intellectual theory that materializes or mechanizes theorists is self-defeating. Behavioral psychology, determinist biology, materialist physics and evolution are all mere gibberish because they explain away the scientist himself and his transcendant pursuit of truth.

The entire history of the twentieth century testifies to the bitter fruits of sacrificing the scientist to the science. Nazism, Marxism, and all the other diseases of politics feed on the belief in some final truth, worthy of some final physical enactment, to which all the mere scientists themselves must succumb. But science is meaningless outside of the unending enterprise of search and discovery, in freedom and faith, by the community of explorers themselves. Outside of the living fabric of knowledge in the minds of believers, free to search for God, beliefs themselves become idols of destruction.

Matter is subordinate to mind and spirit and can only be comprehended by free men. As computers master new levels of the hierarchy of knowledge, human beings can rise to new pinnacles of vision and power and discover new continents of higher truth. As Carver Mead declared in his defense of Boltzmann, the universe offers infinite degrees of freedom. The computer will give mankind new vessels to rule the waves of possibility. Its promise can be betrayed only by the abandonment of the freedom and faith that made it possible.

Wielding the power of knowledge, the human mind is the great and growing disequilibrium at the center of the universe. Made in the image of its creator, human mind wields the power of knowledge against the powers of decay. Conquering the microcosm, mind transcends every entropic trap and overthrows matter itself. This is the true significance of the story of the microcosm and the meaning of the quantum era.

Near the end of his book, Penfield reflects on the further fruits of the overthrow of matter. "What a thrill, then, to discover that the scientist, too, can legitimately believe in the existence of the spirit!" Writing toward the end of his life, he observes that while the brain and body, memory and muscle deteriorate, the mind suffers "no peculiar or inevitable pathology. Late in life it moves to its own fulfillment. As the mind arrives at clearer understanding and better balanced judgment, the other two are beginning to fail. . . ."

Writing "from the physician's point of view," he suggests, "in conformity with the proposition that the mind has a separate existence, it might even be taken as an argument for the feasibility and possibility of immortality!"

In the beginning was the word, the idea. By crashing into the inner sanctums of the material world, into the microcosm, mankind overcame the regnant superstitions of matter and regained contact with the primal powers of mind and spirit. Those new powers have rendered obsolete all the materialist fantasies of the past: the notion that by comprehending things, one could understand thought, and that by controlling things, one could rule the world. The era of the microcosm is the epoch of free men and women scaling the hierarchies of faith and truth seeking the sources of light.

In this unifying search is the secret of reconciliation of science with religion. The quantum vision finds at the very foundations of the material world a cross of light. Combining a particle and a wave, it joins the definite to the infinite, a point of mass to an eternal radiance. In this light, we can comprehend the paradox of the brain and the mind, the temporal and the divine, flesh and the word, freedom and fatality. By this light we can even find the truth. But we cannot see through it. In science and technology, religion and life, we can triumph only by understanding that truth is a paradoxical and re-demptive cross at the heart of light, radiant in the microcosm and in the world.

BIBLIOGRAPHICAL
NOTES

"Tomorrow is going to be wonderful because tonight I do not understand anything."
—Niels Bohr to Abraham Pais.

To the author, researching this book brought a matchless series of dark nights and wonderful tomorrows. Interested readers can share these chiaroscuro pleasures by plunging into the major sources, or they can check details by consulting the more specialized or technical references.

Books and other general sources relevant to each part of *Microcosm* are listed under the part headings; more particular references are described under specific chapter headings. A key further source was conferences and symposia, including classes at Caltech with Carver Mead and at Worcester Polytechnic with Robert Solomon; several seminars on integrated circuit fabrication with Integrated Circuits Engineering Corporation; the ISSCC conferences between 1982 and 1987; and Caltech Conferences on Very Large Scale Integration at MIT and Stanford.

For general coverage of these technologies and their implications, no publication has excelled, for consistency and insight, the weekly output of *Electronic Engineering Times* (CMP Publications, Inc., 600 Community Drive, Manhasset, N.Y. 11030). I relied on their reporters and analysts, from the exemplary Stanley Baker in Silicon Valley to David Lammers in Japan, for much of the background information in the book. Also useful were the San Jose *Mercury, Electronic Business, Electronic News, Release 1.0, Electronics, Solid State Technology, Computerworld, Infoworld, PC Magazine, PC World, Byte, VLSI Update,* Steven Szirom's *SIBS, Technology Research Group Letter, MIPS, The Economist, Harvard Business Review,* and *Japan Economic Journal.*

For coverage of the science, no publication is better written or more scrupulously edited than *Scientific American*. Developing the themes of this book would have been much more difficult without the following issues of this journal:

Scientific American, September 1977, "Microelectronics." The most indispensable publication in the field, useful after more than a decade, this magisterial issue contains articles by Robert Noyce, James Meindl, William C. Holton, William G. Oldham, David A. Hodges, Hoo-Min D. Toong, Lewis M. Terman, Bernard M. Oliver, John S. Mayo, Ivan Sutherland and Carver A. Mead, and Alan C. Kay.

Scientific American, September 1984, "Computer Software." A definitive compendium, led off by Alan Kay.

Scientific American, October 1986, "Materials for Economic Growth." Contains a powerful vision of the future of electronics by John S. Mayo and a good essay on photonics by J. M. Rowell.

Scientific American, October 1987, "The Next Revolution in Computers . . ." With major contributions by James Meindl, Abraham Peled of IBM, Geoffrey C. Fox and Paul C. Messina from Caltech, among others.

Scientific American, Special Issue/vol. 1, 1988, "Trends in Computing." A superb collection of the best computer and semiconductor articles from the magazine, updated by their authors. Includes key essays by Daniel Hillis on the Connection Machine, James Meindl on chips for advanced computing, and Robert T. Bate on Texas Instruments' experiments with quantum devices.

The historic sections of this book were written from a combination of hundreds of interviews with the principals and immersion in the existing literature over a period of seven years. I have tried to select the most important influences, but many necessarily elude my neural net today.

PART ONE: *The Overthrow of Matter*

BASIC TEXTS

FEYNMAN, Richard P., *The Feynman Lectures on Physics*. Vols. 1–3, edited by Robert B. Leighton and Matthew Sands. Reading, Mass.: Addison-Wesley, 1963. Anyone with a serious interest in learning physics cannot do better than these magnificent essays.

PAIS, Abraham, *Inward Bound: Of Matter and Forces in the Physical World*. New York: Oxford University Press, 1986. An often difficult book, full of nuggets of humor and insight and with poignant vignettes of the great figures in the overthrow of matter, written discursively by a onetime close collaborator with Bohr.

PAULING, Linus and Peter, *Chemistry*. San Francisco: W. H. Freeman, 1975. For the interplay between physics, solid state physics, and chemistry, this book is comparable to the Feynman Lectures in lucidity and substance.

WEAVER, Jefferson Hane, ed., *The World of Physics: A Small Library of the Literature of Physics from Antiquity to the Present*. New York: Simon & Schuster, 1988, Vols. 1–3. A cornucopia of writings by leading physical scientists, including Newton, Maxwell, Planck, Einstein, Bohr, Born, Heisenberg, Carnot, Thomson, Boltzmann, Dirac, Wigner, Schwinger, Schrödinger, Delbrück, Bardeen, Hawking, and scores of others, with pithy explanatory notes and biographical materials. An unbeatable way to study the history of the field.

SOME SUPPLEMENTS

DE BROGLIE, Louis, *The Revolution in Physics*. New York: The Noonday Press, 1953. A readable and compelling history of quantum theory by the discoverer of the wave nature of the electron.

―――, *New Perspectives in Physics*. New York: Basic Books, 1962. Fascinating essays by De Broglie struggling with the perplexities of quantum reality and shifting positions over the years.

DELBRUCK, Max, *Mind from Matter?* Palo Alto, Calif.: Blackwell Scientific Publications, 1986. A crisply written and wide-ranging survey of the new science by a man who studied quantum physics with Bohr and went on to win the Nobel Prize in Biology.

FEYNMAN, Richard P., *The Character of Physical Law*. Cambridge, Mass.: The MIT Press, 1967. A splendid set of lectures, extemporized from notes, delivered at Cornell and broadcast by the BBC, and full of Feynman's charm and challenges.

―――, *QED: The Strange Theory of Light and Matter*. Princeton, N.J.: Princeton University Press, 1985. A late book by Feynman exploring the most baffling quantum phenomena, such as electrons momentarily moving backward in time.

―――, Steven Weinberg, *Elementary Particles and the Laws of Physics*, the 1986 Dirac Memorial Lectures. New York: Cambridge University Press, 1987. Two difficult lectures exploring the implications of Dirac's unified theory of special relativity and quantum theory, with his prophecy of antimatter, which Feynman elaborated in his time-reversing positron, and Weinberg developed into the electro-weak force of nuclear radiation.

GAMOW, George, *Thirty Years That Shook Physics: The Story of Quantum Theory*. New York: Dover Publications, 1966. A stirring and readable tale explaining such pivotal events as the Planck "black body" hypotheses as clearly as anyone.

GEORGESCU-ROEGEN, Nicholas, *The Entropy Law and the Economic Process*. Cambridge, Mass.: Harvard University Press, 1971. An elaborately flawed but usefully provocative and ambitious exploration of the role of entropy in economics.

GRIBBIN, John, *In Search of Schrödinger's Cat: Quantum Physics and Reality*. New York: Bantam Books, 1984. One of the most engaging of the popularizers, highly recommended for lay readers.

―――, *In Search of the Double Helix: Quantum Physics and Life*. New York: Bantam Books, 1985. Clear exposition of the links among quantum theory, chemistry, and biology.

HEISENBERG, Werner, *Physics and Beyond, Encounters and Conversations*. New York: Harper & Row, 1971. A fascinating glimpse of the mind and memory of the inventor of the uncertainty principle, full of acute observations from the midst of the action and reports of conversations with Einstein, Bohr, Pauli, and others.

―――, *Tradition in Science*. New York: The Seabury Press, 1983. A series of interesting essays by the pioneer of particle theory, including a telling explanation of the non-existence of so-called elementary particles.

HERBERT, Nick, *Quantum Reality, Beyond the New Physics*. Garden City, N.Y.: Anchor Press/Doubleday, 1985. A wonderfully lucid exposition of quantum theory that is weakened by hyping bizarre possibilities, such as parallel universes, without a skeptical eye.

PAGELS, Heinz, *The Cosmic Code: Quantum Physics as the Language of Nature*. New York: Simon & Schuster, 1984. Probably the most readable and authoritative introduction to the field, by a gifted physicist in his forties who died in 1988 falling off a mountain. The book ends:

> I often dream about falling. Such dreams are commonplace to the ambitious or those who climb mountains. Lately I dreamed I was clutching at the face of a rock, but it would not hold. Gravel gave way. I grasped for a shrub, but it pulled loose, and in cold terror I fell into the abyss. Suddenly I realized that my fall was relative; there was no bottom and no end. A feeling of pleasure overcame me. I realized that what I embody, the principle of life, cannot be destroyed. It is written into the cosmic code, the order of the universe. As I continued to fall in the dark void, embraced by the vault of the heavens, I sang to the beauty of the stars and made my peace with the darkness. . . .

ROTA, Gian-Carlo, Jacob T. Schwartz, and Mark Kac, *Discrete Thoughts: Essays on Mathematics, Science, and Philosophy*. Boston: Birkhauser, 1986. Working a fertile crescent between mathematics, philosophy, and computer science, these lapidary thinkers (at MIT, NYU, and late of Rockefeller U.) can turn a phrase as well as an equation: "The computer is just an instrument for doing faster what we already know how to do slower. . . . Why don't we tell the truth? No one has the slightest idea how the process of scientific induction works."

WILCZEK, Frank, and Betsy Devine, *Longing for the Harmonies: Themes and Variations from Modern Physics*. New York: W. W. Norton, 1988. An often poetic exploration of the world of Maxwell, Fourier, and Feynman.

SOLID STATE

FREDERIKSEN, Thomas M., *Intuitive IC Electronics, A Sophisticated Primer for Engineers and Technicians*. New York: McGraw-Hill, 1982. A prolific writer and analog engineer long at National Semiconductor, he begins with a good exposition of the relevant physics, from band gap theory to quantum tunneling.

KEYES, Robert W., *The Physics of VLSI Systems*. Reading, Mass.: Addison-Wesley, 1987. Written by a researcher at IBM's Watson Laboratories, this short book is comparable to Mead-Conway (below) in its address of the entire field, from the physics to the computer systems. Superb in its distillation of vast quantities of technical material.

MEAD, Carver A., and Lynn Conway, *Introduction to VLSI Systems*. Reading, Mass.: Addison-Wesley, 1980. Still a crucial text in the field. Ends with a discussion of the physics of computation.

QUEISSER, Hans, *The Conquest of the Microchip*. Cambridge, Mass.: Harvard University Press, 1988. The European perspective on the development of the chip, by a director of the Max Planck Institute. Indispensable for its clear explanations of initial ventures in solid state physics.

SHOCKLEY, William, *Electrons and Holes in Semiconductors*. Princeton, N.J.: Van Nostrand, 1953. Difficult to read but full of valuable explanations by the giant in the field.

Chapter 1: The Message from the Microcosm
The thesis of this chapter emerged in part from discussions with David Warsh of the Boston *Globe* on the persistent origin of economic models in the dominant

technology of yesterday. See Warsh, *The Idea of Economic Complexity* (New York: Viking, 1984), pp. 143–53, for examples of Newton's celestial mechanics and economic general equilibrium, and Boyle's law of gases and the quantity theory of money.

I was aided in the shaping of the chapter by Dan Burns of Xpercom, Anya Hurlbert of MIT, Matthew Ridley of *The Economist,* Bo Burlingham of *Inc.,* Robert Noyce, Richard Vigilante, and Walter Tchon of Xicor. The Gordon Moore parable was first told to me by Tchon. A clear early account of the interplay between quantum theory and solid state physics is Gregory H. Wannier, "The Solid State," in Gerard Piel, et al., eds., *Scientific American Reader* (New York: Simon & Schuster, 1953).

A student of Wigner's who later worked at Bell Laboratories, Wannier in the early 1950s did not see any reason to mention silicon. The direct quotations in the chapter come from the essays in Weaver, ed., *The World of Physics,* and from the works listed, by Heisenberg, Weinberg, and Feynman. *The Economist,* long alert to these matters, offers a crisp anonymous presentation of the essential problem in "The Queerness of Quanta" (January 7, 1989): "Many of this century's most familiar technologies come with an odd intellectual price on their heads. The equations of quantum mechanics explain the behavior of sub-atomic particles in nuclear reactors and of electrons in computers and television tubes, the movement of laser light in fibre-optic cables and much else. Yet quantum mechanics itself appears absurd." The problem, however, is in fact the solution.

Chapter 2: The Prophet

This chapter is heavily based on interviews with Carver Mead, his colleagues and his students, and on classes, articles, and other writings of Mead's. The intricacies of tunneling are explained in Keyes. One of Mead's tunneling papers, "Tunneling Physics," delivered at a Caltech Symposium in February 1961, plots the mean free path of tunneling electrons as a function of their speed, with low-energy electrons traveling as far as 1,000 angstroms (one tenth of a micron). Predicting as many as 100 million transistors on a chip a centimeter square, the original papers on the "Fundamental Limitations in Microelectronics—I. MOS Technology," and "II. Bipolar Technology," by Mead and R. Hoeneisen (Mead's graduate student, mentioned in the text), were both published in *Solid State Electronics,* 1972, vol. 15. Mead's speech at Caltech on the prospects for what would be known as personal computers was published in the Caltech magazine, *Engineering & Science,* in 1971 under the title "Computers That Put the Power Where It Belongs." The biographical notes for the article describe Mead's work as a consultant at Lexitron Corporation, helping "design and build a sophisticated electronic typewriter. . . . No paper is necessary until the final letter-perfect document is . . . printed out."

Chapter 3: Wires and Switches

Fractal theory is vividly and stylishly explained in James Gleick, *Chaos: The Making of a New Science* (New York: Viking, 1988), and more pithily in Rudy Rucker, *Mind Tools: The Five Levels of Mathematical Reality* (Boston: Houghton Mifflin, 1987). See also Murray Gell-Mann, "Simplicity and Complexity in the Description of Nature," *Engineering & Science,* Caltech, Spring 1988. The history of Shockley comes from essays by Bardeen and Shockley, interviews with Robert Noyce, Mead, and other industry figures, and copious secondary sources, including the listed histories that follow. Testimonies by Shockley include the essay "Transistors," in Carl F. J. Over-hage, ed., *The Age of Electronics* (New York: McGraw-Hill, 1962, pp. 135–164), and Shockley, "Unipolar 'Field Effect' Transistor" (*Proc. IRE,* vol., 40, 1952, p. 1365). Especially useful also was Jeremy Bernstein, *Three Degrees Above Zero: Bell Labs in the Information Age* (New York: Charles Scribner's Sons, 1984), pp. 73–146, and *Solid State Technology,* December 1987, a commemorative issue on "The Transistor . . . The First 40 Years," including biographical sketches of the co-inventors and recollections by Bardeen, "Solid State Physics—1947," and Ian M. Ross, then president of Bell Laboratories, "Chance Favors the Prepared Mind." For other historical sources, see below, Part Two, Technology of Mind, particularly Braun and Macdonald.

PART TWO: *The Technology of Mind*

HISTORIES

AUGARTEN, Stan, *State of the Art: A Photographic History of the Integrated Circuit,* with a foreword by Ray Bradbury. New Haven: Ticknor & Fields, 1983. Believe it or not, you probably won't be able to put it down. Vivid photographs of all the major chip designs depict the history of the industry in an original and intriguing way.

BRAUN, Ernest, and Stuart Macdonald, *Revolution in Miniature, The History and Impact of Semiconductor Electronics,* 2nd ed. Cambridge, U.K.: Cambridge University Press, 1982. An updated version of the first authoritative history of the industry.

BYLINSKY, Gene, *The Innovation Millionaires.* New York: Charles Scribner's Sons, 1976. Mostly a collection of the author's estimable early *Fortune* articles, including pithy pieces on venture capital, Silicon Valley, and Intel.

———, *Silicon Valley High Tech: Window to the Future.* Hong Kong: Intercontinental Publishing Corp., 1985. A crisply written and spectacularly illustrated survey of the Valley in the early 1980s.

DAVIDOW, William H., *Marketing High Technology: An Insider's View.* New York: The Free Press, Macmillan, 1986. Full of insights and information from the heart of Intel during the battle with Motorola for control of the microprocessor market.

FORESTER, Tom, ed., *The Microelectronics Revolution: The Complete Guide to the New Technology and Its Impact on Society.* Cambridge, Mass.: MIT Press, 1981. A collection of essays, once fresh, now somewhat dated, on the industry and its impact.

HANSEN, Dirk, *The New Alchemists: Silicon Valley and the Microelectronics Revolution.* Boston: Little, Brown, 1982. The first popular survey of the industry, beginning with Edison, and still excellent in its vividness, scope, and theoretical structure.

KIKUCHI, Makoto, *Japanese Electronics, A Worm's-Eye View of Its Evolution.* Tokyo: The Simul Press, 1983. An excellent personal memoir of the rise of Japanese electronics, from an insider at MITI and at Sony, who built the first transistor in Japan. A good antidote to the MITI-did-it theories by American analysts.

LYONS, Nick, *The Sony Vision.* New York: Crown, 1976. An excellent account of the early years at Sony, demonstrating their independent development of powerful manufacturing technology for semiconductors long before MITI's programs.

MALONE, Michael S., *The Big Score: The Billion Dollar Story of Silicon Valley.* Garden City, N.Y.: Doubleday, 1985. Far better than the trite title (what could be less distinctive about Silicon Valley than the money made there?), this book is full of vivid anecdotes and incisive writing about the industry from a Santa Clara perspective (includes Ampex and Hewlett Packard, omits TI).

OKIMOTO, Daniel I., Takuo Sugano, and Franklin B. Weinstein, eds., *Competitive Edge: The Semiconductor Industry in the U.S. and Japan*. Stanford, Calif.: Stanford University Press, 1984. A series of interesting essays on the rise of Japan in electronics, generally disparaging the primacy of industrial policy.

REID, T. R., *The Chip: How Two Americans Invented the Microchip and Launched a Revolution*. New York: Simon & Schuster, 1984. A jewel of a book on the integrated circuit and its invention by Noyce and Kilby, written by a reporter and columnist for *The Washington Post*.

ROGERS, Everett M., and Judith K. Larsen, *Silicon Valley Fever: Growth of High-Technology Culture*. New York: Basic Books, 1984. Often fails to differentiate between what is trivial (jogging habits or gay lifestyle), what is inevitable to achievement in any field ("workaholics"), and what is essential in launching the world's leading technology. But the book also contains many interesting details about the culture and texture of the industry beyond the obvious heroes and breakthroughs, while including some valuable history as well.

ROSTKY, George, *The Thirtieth Anniversary of the Integrated Circuit: Thirty Who Made a Difference. Electronic Engineering Times*, Manhasset, N.Y.: CMP Publications, September 1988. The most authoritative, up-to-date, and technically savvy history of the industry. Readable and comprehensive, including both Japanese and American contributions. See also, *Fifteen Years That Shook the World, E E Times*, November 16, 1987.

Electronics magazine, Special Commemorative Issue, New York: McGraw-Hill, April 17, 1980. Excellent and comprehensive history of the technology until 1980, with inserted biographies of key figures and schematics of key advances.

TECHNICAL GUIDES

ELLIOTT, David J., *Integrated Circuit Fabrication Technology*. New York: McGraw-Hill, 1982. The best introduction to semiconductor process technology that I found, written by an expert on photoresists at Shipley in Newton, Massachusetts.

FREDERIKSEN, Thomas M., *Intuitive IC CMOS Evolution: From Early ICs to microCMOS Technology and CAD for VLSI*. Santa Clara, Calif.: National's Semiconductor Technology Series, 1984. Excellent text with historical perspective as well as technical authority. See also *Intuitive CMOS Electronics: The Revolution in VLSI, Processing, Packaging, and Design*. New York: McGraw-Hill, 1989.

GROVE, Andrew, *The Physics and Technology of Semiconductor Devices*. New York: John Wiley, 1967. The classic text, used by this author in learning the technology and taught before publication by Carver Mead at Caltech.

MORGAN, D. V., K. Board, and R. H. Cocrum, *An Introduction to Microelectronic Technology*. New York: John Wiley, 1985. A good reference, strong in charts and graphs and clear organization; includes gallium arsenide.

ONG, De Witt G., *Modern MOS Technology: Processes, Devices, & Design*. New York: McGraw-Hill, 1984. From a manager of the huge Intel fab in Chandler, Arizona, a valuable new text in the spirit of Grove's original classic, but still too light on CMOS, with little excuse in 1984.

SZE, S. M., *The Physics of Semiconductor Devices*. New York: John Wiley, 1985. Comprehensive and up to date, from IBM, but not for beginners.

Chapter 4: The Silicon Imperative

Patrick E. Haggerty, *Three Lectures at the Salzburg Seminar on Multinational Enterprise* (Dallas, Tex.: Texas Instruments Corp., 1977), and *Management Philosophies and Practices of Texas Instruments* (Dallas, Tex.: Texas Instruments Corp., 1967). Interviews with Mark Shepherd, Fred Bucy, and others, and historic sources, especially Braun and Macdonald. Key elements of the Teal story came from John McDonald, a series in *Fortune* magazine on TI, November and December 1961. Also valuable was Mariann Jelinek, *Institutionalizing Innovation, A Study of Organizational Learning Systems* (New York: Praeger, 1979). This book contains an in-depth analysis of the innovation process at Texas Instruments in the early years. The Texas Instruments Learning Center has consistently published the best instructional books in the field (the "Understanding . . ." series, now available through Tandy Radio Shack). One of their earlier efforts, Robert G. Hibberd, *Integrated Circuits: A Basic Course for Engineers and Technicians* (New York: McGraw-Hill, 1969), gives a window on the state of the technology at TI during the late 1960s. Statistics on energy use and productivity come from Charles A. Berg, "The Use of Electric Power and Growth of Productivity: One Engineer's View" (Department of Mechanical Engineering, Northeastern University, 1988).

Chapter 5: The Monolithic Idea

Interviews with Jack Kilby, Robert Noyce, Victor Grinich, Julius Blank, and others. Also all the listed histories, especially T. R. Reid, *The Chip*.

Chapter 6: Flight Capital

Based on interviews with Andrew Grove, Leslie Vadasz, Frank Wanlass, Tom Longo, and others. The range of Grove's remarkable mind is suggested not only by his technical text (see above) and his business management text, *High Output Management* (New York: Random House, 1986), but also by his wise and savvy column in the San Jose *Mercury* and *EE Times,* excerpted in *One on One with Andy Grove* (New York: G. P. Putnam's Sons, 1987).

Chapter 7: Intel Memories

Interviews with Robert Noyce, Andrew Grove, Federico Faggin, and other principals. Also, Gene Bylinsky, "How Intel Won the Memory Race," in Bylinsky, *Innovation Millionaires* (see above). By far the best article on Noyce and the Fairchild phenomenon is Tom Wolfe, "The Tinkerings of Robert Noyce," *Esquire* magazine, Golden Anniversary Collector's Issue, December 1983.

Chapter 8: Intel Minds

Interviews with Ted Hoff, Federico Faggin, Masatoshi Shima, Stanley Mazor, and others. Essay by Hoff and James Jarrett of Intel (internal publication). Faggin, "Microprocessors in Perspective" (unpublished manuscript available from Synaptics, Inc., Suite 105, 2860 Zanker Rd., San Jose, Calif. 95134).

Chapter 9: The Curve of Declining Entropy

Interviews with Jerry Sanders, William Bain, and others. Also, Bruce D. Henderson, *The Logic of Business Strategy* (Cambridge, Mass.: Ballinger Publishing Co., 1984). For a demolition of the argument that dumping is a significant cause of

Japanese gains in the U.S. market, see Gregory Clark, "Japan's Export Dilemma and the Future of Western Manufacturing," *Japan Economic Journal,* April 5, 1986. For detailed explanation of the experience curve see, Robert H. Hayes and Steven C. Wheelwright, *Restoring Our Competitive Edge: Competing Through Manufacturing* (New York: John Wiley, 1984). For discussion of entropy in economics, see Georgescu-Roegen (above), and Pierce, *Information Theory* (below).

Chapter 10: Japan's Microcosm
Interviews with Shumpei Yamazaki and Matami Yasufucu, Makoto Kikuchi, Robert Watson of National Semiconductor Japan, Keiske Yawata, Nihon-LSI Logic, Daniel Barrett (then of AMD, now of Cypress Semiconductor), Michiyuki Uenohara of NEC Research, and semiconductor executives from Hitachi, Toshiba, and Oki.

Also interviews with Tom Kubo, Tom Kamo, Tom Kodaka, Larry Yoshida. See also, Jun-ichi Nishizawa, "The Transistor: A Look Back and a Look Ahead," *Solid State Technology,* December 1987, p. 72. Among the seventy-one Japanese and twenty-seven foreign patents claimed by Shumpei Yamazaki are patents for silicon-gate, E-Square, EPROM, heterojunction, and silicon nitride technologies.

Chapter 11: The CMOS Slip
Interviews with Grove, Sanders, Frank Wanlass, Tom Longo, T. J. Rodgers, George Huang, John Carey, and others. A good summary of the developments is Alden Hayashi, "CMOS Technology, Once Dismissed as Too Cumbersome, Is Expected to Control Two-Thirds of the Market by 1995," *Electronic Business,* May 1, 1985. For a partisan statement of the case that the differences between the U.S. and Japanese chip markets explain most of the market share gap, see EIAJ Statement on Semiconductors (Electronic Industries Association of Japan, November 21, 1988).

Chapter 12: Mountain of Memories
Interviews with Charles Sporck, Takami Takahashi, Ward and Joseph Parkinson, and others.

PART THREE: *The Transformation of Capital*

Advanced Research in VLSI. Series of volumes published annually by MIT Press, beginning after first conference at Caltech in 1979 and alternating between Caltech and MIT thereafter, until 1987 conference at Stanford. Charles Seitz, Randal Bryant, Charles E. Leiserson, Paul Losleben, et al., eds., containing basic technical papers by Carver Mead, David Johannsen, Seitz, Gordon Moore, Jack Dennis, Richard M. Karp, W. Daniel Hillis, and others, reporting new research in the converging fields of VLSI, parallel computation, systolic arrays, nanostructures, and other developments originating with the Mead-Conway revolution.

Chapter 13: Mead's Theory
Many interviews with Mead and his students, Mead-Conway (see above), and Ivan E. Sutherland and Carver A. Mead, "Microelectronics and Computer Science," *Scientific American,* September 1977. A lucid presentation of the theory and taxonomy of parallel computing and how VLSI affects them is Charles L. Seitz, "Concurrent VLSI Architectures," *IEEE Transactions on Computers,* December 1984. *Proceedings of*

the Caltech Conference on Very Large Scale Integration, January 22–24, 1979, contains historic papers by Mead and by Gordon Moore, "Are We Really Ready for VLSI," Amr Mohsen, "Device and Circuit Design for VLSI," H. T. Kung, "Let's Design Algorithms for VLSI Design," David Johannsen, "Bristle Blocks: A Silicon Compiler," and Ron Ayres, "Silicon Compilation—A Hierarchical Use of PLAs." See also, Mims, Forrest M. III, *Siliconnections: Coming of Age in the Electronic Era.* (New York: McGraw-Hill, 1986), an intriguing personal account of the early era in personal computing, including the saga of MITS and Altair.

Chapter 14: From Boltzmann to Conway

Chapter based on articles in the industry press, on interviews with Mead, Conway, Johannsen, and others, on Moore paper (see above), on Conway, "The MPC Adventures: Experiences with the Generation of VLSI Design and Implementation Methodologies," transcribed from an Invited Lecture at the Second Caltech Conference on Very Large Scale Integration, January 19, 1981 (Palo Alto, Calif.: Xerox PARC, 1981), on Conway and Mark Stefik, "Toward the Principled Engineering of Knowledge" (*The AI Magazine,* Summer 1982), and on Conway, Lynn, Alan Bell, and Martin Newell, "MPC 79: A Large-Scale Demonstration of a New Way to Create Systems in Silicon" (*LAMDA* magazine, second quarter, 1980). All of these Conway papers offer clear expositions, from the perspective of Xerox Palo Alto Research Center (PARC), of the development of the first projects in hierarchical design of VLSI systems. The *Electronics* magazine cover story announcing the 1981 Award for Achievement to Lynn Conway and Carver Mead appeared October 20, 1981, pp. 102–105.

Chapter 15: The Silicon Compiler

Proceedings of 1979 Caltech Conference on VLSI (see above). Interviews with Mead, Johannsen, Hal Alles, Edmund Cheng, Misha Burich, Thomas Matheson, and others, and many publications, including Daniel D. Gajski, "Silicon Compilation" (*VLSI Systems Design,* November 1985), and Stephen C. Johnson, "Top Down Systems Design Through Silicon Compilation" (*Electronics* magazine, May 3, 1984, pp. 120–128).

Chapter 16: Competing Visions

Interviews with the principals and articles by them, including Misha R. Burich and Thomas G. Matheson, "Environments for Silicon Compilation" (*Proceedings,* Custom Integrated Circuits Conference, May 20–22, 1985, Portland, Oreg.). Also, Richard Gossen interviews and videocassette on the flight of the Cristen Eagle.

Chapter 17: The New Balance of Power

Valuable interviews with Andrew Rappaport and issues of his *Technology Research Group Letter.* Interviews with Gordon Campbell and Edmund Cheng.

PART FOUR: *The Imperial Computer*

AGHA, Gul, *Actors: A Model of Concurrent Computation in Distributed Systems.* Cambridge, Mass.: MIT Press, 1986. An interesting theory of parallel computation.
BERNSTEIN, Jeremy, *The Analytical Engine: Computers Past Present and Future.*

New York: William Morrow, 1981. A perceptive and elegantly written survey by the physicist at *The New Yorker*.

BISHOP, Peter, *Fifth Generation Computers: Concepts, Implementations and Uses*. New York: Halstead Press, a division of John Wiley, 1986. A broad survey of the field, from the conceptual foundations to the various computer architectures, by a British consultant.

BOLTER, J. David, *Turing's Man: Western Culture in the Computer Age*. Chapel Hill, N.C.: University of North Carolina Press, 1984. A readable introduction to the subject.

BRAND, Stewart, *The Media Lab: Inventing the Future at MIT*. New York: Viking, 1987. An intriguing exploration of the future of television and almost everything else.

BRISTOW, Geoff, ed., *Electronic Speech Recognition: Techniques, Technology, and Applications*. New York: McGraw-Hill, 1986. A valuable collection of articles on the subject, first published in Great Britain.

CHAMBERLIN, Hal, *Musical Applications of Microprocessors*, 2nd. ed. Hasbrouck Heights, N.J.: Hayden Book Co., 1985. The definitive work. Chamberlin is now at Kurzweil Music Systems.

FREDERIKSEN, Thomas M., *Intuitive Analog Electronics: From Electron to Op Amp*. New York: McGraw-Hill, 1988. Detailed and mostly technical explanation of analog devices.

———, *Intuitive IC Op Amps: From Basics to Useful Applications*. Santa Clara, Calif.: National Semiconductor Technology Series, 1984. A comprehensive explanation of op amp technology.

GOLDBERG, Adele, ed., *A History of Personal Workstations*. New York: ACM Press, Addison-Wesley, 1988. A fascinating compendium, led off by the authoritative Gordon Bell, "Toward a History of (Personal) Workstations," which tells you almost all you need to know about the evolution of computers.

GRAUBARD, Stephen B., ed., *Artificial Intelligence*, issue of *Daedalus, Journal of the American Academy of Arts and Sciences*, Winter 1988. Contains an array of fascinating essays by leading figures in the field from Seymour Papert to John McCarthy. Particularly pertinent for this book were contributions by Anya Hurburt and Tomaso Poggio, W. Daniel Hillis, Jacob T. Schwartz, and Hilary Putnam.

GREBENE, Alan B., ed., *Analog Integrated Circuits*. New York: The Institute of Electrical and Electronics Engineers, 1978. Contains historic papers by Widlar, Frederiksen, Brokaw, Dobkin, Gilbert, Hoff, Erdi, and others.

HAUGELAND, John, *Artificial Intelligence: The Very Idea*. Cambridge, Mass.: MIT Press, 1985. A lucid account of the history and the techniques.

HEIMS, Steve J., *John Von Neumann and Norbert Wiener: From Mathematics to the Technologies of Life and Death*. Cambridge, Mass.: MIT Press, 1984. Interesting biographical materials about both these titans of computer science.

HILLIS, W. Daniel, *The Connection Machine*. Cambridge, Mass.: MIT Press, 1985. Excellent explanation of the need to transcend the Von Neumann model of computation and good description of Hillis's connection machine, now with 64,000 processors working in parallel.

HOROWITZ, Paul, and Winfield Hill, *The Art of Electronics*. Cambridge, U.K.: Cambridge University Press, 1980. An encyclopedia of circuitry, especially good on analog devices such as the op amp.

KARIN, Sidney, and Norris Parker Smith, *The Supercomputer Era*. New York: Harcourt Brace Jovanovich, 1987. A good exposition of the uses and technologies of supercomputers.

KURZWEIL, Raymond, *The Age of Intelligent Machines*. Cambridge, Mass.: MIT Press, 1989. Superb survey of the field of artificial intelligence, from the theory to the technology, by the pioneer of the field.

LEISERSON, Charles E., *Area-Efficient VLSI Computation*. Cambridge, Mass.: MIT Press, 1983. A theory of parallel computation through systolic arrays.

LUTHER, Arch C., *Digital Video in the PC Environment*. New York: McGraw-Hill, 1989. Excellent text on the increasing digitization of video technology, with a focus on Intel's Digital Video Interactive system.

MEAD, Carver A., *Analog VLSI and Neural Systems*. Reading, Mass.: Addison Wesley, 1989. A new language and technical repertory for analog devices.

PAYNTER, H. M., ed., *Palimpsest on the Electronic Analog Art: A Collection of Reprints and Papers and Other Writings Which Have Been in Demand*. Boston: George A. Philbrick Researches, 1955. A fascinating collection of readings relating to the early analog computers.

PENZIAS, Arno, *Ideas and Information: Managing in a High-Tech World*. New York: W. W. Norton, 1989. A splendid short text by the director of Bell Labs. A valuable antidote for any unclear technological explanations in this book, from computer software code to neural networks.

PHILBRICK, George A., *Applications Manual for Computing Amplifiers, for Modelling, Measuring, Manipulating, and Much Else*. Boston: George A. Philbrick Researches, 1966. Designs and essays in analog computing, this book has been distributed by the scores of thousands.

PIERCE, John R., *An Introduction to Information Theory: Symbols, Signals and Noise*. New York: Dover Publications, 1980. 2nd rev. ed. A definitive introduction to Shannon's Information Theory by a reverent student from Bell Labs, now teaching at Caltech.

RUMELHART, David E., James L. McClelland, and the PDP Research Group, *Parallel Distributed Processing: Explorations in the Microstructures of Cognition*, Vol. 1, *Foundations*, and Vol. 2, *Psychological and Biological Models*. Difficult but important articles on the origins of parallel processing, from the biological and psychological models to the computational schemes.

WIENER, Norbert, *Cybernetics: Control and Communication in the Animal and the Machine*. Cambridge, Mass.: MIT Press, 1961. An ambitious and finally flawed but widely inspirational case for a new scientific discipline.

WILDES, Karl L., and Nilo A. Lindgren, *A Century of Electrical Engineering and Computer Science at MIT, 1882–1982*. Cambridge, Mass.: MIT Press, 1985. Contains good accounts of the early analog computers made by Vannevar Bush at MIT.

Chapter 18: Patterns and Analogies

I was introduced to the charms and continuing importance of analog electronics by Daniel Sheingold of Analog Devices, formerly with George A. Philbrick Researches. Sheingold took over as editor of *The Lightening Empiricist,* the elegantly written philosophical house organ of Philbrick, Inc., after George A. Philbrick gave up the job in 1957. My sources for this chapter are mostly Sheingold and the volumes by Paynter and Philbrick (see above). Also useful was Sheingold, introductory chapter, in Roger Thielking, ed., *Operational Amplifiers Applications Handbook* (New York: John Wiley, forthcoming). An eloquent essay on the perplexities of analog and digital electronics as applied to high-fidelity audio is Edward Rothstein, "The Quest for Perfect Sound," *The New Republic,* December 30, 1985. For a lucid exploration of feedback, see Peter Beckmann, *Feedback and Stability in Science and Society* (Boulder, Col.: The Golem Press, 1978).

Chapter 19: Analog People
Based on interviews with the principals and *The Collected Writings of Ambrose Bierce* (Secaucus, N.J.: The Citadel Press, 1946).

Chapter 20: The Age of Intelligent Machines
Based on interviews with the principals, with Richard Lyons of Schlumberger Research in Palo Alto, with James and Janet Baker of Dragon Systems, and others.

Chapter 21: The IBM Machine
Interviews with the principals and Pierce, *Information Theory* (see above).

Chapter 22: The Neural Computer
Interviews with the principals. Mead, *Analog VLSI* (see above) presents the entire theory and technology. Mead, "Neural Hardware for Vision," *Engineering & Science,* Caltech, June 1987, is an abridged version of the lecture to the Feynman class. Also, David H. Hubel, "The Brain," and Francis H. C. Crick, "Thinking about the Brain," *Scientific American,* offprint, 1984; and Max V. Mathews and John R. Pierce, "The Computer as Musical Instrument," *Scientific American,* February 1987.

Chapter 23: The Death of Television
The essentials of this argument appeared in *Forbes,* "IBM-TV," February 20, 1989, by the author. For the background of the technical argument and a detailed explanation of DVI, see Luther, *Digital Video* (see above). For an effective presentation of the case and technical requirements for HDTV, assuming the continued desirability of over-the-air broadcast television, see William F. Schreiber of the MIT Media Lab, "Advanced Television Systems and Their Impact on the Existing Television Broadcasting Service," a paper presented to the Federal Communications Commission, November 17, 1987.

For the layman, Norm Alster, "TV's High-Stakes, High-Tech Battle" (*Fortune,* October 24, 1988), offers the most comprehensive view of the argument. The historical background is presented in "The Decline of U.S. Consumer Electronics Manufacturing: History, Hypotheses, and Remedies," by David H. Staelin (Chairman), Charles Ferguson, Robert Solow, et al. (Consumer Electronics Sector Working Group: MIT Commission on Industrial Productivity, Cambridge, Mass., 1988). For technical references, see K. Blair Benson, ed., *Television Engineering Handbook* (New York: McGraw-Hill, 1986), and John D. Lenk, *Complete Guide to Digital Television: Troubleshooting and Repair* (Englewood Cliffs, N.J.: Prentice Hall, 1989). For evidence that fiber-optics technology is ready, see Paul W. Schumate, "Outlook for Fiber-to-the-Home: Healthy But Cloudy" (*Lightwave,* February 1989), and with J. E. Berthold, "Progress in Switching Technology for the Emerging Broadband

Network," in J. Raviv, ed., *Computer Communication Technologies for the '90s*. Amsterdam: Elsevier Science Publishers, 1988.

PART FIVE: *The Quantum Economy*

ECONOMICS AND POLICY

BELL, Daniel, *The Coming of Post-Industrial Society: A Venture in Social Forecasting*. New York: Basic Books, 1973. An authoritative prophecy of the new era of information, from a learned social scientist who missed the erosive impact of microelectronics on large bureaucracies.

BORRUS, Michael G., *Competing for Control: America's Stake in Microelectronics*. Cambridge, Mass.: Ballinger Publishing Co., 1988. A good statement of the case for governmental action by an estimable member of the Berkeley Roundtable on International Economics.

BROOKES, Warren T., *The Economy in Mind*. New York: Universe Books, 1982. An early and exciting statement of many of the themes of this book.

CLEVELAND, Harlan, *The Knowledge Executive: Leadership in an Information Society*. New York: Truman Talley Books/E. P. Dutton, 1985. Full of interesting insights about the differences between information and other industrial resources.

COHEN, Stephen S., and John Zysman, *Manufacturing Matters: The Myth of the Post-Industrial Economy*. New York: Basic Books, 1987. Effective presentation of the argument that manufacturing is still central to economic growth and power, by two members of the Berkeley Roundtable on International Economics.

FEIGENBAUM, Edward A., and Pamela McCorduck, *The Fifth Generation: Artificial Intelligence and Japan's Computer Challenge to the World*. Reading, Mass.: Addison-Wesley, 1983. Tired of panicking over Japanese microchips? Now you can panic over Japanese AI. An effective polemic.

FERGUSON, Charles, *Neoclassical Economics and Strategic Evolution: International Competition and Its Theoretical Lessons*. MIT: unpublished paper, 1986. Strong statement of the case against classical trade theory.

———, *American Microelectronics in Decline: Evidence, Analysis, and Alternatives*. MIT: Department of Political Science and Program in Science, Technology and Society, 1985. This paper comprises approximately one third of the author's Ph.D. thesis, *International Competition, Strategic Behavior and Government Policy in Information Technology Industries*, finished in 1988, and due for publication in a different form by Random House.

FLAMM, Kenneth, *Targeting the Computer: Government Support and International Competition*. Washington, D.C.: The Brookings Institution, 1987. Demonstrates that government was the central force in developing and purchasing early U.S. computers.

———, *Creating the Computer: Government, Industry and High Technology*. Washington, D.C.: The Brookings Institution, 1988. A further development of the argument above.

HUBER, Peter W., *The Geodesic Network: 1985 Report on Competition in the Television Industry*, for the Anti-Trust Division of the U.S. Department of Justice. Washington, D.C.: Government Printing Office, 1987. A brilliant treatise by the best mind in the field.

JACOBS, Jane, *Cities and the Wealth of Nations: Principles of Economic Life*. New

York: Random House, 1984. Jacobs's most recent and difficult book, but it is full of entirely original insights, especially on trade. Her chapter on exchange rates has been ignored because of its outlandish implication of the need for separate currencies for every city. But she puts her finger on a crucial problem ignored by all other economists and deeply relevant to the current predicament of U.S. high technology while the value of the dollar is established by political forces shaped by declining heavy industries.

MACRAE, Norman, *The 2025 Report: A Concise History of the Future, 1975–2025*. New York: Macmillan, 1984. For readers unwilling to pore through back issues of *The Economist,* this somewhat ill-conceived volume is the only available text espousing the powerful vision of the future of technology and entrepreneurship by a writer who predicted a coming glut of oil in 1974.

McKENNA, Regis, *Who's Afraid of Big Blue: How Companies are Challenging IBM —and Winning.* Reading, Mass.: Addison-Wesley, 1989. Presents case for small computer, digital everything, in networks, with entrepreneurial economy.

McKENZIE, Richard B., and Dwight R. Lee, *The Twilight of Government Power: How the Global Economy Is Limiting the Ability of Governments to Tax and Regulate.* Washington, D.C.: Cato Institute, 1989. A rigorous and sophisticated presentation of the economic argument of Part V of *Microcosm.*

MAGAZINER, Ira C., and Robert B. Reich, *Minding America's Business: The Decline and Rise of the American Economy.* New York: Harcourt Brace Jovanovich, 1982. A shrewdly argued case for industrial policy, based on a sophisticated understanding of the strategic conditions of real companies. It is marred only by skewed statistics on the American failure and a gullible mercantilism that assumes a trade gap is always bad. Magaziner's 1989 publication, *The Silent War* (New York: Random House), shows that this brilliant analyst has learned little since 1982 but still knows more about business than anyone else on the left.

NESHEIM, John L., *Starting a High-Tech Company and Securing Multi-Round Financing: A Strategic Report Analyzing the Successful Techniques Used to Start, Finance, Value, and Launch a New High-Tech Business from the Seed Round Through the Initial Public Offering.* Saratoga, Calif.: Electronic Trend Publications, 1988. A definitive study of the entrepreneurial process in Silicon Valley high-technology startups.

OHMAE, Kenichi, *Triad Power: The Coming Shape of Global Competition.* New York: The Free Press, Macmillan, 1985. The single best analyst of the new global economy.

PETERS, TOM, *Thriving on Chaos.* New York: Alfred A. Knopf, 1987. The bottom line on microcosmic management: "Learning to Love Change." An encyclopedia of insights for a new era.

PRESTOWITZ, Clyde V., Jr., *Trading Places: How We Allowed Japan to Take the Lead.* New York: Basic Books, 1988. Performs the valuable service of taking us deep into the heart of darkness of the U.S. trade policy apparatus that gave us the Semiconductor Trade Agreement with Japan. However, Prestowitz is hopelessly mercantilist on trade.

REICH, Robert B., *Tales of a New America.* New York: Times Books, 1987. The latest and most eloquent statement of Reich's case for industrial policy.

SUMMING UP

DAVIES, Paul, *God and the New Physics*. New York: Simon & Schuster, 1983. A broad survey of the various theories of God and the quantum from a routine scientific perspective.

ECCLES, Sir John, ed., *Mind and Brain: The Many-Faceted Problems*. New York: Paragon House, 1985. An excellent symposium on the current state of the argument on mind and brain, including memorable essays by Louis de Broglie, Gunther S. Stent, Karl H. Pribam, W. H. Thorp, and the editor himself, among others.

GARDNER, Howard, *The Mind's New Science: A History of the Cognitive Revolution*. New York: Basic Books, 1985. An ambitious and comprehensive survey, but it fails to demonstrate the revolution that it proclaims.

GELWICK, Richard, *The Way of Discovery: An Introduction to the Thought of Michael Polanyi*. New York: Oxford University Press, 1977. Does the job well.

JAKI, Stanley L., *Brain, Mind and Computers*. South Bend, Ind.: Gateway Editions, 1969. An early presentation of the essential arguments, by a Roman Catholic theologian and physicist. Quotes Sir Arthur Eddington to the effect that to explain thought in terms of physics is "like trying to find the square root of a sonnet."

KOESTLER, Arthur, *The Ghost in the Machine*. London: Pan Books, 1975. The most thoroughly argued of Koestler's ardent polemics against the prevailing materialism in science and technology.

PAGELS, Heinz R., *The Dreams of Reason: The Computer and the Rise of the Sciences of Complexity*. New York: Simon & Schuster, 1988. An intellectual peregrination on the issues of mind, brain, and computer by one of the best writers in science, who died in 1988.

PENFIELD, Wilder, *The Mystery of the Mind*. Princeton, N.J.: Princeton University Press, 1973. An indispensable work based on unique research findings and bold speculative thinking.

POLANYI, Michael, *Knowing and Being*. London: Routledge & Kegan, Paul, 1969. A good sample of Polanyi's trenchant thought on the epistemology of science.

POPPER, Karl, *Unended Quest, An Intellectual Autobiography*. London: Fontana/Collins, 1976. A stirring series of reflections on the great issues of twentieth-century science and philosophy by a great modern thinker.

SCHAEFFER, Francis A., *How Should We Then Live?: The Rise and Decline of Western Thought and Culture*. Westchester, Ill.: Crossway Books, 1976. A powerful Christian case against scientistic and humanistic culture.

SHAFTO, Michael ed., *How We Know: Nobel Conference 20, Gustavus Adolphus College, St. Peter, Minnesota*. San Francisco: Harper & Row, 1985. With contributions by Gerald M. Edelman, Brenda Milner, Roger C. Schank, Herbert A. Simon, Daniel Dennett, and Arthur Peacocke.

TOFFLER, Alvin, *The Third Wave*. New York: William Morrow, 1980. Okay, okay, I know it is hokey, stilted, and plain wrong in parts, but he got the essence of the changes years before the rest of us. It's a rare best-seller that beat all the academics to the punch.

Chapter 24: The New American Challenge

Charles A. Berg, "The Use of Electric Power and Growth of Productivity: One Engineer's View" (Department of Mechanical Engineering, Northeastern University,

1988), presents a powerful argument for the theory that the use of microcosmic forms of power is critical to the advance of productivity. For a detailed discussion of the failure of European industrial policy for semiconductors, see Franco Malerba, *The Semiconductor Business: The Economics of Rapid Growth and Decline* (Madison, Wis.: University of Wisconsin Press, 1985). The best compendium of statistics on the U.S. and global industry is William J. McClean, ed., *Status 1989: A Report on the Integrated Circuits Industry* (Scottsdale, Ariz.: Integrated Circuit Engineering Corp., 1989). See also *Status* for 1988 and 1987. These reports demonstrate the devastating impact of the rising value of the yen on market share statistics in semiconductors and on the profits and investment of Japanese firms (also massively enhanced by the Semiconductor Trade Agreement). The statistics of the U.S. lead in computer applications can be found in Egil and Karen Juliussen, *The Computer Industry Almanac 1989* (New York: Brady, 1988). For tables showing that for industrial countries the more government aid in semiconductors the less national advance, see Dmitri Ypsilanti, ed., *The Semiconductor Industry: Trade Related Issues,* a report for the Organization for Economic Cooperation and Development (Paris: OECD Press, 1985).

Chapter 25: Revolt against the Microcosm

This chapter and the following are heavily based on the author's "The Revitalization of Everything: The Law of the Microcosm," *Harvard Business Review,* March–April 1988. Charles Ferguson replied in the May–June issue, "From the Folks Who Brought You Voodoo Economics . . ." and this issue and the next contained a long series of valuable correspondence on the subject. Also relevant is the author's "The DRAM Crisis," *Forbes,* June 13, 1988, and responses from Robert Noyce and Jerry Sanders and a reply by the author. For a detailed presentation of the devastating impact of the Semiconductor Trade Agreement on U.S. computer and semiconductor firms, see Kenneth Flamm, "Policy and Politics in the International Semiconductor Industry," a presentation to the SEMI-ISS Seminar, Newport Beach, California, January 16, 1989.

See Jacobs, *Cities* (above), for a presentation of the theoretical basis for the argument that currency values established by ascendant industries such as semiconductors and computers are damaging to less progressive economic sectors and lead to powerful political forces for devaluation. In the early 1980s the upsurge of the computer industry, together with major tax cuts, led to a strong dollar and a rebellion by major U.S.–based commodity producers, which induced the U.S. government to push for a 50 percent depreciation of the dollar against the yen. This devaluation was crucial to the Japanese comeback in the late 1980s. For a close empirical study of high-technology startups, see Nesheim (above).

Chapter 26: The Law of the Microcosm

Charles J. Bashe, Lyle R. Johnson, John H. Palmer, and Emerson W. Pugh describe the ferrite core memories in early IBM machines in *IBM's Early Computers* (Cambridge, Mass.: MIT Press, 1986), pp. 231–272. Meindl's projections of a billion transistors on a chip by the end of the century appeared in *Scientific American,* October 1987 (see above), and the estimated rise in costs of interconnections as they move from chips to printed circuit boards and to computer networks appeared in John Mayo, *Scientific American,* October 1986 (see above).

Chapter 27: The Eclipse of Geopolitics

Interviews with Charles Seitz of Caltech and others on the flaws of the prevailing computer strategies for strategic defense. Robert Jastrow, *How to Make Nuclear Weapons Obsolete* (Boston: Little, Brown, 1985), offers a pithy, short presentation of the case for anti-missile systems. See also John Gardner, Edward Gerry, Robert Jastrow, William Nierenberg, and Frederic Seitz, *Missile Defense in the 1990s* (Washington, D.C.: George C. Marshall Institute, 1987). *Balancing the National Interest,* Report of the Panel on the Impact of National Security Controls . . . (Washington, D.C.: National Academy of Sciences, 1987), presents an effective case against drastic measures to inhibit technology transfers except in the case of the most specialized military codes and systems. The leading voice of the software catastrophists is David Lorge Parnas, "Software Aspects of Strategic Defense Systems: A Former Member of an SDIO Advisory Panel Explains Why He Believes We Can Never Be Sure SDI Software Will Work" (*Communications of the ACM:* A publication of the Association for Computing Machinery, December 1985). For an especially pithy presentation of some of the economic argument, see Alan Greenspan, "Goods Shrink and Trade Grows" (*The Wall Street Journal,* October 24, 1988, p. A12). In all these writings I benefit from the regular guidance of Alan Reynolds of Polyconomics, Morristown, N.J., who is the world's leading analyst of international economic developments.

Chapter 28: Triumph Over Materialism

References include Penfield, Kurzweil, Polanyi, Koestler, Schaeffer, and Popper (see above). For an excellent presentation of the crisis in the contemporary philosophy of knowledge, see Frederick Turner, "Escape from Modernism: Technology and the Future of the Imagination," *Harper's* magazine, November 1984. Schaeffer also sums up the flaw of scientist and humanistic culture: ". . . the individual things, the particulars, tended to be made independent, autonomous, and consequently the meaning of the particulars began to be lost. We can think of it as the individual things, the particulars, gradually and increasingly becoming everything and thus devouring all meaning until meaning disappears." But Steven Weinberg, Nobel Laureate for pathbreaking discoveries in particle physics, puts the problem more succinctly: "The more the universe becomes comprehensible the more it also seems pointless."

Acknowledgments

This book explains and celebrates this fourfold transformation of science, technology, business, and politics. Obviously, I could not have written it without extraordinary help and support from many friends, associates, teachers, and editors.

Robert Asahina of Simon & Schuster was central in developing the manuscript from a history of the semiconductor industry into a theory of the microcosm, on the order of *Wealth and Poverty*. From the beginning, his persistent guidance and inspiration—his refusal to accept any compromise of his own sense of the book's possibilities—was vital to the concept and structure and emerging vision of *Microcosm*.

Caltech Professor Carver Mead's role in the book will be evident to any reader. But also important was the contribution of his students, including Massimo Civilotti, Mary Ann Maher, Mischa Mahowald, Telle Whitney, and John Tanner, and Mead's assistant Helen Derevan, who makes it possible to communicate with Carver Mead.

The contributions of entrepreneurs and engineers in the semiconductor and computer industries will also be evident in reading the book. However, some important guides in my education in microelectronics had to be cut for structural reasons. My gratitude is no less

great and the book could not have been written without the generous
assistance of Fred Bucy, Mohon Rao, and George Heilmeir of Texas
Instruments (particularly for Heilmeir's prophetic papers on the industry), Dr. Phillip Salsbury of SEEQ, Richard Pashley of Intel, the
encyclopedic Thomas M. Frederiksen of National, John Lazlo and
Stephen Roach of Morgan Stanley, Mel Phelps of Hambrecht &
Quist, Daniel Sheingold of Analog Devices, and especially, on the
physics, Walter Tchon of Xicor, among many others who lavished
time and insights on me.

Among my most important teachers were Ward and Joe Parkinson,
Juan Benitez, and my many other instructors at Micron Technology
(a firm whose incredible history I tell in my earlier work, *The Spirit of
Enterprise*). They gave me a vivid sense of the perils and possibilities
of entrepreneurship in technology.

I am deeply indebted to Peter Sprague and Nick Kelley for introducing me to this stirring and crucial subject, to Ben Rosen whose
superb newsletter taught me how to write about it, and most of all to
Esther Dyson, the splendid writer and analyst who hired me to cover
the microchip industry and plunged me into the educational crucible
that produced this book. I am especially grateful to Nick Kelley for
bringing me into his company, Berkshire Corporation—a supplier to
semiconductor and disk drive producers—and thus giving me an inside view into the commercial dynamics of the industry.

Through long winter evenings in Tyringham and a crucial summer
afternoon, screenwriter-novelist Deric Washburn persuaded me that
the heart of the book must be the physics and philosophy, as well as
the technology, of the microcosm. For many other vivacious conversations on the subjects of this book, I could mention many friends
and family, but two uncles put up with most: I am deeply indebted
to Rodman Gilder and Reese Alsop (and Mellie and Lee) for minis-
trations to mind, body, and spirit.

Although some authors imagine otherwise, a writer can never know
what he has written without readers who are willing to labor through
unfinished work and dare to offend him by delivering bad news. That
elegant stylist and economic philosopher Tom Bethell gave me an
early sense that explaining this technology requires far more effort
and theoretical scaffolding than I was then providing. At the end,
Mark Stahlman of Sanford C. Bernstein gave me the benefit of his
sophisticated insights on information technology and his broader
sense of the philosophical issues of the microcosm.

Taking time off from vital chores as a presidential speechwriter,

Joshua Gilder did heroic work as the first reader to emerge from the swamps of a much longer and less intelligible version of this manuscript with the happy news that it might someday be a book, if I took the time to write it.

Bill Hammett frequently encouraged me in this project and boldly embraced Carver Mead as the Tenth Anniversary Speaker and Walter Wriston Award Lecturer for the Manhattan Institute. Hammett's entrepreneurial vision and Mead's mesmerizing speech in November 1987 thus gave some of the themes of the book an early exposure to a key New York audience. Followed up by *Forbes*'s superb editor Lawrence Minard, Hammett's vision also led to a *Forbes* cover story on Mead in 1988.

David and Peggy Rockefeller once again offered their generous hospitality for long stints of uninterrupted writing and editing, and gave me a vivid sense of the presence of my father in my life and in theirs.

My mother and Gilly not only read and criticized important parts of the manuscript but also provided me with their unswerving support, including an office for the end game of this long project, protecting me from creditors, visitors, callers, predators, and many friends who could not be expected to comprehend that I was off in the microcosm for the duration.

For many services beyond the call of duty, wrenching these ideas from the microcosm into the world, I dedicate the book to Richard Vigilante.

My wife Nini is co-author and editor of all I am and do.

George Gilder
Tyringham, Mass.
April 5, 1989

INDEX

ABOUT THE AUTHOR

George Gilder lives in Tyringham, Massachusetts, with his wife, Nini, and four children, Louisa, Mellie, Richard, and Nannina, and five computers. He is an associate of the Manhattan Institute and the Tyringham Union Church, and author of seven previous books.